Approaches to Videogame Discourse

Approaches to Videogame Discourse

Lexis, Interaction, Textuality

Edited by
Astrid Ensslin and Isabel Balteiro

BLOOMSBURY ACADEMIC
NEW YORK • LONDON • OXFORD • NEW DELHI • SYDNEY

BLOOMSBURY ACADEMIC
Bloomsbury Publishing Inc
1385 Broadway, New York, NY 10018, USA
50 Bedford Square, London, WC1B 3DP, UK

BLOOMSBURY, BLOOMSBURY ACADEMIC and the Diana logo are trademarks of
Bloomsbury Publishing Plc

First published in the United States of America 2019
This paperback edition published in 2021

Cover design by Louise Dugdale
Cover image © iStock.com/Bigmouse108

A catalog record for this book is available from the Library of Congress.

ISBN: HB: 978-1-5013-3845-8
PB: 978-1-5013-7544-6
ePDF: 978-1-5013-3847-2
eBook: 978-1-5013-3846-5

Typeset by Deanta Global Publishing Services, Chennai, India

To find out more about our authors and books visit www.bloomsbury.com
and sign up for our newsletters.

CONTENTS

List of Illustrations vii
Acknowledgments x

Locating Videogames in Medium-specific,
Multilingual Discourse Analyses *Astrid Ensslin and
Isabel Balteiro* 1

Part One Lexicology, Localization, Variation 11

1 Videogames: A Lexical Approach *Carola Álvarez-Bolado
   and Inmaculada Álvarez de Mon* 13

2 Lexical and Morphological Devices in Gamer Language in
   Fora *Isabel Balteiro* 39

3 Phraseology and Lexico-grammatical Patterns in Two
   Emergent Paragame Genres: Videogame Tutorials and
   Walkthroughs *Christopher Gledhill* 58

4 Playing with the Language of the Future: The Localization
   of Science-fiction Terms in Videogames *Alice Ray* 87

5 End-user Agreements in Videogames: Plain English at Work
   in an Ideal Setting *Miguel Ángel Campos-Pardillos* 116

Part Two Player Interactions: (Un)Collaboration,
   (Im)Politeness, Power 137

6 Bad Language and Bro-up Cooperation in Co-sit
   Gaming *Astrid Ensslin and John Finnegan* 139

7   "Shut the Fuck up Re! Plant the Bomb Fast!":
    Reconstructing Language and Identity in First-person
    Shooter Games *Elisavet Kiourti* 157

8   "I Cut It and I ... Well Now What?": (Un)Collaborative
    Language in Timed Puzzle Games *Luke A. Rudge* 178

9   "Watch the Potty Mouth": Negotiating Impoliteness in
    Online Gaming *Sage L. Graham and Scott Dutt* 201

Part Three Beyond the "Text": Multimodality, Paratextuality,
    Transmediality 225

10 On the Procedural Mode *Jason Hawreliak* 227

11 The Player Experience of *BioShock*: A Theory of
    Ludonarrative Relationships *Weimin Toh* 247

12 Language Ideologies in Videogame Discourse: Forms
    of Sociophonetic Othering in Accented Character
    Speech *Tejasvi Goorimoorthee, Adrianna Csipo,
    Shelby Carleton, and Astrid Ensslin* 269

13 Playing It By the Book: Instructing and Constructing the
    Player in the Videogame Manual Paratext
    Michael Hancock 288

Afterword *James Paul Gee* 305

Notes on Contributors 311
Index 315

ILLUSTRATIONS

Figures

1.1 Videogame frames 22

4.1 Tech purchase menu for Terran units. Screenshot by Alice Ray, © Blizzard Entertainment 93

4.2 Ion torch. Screenshot by Alice Ray, © The Creative Assembly Limited and Sega Corporation 99

4.3 Hellion unit. Screenshot by Alice Ray, © Blizzard Entertainment 102

4.4 Zerg tech tree. Screenshot by Alice Ray, © Blizzard Entertainment 102

4.5 Working Joe. Screenshot by Alice Ray, © The Creative Assembly Limited and Sega Corporation 103

4.6 Vidicom. Screenshot by Alice Ray, © The Creative Assembly Limited and Sega Corporation 105

6.1 Relative distribution of BLEs across corpus 145

7.1 *CS:GO* as a multimodal environment 165

7.2 Hookah is changing his gun to a knife. Screenshot by Elisavet Kiourti, © Valve Corporation 173

8.1 A side view of the experimental setup 186

8.2 Mean values of clauses used per round, split for task success 188

9.1 Twitch stream screen layout using *League of Legends* as an example, © by Riot Games 206

11.1 Ludonarrative relationships and the players' experiences in gameplay videos 249

11.2   The *Let's Play Onion* and its different layers
       (reproduced from Recktenwald 2014) 250
12.1   Race and accent distribution in *Dragon Age:
       Origins* 275
12.2   Race ratio in *Dragon Age: Origins* 276
12.3   Gender and accent distribution in *Dragon Age:
       Origins* 277

Tables

1.1   Content words in the CVC keyword list 17
1.2   Keyword list of nouns in CVC before processing 18
1.3   Videogame frames and associated specific
       vocabulary 23
1.4   Semantic neologisms in the CVC corpus 24
1.5   Examples of use of *arcade* 29
1.6   Examples of use of *combo* 30
1.7   Examples of use of *disco* 31
1.8   Examples of use of *escenario* 32
1.9   Examples of use of *expansión* 33
1.10   Examples of use of *jefe* 34
1.11   Examples of use of *jugabilidad* 35
3.1   Key grammatical items in a sample of four technical
       registers 68
3.2   Semantic zones and grammatical keywords in the
       VGT and VGW corpora 70
4.1   Analysis of N + N compounds 95
4.2   Analysis of Adj + N structure 101
4.3   Analysis of derived nouns 104
4.4   Analysis of fictive meanings 107
5.1   Word, syllable, and sentence count for EULAs 122
5.2   Readability indexes for EULAs 123

5.3   Compound adverbs in EULAs analyzed 126

5.4   Use of archaisms in EULAs analyzed 128

6.1   Normalized frequencies of BLE categories with types occurring in corpus 146

8.1   The organization of participants in the ten rounds played 187

8.2   Number of clauses used (with mean values per round in parentheses), split by task success 189

9.1   Types of rules 208

9.2   Total sanctions by rule category 214

11.1   Participants' profiles 253

11.2   Relevant subcategories of ludonarrative dissonance (Toh 2015, 2018) 256

11.3   Relevant subcategories of ludonarrative resonance (Toh 2015, 2018) 256

11.4   Relevant subcategories of ludonarrative irrelevance (Toh 2015, 2018) 257

ACKNOWLEDGMENTS

First and foremost, we would like to thank our respective (young and extended) families for their ongoing support and understanding, especially during crunch-time copyediting. We hope that the insights gained from this book will help reflect and shape the academic and playful futures of our children, Anton, Leo, Carolina, and Guillermo. They are representatives of a whole generation of young people growing up to be natives of diverse videogame discourses. Giving them the tools to understand and critically evaluate the ludic languages they speak will be key to their socialization and their digital-born identities.

We would like to thank the Department of English Philology as well as the Faculty of Arts of the University of Alicante for their financial support, as well as the members of the LexESP Research Group, especially José Ramón Calvo-Ferrer, for their support during the LexESP IV Seminar and also for acting as one of the reviewers of this manuscript. We received a lot of in-kind and financial support from the Departments of Modern Languages and Cultural Studies and Humanities Computing at the University of Alberta, for which we are immensely grateful. We would like to thank Jennie Dailey-O'Cain in particular for her constructive comments during the redrafting process, and Elizaveta Tarnarutckaia for her work on the index. Finally, our thanks go to the editorial and production teams at Bloomsbury and Deanta, who were unfailingly helpful, generous, and accommodating throughout the reviewing and publication cycle.

Locating Videogames in Medium-specific, Multilingual Discourse Analyses

Astrid Ensslin and Isabel Balteiro

Videogames, gamification, game culture, and cultures of play have become a global, ethnographically and culturally diverse paradigm of our hypermediated everyday lives. Games—digital and analog—are played and revered by countless player communities worldwide, whether for leisure and entertainment or professionally, as, for example, in e-sports. Videogames are vilified, pathologized, or even banned by politicians and health organizations. They are critiqued, analyzed, and studied by journalists, bloggers, and scholars alike and are utilized by marketing campaigns and in occupational and academic education across public and private sectors. As "affinity spaces," or "loosely organized social and cultural settings in which the work of teaching tends to be shared by many people, in many locations, who are connected by a shared interest or passion" (Gee 2018: 8; Gee 2007a, b), games and their paratextual environments serve their players and their communities of practice (Lave and Wenger 1991) as personalizable, experiential tools and objects of learning, communication, and the promotion of values. Yet, as "social semiotic spaces" (Gee 2005), they also carry enormous ideological weight that can inform people's views and behaviors inside and outside the fictional gameworlds they inhabit (see Goorimoorthee et al., this volume).

In this context, it is perhaps surprising that little comprehensive work exists to date that examines digital games as diverse, medium-specific objects and tools of language studies and discourse analysis more specifically. With the exception of broader educational, rhetorical, and introductory discourse-analytical work done by James Paul Gee (2003, 2007a, b, 2013, 2018),

Christopher Paul (2012), and Astrid Ensslin (2012), there has not been any systematic, book-length attempt at bringing together specific areas within discourse analysis that examine in detail how videogames function as means and objects of communication; how they give rise to new vocabularies, meanings, textual genres, and discourse practices; and how they serve as rich vehicles of ideological signification and social engagement.

That said, recent years have seen a sharp increase in academic interest in studying videogames as medium-specific platforms and multidimensional communicative objects that give rise to a plethora of paratextual phenomena across social media, fora, streaming platforms, and other contemporary online platforms. In 2018, for example, the area of multimodal discourse analysis saw the publication of several book-length applications to digital ludonarrativity and procedurality (Toh 2018; Hawreliak 2018). Similarly, an increasing number of articles in leading media and communication journals and edited collections have emerged over the past decade that deal with highly specialized, cutting-edge discourse-analytical research into games and gamer language. This work includes for example an examination of the role of pronouns in the construction of gendered players and videogame characters (Carrillo Masso 2011); an investigation of Blizzard's (2004) *World of Warcraft* guild members' use of online conversational turn-taking in the performative construction of "identities of expertise" (Newon 2011); an analysis of the multimodal persuasive design strategies in Hideo Kojima's (1998) *Metal Gear Solid* (Stamenković et al. 2017); a study into the role of swearing in the creation of celebrity YouTube Let's Player PewDiePie's online persona (Fägersten 2017); and an exploration of immersed digital game players' multifaceted response cries (Conway 2013).

As the above examples demonstrate, the discourse of games involves various layers of communicative interaction and multiple types of social actors (Ensslin 2015). These include, of course, the players themselves and the ways in which they negotiate meanings about games and gaming, for instance on specialized subreddits and Twitch chat logs and through paratextual Let's Plays and walkthroughs (see Gledhill, this volume), but also in couch co-op and other forms of co-situated gaming (see Kiourti, this volume; Ensslin and Finnegan, this volume) and online game chats. A second group of actors participating in the discourse of games and naturally overlapping with the former group are industry professionals— people who develop, produce, publish, and disseminate games, including those that create rules, instructions (see Hancock, this volume), restrictions, and legislation for player interaction (see Campos-Pardillos, this volume), as well as those that translate the language used in videogame interfaces (such as menu items and character dialogue) for other, "localized" player communities (see Ray, this volume). Third, there are journalists, politicians, educators, parents, activists, and other (media) stakeholders who engage in debates about games, gameplay, and the alleged effects of gameplay on people's health and behavior. All these social actors are deeply invested and engaged in the construction and evolution of lexical and phraseological items that form the building blocks of videogame lexicons across languages (see Álvarez-Bolado and Álvarez de Mon, this volume).

Viewed from a more representational angle, the discourse of games relates to the language and multimodal designs of games themselves, as well as to paratexts such as instruction manuals, end-user license agreements (EULAs), fora, blurbs, and games advertising. Games as cultural artifacts communicate meanings via user interfaces, audiovisual character design, backstories, instructions, and scripted dialogues; and, in order to analyze these diverse modes of representation appropriately, a wide range of discourse-analytical methods can and need to be used in order to address the full range of lexical and phraseological elements as well as con-, hyper-, sub-, inter-, and paratextual elements that constitute videogame discourse.

The idea to address this need in the form of the present volume arose in the context of the Fourth International Seminar on English and ESP Lexicology and Lexicography (LexESP IV) on Video Games and Language, held at the University of Alicante in May 2016. In our respective roles as keynote speaker and conference chair, we became aware of delegates' keen interest in this evolving subject area, coupled with a strongly felt demand for leadership and consolidated, collaborative scholarship to drive the field forward. Inspired by this insight, we decided to solicit contributions to what is the first significant collection of international, cutting-edge research in videogame linguistics (understood as a subarea of media linguistics), performed by linguists and media and communication scientists and scholars from around the world, in and about multiple languages. A number of contributors to this book presented earlier drafts of their chapters at LexESP IV. Others followed the ensuing call for papers. The result is a refreshing mixture of cultural, linguistic, and disciplinary backgrounds and career stages, and it was particularly exciting for us to see the future of the field heralded in the work of our more junior contributors. By the same token, we are honored and pleased to have been able to bring on board the perhaps most eminent international scholar in the field of videogame literacy, Prof. James Paul Gee, who kindly provided a thought-provoking afterword for the volume.

Taking advantage of the unique lexicological and LSP (Language for specific purposes) expertise of the LexESP research group in and around Drs Isabel Balteiro, Miguel Ángel Campos-Pardillos, and José Ramón Calvo-Ferrer at Alicante, we have structured this book in such a way as to highlight the importance of micro- as well as macrostructural phenomena. Thus, we have singled out a specific focus on videogame lexicology and how it applies to language- and industry-specific jargon, slang, and localization processes. Lexical processes surrounding ludolectal morphology, creativity and productivity, stylistic choices, and borrowing across languages are key to understanding the linguistic economies and ecologies of games. Issues relating to lexicology, localization, and variation therefore form one of three main pillars of videogame discourse analysis reflected in the structure of this book, and we are delighted to offer analyses of videogame lexis in Spanish and French as well as English. The other two pillars, or parts, focus on linguistic and pragmatic nuances of "player interactions" in various parts of the world, from Cyprus through Wales and the United States, on the one hand, and on discursive phenomena "beyond the text" of the game itself, on the other. In what follows, we shall outline the individual contributions to this volume and how they map onto these three pillars, often in overlapping ways.

Part One, "Lexicology, Localization, Variation," begins with a case study by Carola Álvarez-Bolado Sánchez and Inmaculada Álvarez de Mon exploring what a "lexical approach" to videogames—or, more precisely, to a specific form of journalistic videogame metadiscourse—might look like. Using a review corpus from the Spanish technology weekly *CiberPaís*, they offer a corpus-driven analysis of Spanish keyword nouns relating to videogames and their collocates for contextualization. They group their results into four thematic areas associated with videogame production and use and subsequently perform an analysis of neologisms in the corpus, adopting an inclusive concept of neology that combines new formation, borrowing, and semantic shift. Their findings suggest that semantic neologisms—that is, shifts in the meaning of existing words in Spanish—are dominant in the data, as opposed to a very small number of borrowings from English. Finally, the authors are careful to remind us about the medium-specificity of the chosen corpus material, which may yield highly idiosyncratic results depending on genre, platform, and questions of authorship.

Chapter 2, by Isabel Balteiro, takes us into the world of videogame fora and the lexical and morphological devices used and developed by players inhabiting the popular NeoGaf gamer forum. More specifically, using a corpus-driven analysis, Balteiro explores gamers' language choices within the NeoGaf community or group. The author suggests that the interactions of gamers in fora through posts have their own register which presents a balance between specialized/technical jargon and relaxed slang. Accordingly, Balteiro's findings include the identification of specific and highly creative lexical units which seem exclusive to this group and, in general, to gamers' lexical stock in fora. As expected and hypothesized in this study, among the word-formation mechanisms employed, gamers' lexis and terminology are highly prolific in abbreviations (mainly acronyms and initialisms), which the author considers as a consequence of the conditions imposed by the online written medium where the interactions take place, and which make the register analyzed lexically and stylistically closer to spoken discourse.

Moving from lexis to phraseology in Chapter 3, Christopher Gledhill looks at the lexico-grammatical patterns found "in two paragame [paratextual game] genres: videogame tutorials and walkthroughs." Adopting a "contextualist" (Firth 1957) approach, he posits that these two genres are both well-defined and phraseologically distinct, and he demonstrates this through a corpus-driven, phraseological analysis of "key" grammatical items. His analysis allows insight into statistically more likely occurring particles, adverbs, conjunctions, and pronouns in these genres, as well as three main sets of patterns relating to (1) the management of the player's moves within the imaginary space of the game, (2) the framing of advice, and (3) the tracing of relationships between individual discourse referents, for example, through the use of pronouns. Adopting a systemic functional approach to discussing his results, Gledhill concludes that it is erroneous to assume the existence of LGP (language for general purposes) because every language event is contextualized and codified accordingly, and that videogame tutorials and walkthroughs are also to be seen as a form of LSP.

In Chapter 4, "Playing with the Language of the Future: The Localization of Science Fiction Terms in Videogames," Alice Ray focuses on lexico-translational complexities by looking at a particular case study of localization from English to French. She chooses the science-fiction videogame as an idiosyncratic vignette for how games from one and the same genre might display very different lexical and semantic strategies to naming fictional in-game objects. While *Starcraft II: Wings of Liberty* (Blizzard 2010) tends to invent completely new expressions that only make sense in the immediate context of the game world, *Alien: Isolation* (Creative Assembly 2014) uses lexical items already familiar to players and adapts them semantically to the game. These diverse approaches to lexical productivity directly affect translators' levels of creativity in localizing science-fiction games as either more recognizable and mimetic or more visionary and fantasy-like.

Another important area of professional videogame discourse is the legal language surrounding the products and their users. This is tackled by Miguel Ángel Campos-Pardillos in the final contribution to Part One, where he looks at end-user agreements and the use of plain English "in an ideal setting" that not only improves comprehensibility of legal jargon but also operates as a community-building tool and promotion strategy. By comparing *Minecraft*'s EULA with those of other videogames, the author highlights a number of specific devices that successfully present the game as an "indie" product, where developers and gamers participate in a friendly, relaxed atmosphere. The gamers' hearts are won by avoiding legalese but also through colloquialisms, contractions, and side remarks that make this EULA an interesting example, which may perhaps pave the way for the future.

The second thematic pillar of this book is interaction between players, which, as various contributions to Part Two demonstrate, shows some intriguing commonalities between online and offline, physically co-located and remote communication—such as bad language, swearing, discourses of "cool" and fun, and the strangely paradoxical phenomenon of cooperative impoliteness (see also Ensslin 2012). These commonalities are partly due to players' heightened emotional stance during gameplay, but also to the specific social rules *at* play *during* play. In their chapter on "Bad Language and Bro-up Cooperation in Co-sit Gaming," Astrid Ensslin and John Finnegan offer a corpus-driven analysis of co-situated gamers' prolific use of "Bad-Language Expressions" (McEnery 2006), the vast majority of which turn out to be religious terms of abuse and actual expletives and which they observe to be part of a general "bro-up" tendency that involves high levels of polite and mutually supportive behavior. Interviews with players in their study suggest that this overuse (compared to the BNC Spoken) is not primarily intended as subversive behavior vis-à-vis perceived social norms, as McEnery (2006) suggests, but a symptom of deep immersion that removes verbal inhibitions, of extreme levels of affect and emotional investedness, and of a degree of performative pressure to use bad language, imposed by the community of practice themselves.

A specific ethnographic lens through which to view bad language in the form of dysphemisms (swearing, expletives, and irony) in co-sit gameplay is offered by Elizavet Kiourti in Chapter 7. Her chapter focuses on the ways in which bad language serves as a means of linguistic identity (re-)construction. It zooms in on a small group of young Cypriot gamers playing the multiplayer first-person shooter *Counter Strike: Global Offensive* (Valve 2018) on a regular basis. In line with Ensslin and Finnegan (this volume), Kiourti's analysis of participants' turn-taking in the Greek-Cypriot dialect suggests that the use of swearing and bad language is socially functionalized. More specifically, dysphemisms serve as situationally codified, communicative short-cuts to prevent team-based performative face-loss, to ease stressful situations, and to ensure in-group bonding.

Chapter 8 by Luke A. Rudge addresses the idea of (un)collaborative gamer language and examines it in relation to player cooperation in timed puzzle games. Combining elements of Systemic Functional Linguistics and Conversation Analysis, he highlights the importance of effective verbal communication in situations where players engaged in collaborative play are confronted with limited time, limited information, and limited communicative capabilities. Among his findings is the observation that the use of complete rather than elliptical clauses and mutual completions of adjacency pairs is associated with successful task completion, thus suggesting that it is worth players' time to invest in effective communication despite, or indeed because of, a race against time.

A specific take on impoliteness as a common phenomenon in online discourse is offered by Sage L. Graham and Scott Dutt. In Chapter 9, "'Watch the Potty Mouth:' Negotiating Impoliteness in Online Gaming," the authors examine emergent interactions in online gaming streams to illuminate the ways that the posted rules of interaction intersect with communicative practice. Using a corpus of video streams and online chats from Twitch.tv, this chapter investigates how rules and guidelines for appropriate behavior are understood by participants in synchronous online chat and then explores how these same rules are understood and enforced by both botmods and human moderators who are responsible for identifying and then controlling impolite, disruptive, and/or aggressive behavior. The authors observe how competing interests—the duty to maintain order in the chatroom and the need to avoid ostracizing participants and viewers by overly harsh rule enforcement—have to be balanced by streamers, particularly in cases of spam and copypasta. They also highlight the dilemma posed by automated regulatory behavior resulting in out-of-sync punishment.

Part Three of this book, then, tackles the complex, multilayered discoursal ramifications of game design, from mechanics to storytelling, voice acting, and player instructions. It moves beyond existing "textual" approaches to game analysis by offering new perspectives on videogame multimodality, narrativity, and paratextuality. In Chapter 10, "On the Procedural Mode," Jason Hawreliak examines procedurality as a medium-specific mode of expression that has hitherto been largely neglected by frameworks of multimodal discourse analysis. Using close-play as analytical method, he demonstrates how the procedural mode dynamically influences the meaning of other communicative modes often found in video games, such as moving images, music, and haptics, and that it can either consonantly align with or dissonantly work against the informational content produced by other modes, thus creating powerful aesthetic effects such as ludonarrative dissonance (Hocking 2007).

In Chapter 11, "The Player Experience of *BioShock*: A theory of ludonarrative relationships," Weimin Toh examines medium-specific ludonarrative relationships as a key element of videogame discursivity. The author examines how players' interactions and gameplay performances in gameplay videos are actualized based on players' understandings of the relationships between the games' mechanics and their storytelling elements. Furthermore, Toh analyzes *how* the interaction of the player in the videogame with the player's simultaneous gameplay performance in the video contributes to the actualization of the player's narration-commentary. His findings are intended to help videogame developers and researchers obtain a better understanding of how players make meaning and interact with the videogame content (ludonarrative) to produce their gameplay videos, which are, after all, a key factor of commercial and noncommercial player engagement and thus a symptom of a game's popularity.

Venturing a hitherto rare look into the audiodiscursive landscape of game design, in Chapter 12, Tejasvi Goorimoorthee, Adrianna Csipo, Shelby Carleton, and Astrid Ensslin examine the voiced-over speech accents of nonplayer characters (NPCs) in BioWare's blockbuster fantasy role-playing game *Dragon Age: Origins* (*DAO* 2009) as a form of "sociophonetic othering." The authors present a detailed view of the speech accent distribution in the game during a typical, full playthrough and analyze how native and foreign accents of English are used as sites of othering and of perpetuating standard language ideology. They identify patterns of both intentional and unintentional othering that occurs based on specific voice-acting choices and practices. They discuss how sociophonetic representations function politically in *DAO* by examining how accents delineate class structure and social hierarchy among races. Their findings show that *DAO* lacks cultural diversity in the speech accents assigned to its characters.

The final chapter of the book offers a diachronic look at a paratextual genre that seems key to games in general but has essentially faded as a textual phenomenon in the videogame industry in the past decade: the instruction manual. In "Playing it By the Book: Instructing and Constructing the Player in the Videogame Manual Paratext," Michael Hancock conceptualizes the videogame instruction manual as a paratext with a view to forming an understanding of how the manual constructs the ideal gamer, by pushing the player toward the new cool. He takes *the Donkey Kong Country* manual's (Rare 1994) lead character Cranky as his main case study, addressing the question of why a textual genre meant to inform players about a game features a character who mocks the player for seeking that information, contradicts the manual's descriptions of the game, and opposes the very notion that the manual should exist. As well as heralding the surmise of the genre as a separate paratextual item in what is by now also an almost dated form of distribution, the game box, the *Donkey Kong Country* manual presents its audience with a very specific message about game culture and where it is going.

The "Where is it going?" question is again addressed by James Paul Gee in the Afterword to this book. He announces the discourse analysis of gaming as a "new field of inquiry" that examines a "domain of human activity" that is language in use and thus highly contextualized. He emphasizes how the contributions to this book set the scene for a new area of linguistic research and that "the deeply important study of how language helps form and, in turn, is changed by gaming as a distinctive human activity." He also highlights that the research presented in this book offers methodological and conceptual insights that may become applicable to a variety of other areas of human life. Gee's vision of a possible future for (videogame) discourse analysis is a neostructuralist return to grammar as "anchor and choice maker" in people's ongoing endeavor to produce and process situational meaning. The point of this "linguistic" approach to discourse would provide a lens through which to view human nature and social interactions on a more fundamental level. Videogames, in turn, will enrich this prospect with a medium-specific mirror of humanity's "shared imagination."

Let us close with a methodological note. A book that bears within its title one of the most diverse, multi- and interdisciplinary analytical approaches in use across humanities and social sciences needs to map its own tools carefully, not least because its media focus is so diverse and fast changing that it requires constant scholarly innovation. It is thus not surprising that the contributions to this volume showcase a broad, yet by no means exhaustive, range of possible theoretical and methodological combinations and triangulations. While various chapters follow a mostly corpus-driven, quantitative approach (e.g., Álvarez-Bolado Sánchez and Álvarez de Mon; Gledhill), enriched by descriptive analysis (e.g., Balteiro), others demonstrate the importance of complementing quantitative with qualitative research for their goals (e.g., Campos-Pardillos; Ensslin and Finnegan; Goorimoorthee et al.). Some choose case studies to illustrate theoretical innovation (e.g., Hawreliak), professional-creative concerns (e.g., Ray), and historical developments (e.g., Hancock), while others examine conversational and/ or grammatical patterns and cooperation in ethnographical studies (e.g., Graham and Dutt; Kiourti; Rudge; Toh). Together, these methodological choices constitute the foundation of a new and expanding repository for the medium-specific discourse analysis of videogames, which we hope will be an inspiration as well as a critical starting point for future research.

References

2K Boston (2007), *Bioshock*, Quincy, MA: 2K Games.

BioWare (2009), *Dragon Age: Origins*, Redwood City: Electronic Arts.

Blizzard Entertainment (2004), *World of Warcraft*, Irvine, CA: Blizzard Entertainment.

Blizzard Entertainment (2010), *Starcraft II: Wings of Liberty*, Irvine, CA: Blizzard Entertainment.

Carillo Masso, I. (2011), "The grips of fantasy: The construction of female characters in and beyond virtual game worlds," in A. Ensslin and E. Muse (eds.), *Creating Second Lives: Community, Identity and Spatiality as Constructions of the Virtual*, 113–42, New York, NY: Routledge.

Conway, S. (2013), "Argh! An exploration of the response cries of digital game players," *Journal of Gaming & Virtual Worlds*, 5 (2): 131–46.

Creative Assembly (2014), *Alien: Isolation*, Tokyo: Sega.

Ensslin, A. (2012), *The Language of Gaming*, Basingstoke: Palgrave Macmillan.

Ensslin, A. (2015), "Discourse of games," in C. Ilie and K. Tracy (eds.), *International Encyclopedia of Language and Social Interaction*, Hoboken, NJ: Wiley-Blackwell, doi:10.1002/9781118611463.wbielsi154.

Fägersten, K.B. (2017), "The role of swearing in creating an online persona: The case of YouTuber PewDiePie," *Discourse, Context & Media*, 18: 1–10.

Firth, J.R. (1957), *Modes of Meaning: Papers in Linguistics, 1934–1951*, Oxford: Oxford University Press.

Gee, J.P. (2003), *What Video Games Have to Teach Us about Learning and Literacy*, New York: Palgrave Macmillan.
Gee, J.P. (2005), "Social semiotic spaces and affinity spaces: From *The Age of Mythology* to today's schools," in D. Barton and K. Tusting (eds.), *Beyond Communities of Practice*, New York: Cambridge University Press.
Gee, J.P. (2007a), *What Video Games Have to Teach Us about Learning and Literacy*, 2nd edn., New York: Palgrave Macmillan.
Gee, J.P. (2007b), *Good Video Games and Good Learning: Collected Essays on Video Games, Learning and Literacy*, New York: Peter Lang.
Gee, J.P. (2013), *Good Video Games and Good Learning: Collected Essays on Video Games, Learning and Literacy*, 2nd edn., New York: Peter Lang.
Gee, J.P. (2018), "Affinity spaces: How young people live and learn on line and out of school," *Phi Delta Kappan*, 99 (6): 8–13.
Hawreliak, J. (2018), *Multimodal Semiotics and Rhetoric in Videogames*, New York: Routledge.
Hocking, C. (2007), "Ludonarrative dissonance in Bioshock," *Click Nothing*, October 7. Available online: http://clicknothing.typepad.com/click_nothing/2007/10/ludonarrative-d.html.
Kojima, H. (1998), *Metal Gear Solid*, Japan, Tokyo: Konami Computer Entertainment.
Lave, J. and E. Wenger (1991), *Situated Learning: Legitimate Peripheral Participation*, Cambridge: Cambridge University Press.
McEnery, T. (2006), *Swearing in English: Bad Language, Purity and Power from 1586 to the Present*, London: Routledge.
Mojang (2011), *Minecraft*, Sweden: Mojang AB.
Newon, L. (2011), "Multimodal creativity and identities of expertise in the digital ecology of a World of Warcraft guild," in C. Thurlow and K. Mroczek (eds.), *Digital Discourse: Language in the New Media*, 131–53, Oxford: Oxford University Press.
Paul, C.A. (2012), *Wordplay and the Discourse of Video Games: Analyzing Words, Design, and Play*, New York, NY: Routledge.
Rare (1994), *Donkey Kong Country Instructional Booklet*, Kyoto: Nintendo.
Stamenković, D., M. Jaćević, and J. Wildfeuer (2017), "The persuasive aims of *Metal Gear Solid*: A discourse theoretical approach to the study of argumentation in video games," *Discourse, Context & Media*, 15: 11–23.
Toh, W. (2018), *A Multimodal Approach to Video Games and the Player Experience*, New York: Routledge.
Valve Corporation (2018), *Counter Strike: Global Offensive*, Bellevue, WA: Valve Corporation.

PART ONE

Lexicology, Localization, Variation

1

Videogames

A Lexical Approach

Carola Álvarez-Bolado and Inmaculada Álvarez de Mon

1 Introduction

When we first came across the lexis of videogames years ago, we were actually looking for specialized texts on new technologies and the internet to be used in the ESL (English as a second language) classroom with our engineering students at Universidad Politécnica de Madrid (UPM). On that occasion, we were trying to select technical texts in Spanish that could serve as a reference of the kind of vocabulary the students would encounter in similar texts in English. Among the different kinds of passages selected with that teaching purpose in mind, we found videogame reviews published in *CiberPaís*, the supplement on new technologies and the internet of *EL PAÍS*, Spain's largest newspaper. When trying to decide whether to include some videogame technical vocabulary in our selection, the lexis we found in those reviews surprised us. We had expected to find abundant computer terminology, but our findings were rather unexpected instead: the texts did certainly include specialized terms, but there were also characters, stories, missions, enemies, adventures, and avatars. In addition to videogame engines, polygons, and graphics, we could read about spells and secret codes, designers and artists. That mixture of ludic, narrative, industrial, and technological elements made the lexis of videogames in Spanish one that was both surprising and attractive, and one that, to our view, deserved some academic attention.

For this reason, we decided to analyze videogame reviews with the aim of finding out more about the lexis used in them. We tried to establish if they were new words or already existing words in Spanish acquiring a different meaning, and if so, how the new meanings had originated and what kind of reality they described. In addition, the fact that *CiberPaís* included reports and analyses of new technologies other than videogame reviews allowed us to determine which words were exclusive to the videogame lexical domain by comparing the different genres present in the supplement. Fortunately, *EL PAÍS* was helpful enough to provide us with the corpus of *CiberPaís* and, as will be explained later, it also gave us the possibility of using a collection of reviews large enough to produce relevant results as well as consistent data on the lexis of videogames as a specific domain. This chapter shows the results of analyzing the aforementioned corpus over a period of 10 years.

When facing the analysis of the lexis of a domain from the point of view both of its specificity and its formal as well as semantic features, as it is the case with videogames, it seems relevant to clarify first the notion of *neologism*. This can refer to a newly created word; a word coming from another language, such as English; or one already existing in the language under analysis but with a different meaning. Since new advances require new denominations, the study of neologisms has been of special interest when it comes to the terminology of science and technology. It acquired great importance at the end of the twentieth century given the explosion of technological advances taking place in the fields of electronics and telecommunications (Álvarez de Mon 2006: 241–62), which made especially relevant the analysis of the creation of terminology in Spanish, as Aguado de Cea's work in the field of computer science reveals (1995, 2006). Although traditionally any new word widely used is considered a neologism, neology is not a clear-cut concept, as evidenced by different authors (Rey 2005: 311–33; Cabré 1993: 450). In specialized terminology (Cabré 1993: 446), neology results in a new term, *neonym*, which refers to new denominations in specialized languages.

Algeo (1991: 2) refers to the question of identifying neologisms by stating that a new word can be either a word with a form that has never been seen or heard before or perhaps a preexisting form showing a new use:

> The form of the word itself may be novel, a form that has not been seen in English … or the newness may lie in a novel use of an existing form. In the latter case, the novelty may be in what the word refers to … the word's grammar … or even its relationship to those who use it.

As will be explained here, it is precisely the "newness" in the use of an existing form, that is, a word used with a new meaning in a different context, that we consider relevant for clarification. According to Rey (2005: 312), those words that are not strictly new but have acquired a new meaning are to be considered neologisms of meaning or *neosemanticisms*, as new meanings develop for already existing words in the language. This type of neologism, which is based on the distinction between formal and semantic neology, has also been applied to studies of neology in nonspecialized Spanish (Guerrero Ramos 1995: 24) and is present in the *CiberPaís* corpus designating some of the key concepts of the videogame domain. More recently, other concepts such as pseudo-anglicism, hybrid anglicism, and false anglicism have been used when studying the lexical input of English into European languages (Görlach 2001; Onysko 2007; Furiassi 2010).

Another crucial notion of our study is *context*, because it has allowed us to identify new meanings of the words under study and to determine the characteristics of the new concepts to which they refer. From a linguistic point of view, context has many features. In order to reach a definition and describe its importance in the identification of new meanings in the specialized corpus, this research has taken into account different approaches. Coseriu (1973: 230–33) defines context as the reality surrounding a linguistic sign, an act of speech, or a text. In British linguistics, the work of the anthropologist Malinowski led to Halliday's (1978: 143) description of context as a situation and its components: field (content), tenor (relationship between speakers), and mode (channel or medium of communication). On the other hand, Miller (1999: 15) defines the situational context as the information "about the purposes and goals of the communicative interaction," the topic context as "that depending on the domain of discourse," and the local context as "information provided by words in the immediate neighborhood." Context is also taken into account by relevance theory (Sperber and Wilson 1986: 15), which describes it as "the set of premises used in interpreting an utterance." More recently, Van Dijk's (2009: 165) comments on context have added value to the debate. His view of contexts as "mental constructs of relevant aspects of social situations [that] influence what people say and especially how they do so," thus determining lexical choice and syntax, is of special interest regarding the gaming experience and the circumstances which surround it. Notwithstanding their role as journalists, videogame reviewers are usually accomplished players who analyze and write about their own practice and skill when playing, which can have an impact on their selection of words when writing.

Finally, and still in connection with context, there is another key concept that was particularly convenient for our study of videogame lexis, that of *collocation*. Since Firth's often-quoted statement "you shall know a word by the company it keeps" (Firth 1957: 11), usual combinations of words or lexical solidarities have remained a controversial idea. For the structural approach (Sinclair 1991: 65; Hunston 2002: 142; Francis 1993: 147), a collocation is determined by its structure and occurs in patterns, so its study should include grammar. However, lexical combinations are also an object of study for lexicographers (L'Homme 2009: 238; Bergenhholtz and Tarp 2010: 33), who differ on the way collocations should be grouped, codified, and organized in dictionaries. The lexical approach, whose main advocates are Halliday (1996) and Sinclair (1991), agrees with Firth that the words that surround a term determine its meaning, and that part of the meaning is consequently the result of the terms occurring next to one or several specific words. For our study, we adopted a generic concept of collocation: a frequent combination of words, which allowed us to analyze *concordances*, that is, cooccurrences of keywords, as well as obtain data for the disambiguation of new meanings.

2 Corpus description and analysis

Years before they were officially established as a cultural reality, the need arose in the press for a specialized section analyzing videogames, thus adding a new genre to the already traditional reviews of books and films.[1] Published in the *CiberPaís* digital edition for the first time in 1996, the singularity of videogame reviews is the mixture of artistic and technological features present in them in order to guide the readers. Authored by specialized journalists who master the language of so-called electronic entertainment, they soon became regular articles in the written press. Therefore, videogame reviews seemed the perfect context for a study of lexis. After all, they are written by experts whose purpose is to inform the reader about the game being reviewed and therefore include all the information relevant for that purpose. The reviews in the corpus include the following information: the genre, the game developers, the type of device required, the suitable age range of its players, and its price. From a technical and artistic perspective, many details are added to inform the reader on how the game was developed, its difficulty, and the degree of entertainment provided. Usually, an overall evaluation of the videogame is included together with some information on localization, dubbing, and release date.

For our study, the initial corpus, including every article published in *CiberPaís*, was divided into two smaller corpora: one consisting of 983 videogame reviews, called the CVC (*Corpus de Videojuegos CiberPaís*); and a reference corpus, CRC (*Corpus de Referencia CiberPaís*), made up of the rest of articles published in the supplement between 1998, the publication date of the first videogame review in the written edition, and 2008. Our choice of that particular span was driven by the fact that *CiberPaís* ceased to be issued as such in 2010, when information on technology and the internet began to be part of *EL PAÍS*. We decided to stick to a span of 10 years and considered it a happy coincidence that the period covered by those reviews is one of special relevance for videogame development. As it happens, in

1998 a new generation of consoles was born with the launch of the first one equipped with a modem for surfing the internet and online gaming. Such advances in technology led to some of the most classic and well-known videogames (*The Legend of Zelda "Ocarina of Time," Fallout 2,* and *Parasite Eve,* among others). Moreover, once the millennium virus of the year 2000 was overcome, portable consoles appeared, and three-dimensional games revolutionized the industry, changing the concept of the videogame. These trends later promoted the extension of videogames to an enormous global market in terms of target age groups and technologies, proliferating new genres and subgenres such as exergames and casual and social network games.

From the point of view of content, both corpora can be considered specialized, but CVC is also a genre-specific corpus as it only consists of videogame reviews. The total number of tokens in CVC is 370,138, and there are 235,908 tokens in CRC.

The initial analysis of the corpus was carried out using *Wordsmith Tools 4.0* (Scott 2004). By means of the *Wordlist* tool, two lists of words were produced: a list for the CVC corpus and another for the CRC corpus. The comparison between these two lists made using the *Keyword* tool (Scott 2004) resulted in the list of CVC keywords consisting of 500 lexical units, mostly nouns, but also verbs, adjectives, prepositions, articles, and pronouns. From this keyword list, we discarded proper names and functional words such as conjunctions, prepositions, pronouns, and quantifiers. Proper names are highly frequent for referring to the videogame manufacturers or the journalists authoring the reviews, but since they do not belong to the Spanish word stock but rather to what is considered encyclopedic knowledge, they are not the focus of our study.

The keyword list served to identify the most relevant concepts in the thematic domain. Not surprisingly, the first word in the list is *juego* (game). We decided to focus on the study on nouns, as they refer to the concepts in the domain, and to analyze adjectives and verbs as collocates of those nouns. The number of lexical units in CVC for each word class is shown in Table 1.1.

After discarding both grammatical words and proper names, the list was reduced to 150 nouns. Table 1.2 shows this preliminary list of nouns, which

TABLE 1.1 *Content words in the CVC keyword list*

Nouns	319
Adjectives	25
Verbs	46

TABLE 1.2 *Keyword list of nouns in CVC before processing*

acción ("action")	desarrollador ("developer")	hechizo ("spell")	ocasión ("occasion")	secuela ("sequel")
ambientación ("setting")	dificultad ("difficulty")	historia ("story")	oro ("gold")	seguidor ("follower")
arcade ("arcade")	dimension ("dimension")	infiltración ("infiltration")	palanca ("lever")	sensación ("sensation")
arena ("arena")	dios ("god")	ingredientes ("ingredients")	partida ("game")	serie ("series")
argumento ("plot")	disparos ("shots")	inventario ("inventory")	patada ("kick")	shift
arma ("gun")	diversion ("entertainment")	isla ("island")	película ("film")	simulador ("simulator")
armamento ("armament")	doblaje ("dubbing")	joven ("youngster")	pelota ("ball")	skater
aspecto ("look")	dosis ("dose")	juego ("game")	perfección ("perfection")	soldado ("soldier")
ataque ("attack")	edad ("age")	jugador ("player")	personaje ("character")	sombra ("shadow")
aventura ("adventure")	editor ("editor")	lanzamiento ("launch")	piloto ("driver")	tablero ("board")

bonus ("bonus")	*ejército* ("army")	*lucha* ("fight")	*pista* ("track")	*táctica* ("tactic")
botones ("buttons")	*enemigo* ("enemy")	*luchador* ("fighter")	*pistola* ("pistol")	*tecla* ("key")
campeonato ("championship")	*entrega* ("instalment")	*mando* ("control")	*plataforma* ("platform")	*terror* ("terror")
carrera ("race")	*entrenamiento* ("training")	*máquina* ("machine")	*poder* ("power")	*textura* ("texture")
chica ("girl")	*entretenimiento* ("entertainment")	*metal* ("metal")	*polígono* ("polygon")	*tiempo* ("time")
chico ("boy")	*equipo* ("team")	*mezcla* ("mixture")	*precio* ("price")	*título* ("title")
ciudad ("city")	*escenario* ("background")	*minijuego* ("minigame")	*princesa* ("princess")	*todo* ("everything")
código ("code")	*escenas* ("cutscenes")	*misión* ("mission")	*prisión* ("prison")	*trama* ("storyline")
combate ("combat")	*estrategia* ("strategy")	*modo* ("mode")	*protagonista* ("protagonist")	*trampa* ("trap")
combo ("combo")	*experiencia* ("experience")	*monstruo* ("monster")	*puntería* ("aim")	*triángulo* ("triangle")

(Continued)

TABLE 1.2 (*Continued*)

compañía ("company")	*fabricante* ("manufacturer")	*motor* ("motor")	*puntuación* ("punctuation")	*trono* ("throne")
competición ("competition")	*fase* ("phase")	*movimientos* ("movements")	*puzle* ("puzzle")	*truco* ("trick")
conducción ("driving")	*frienduare*	*multijugador* ("multiplayer")	*realismo* ("realism")	*turno* ("turn")
consola ("console")	*género* ("genre")	*mundo* ("world")	*recomendación* ("recommendation")	*unidad* ("unit")
continuación ("continuation")	*golf* ("golf")	*munición* ("ammunition")	*reto* ("challenge")	*usuario* ("user")
contrincante ("opponent")	*golpe* ("blow")	*nena* ("little girl")	*rol* ("role")	*vehículo* ("vehicle")
control ("control")	*gracias* ("thanks")	*ninja* ("ninja")	*rompecabezas* ("puzzle")	*victoria* ("victory")
creador ("author")	*gráfica* ("graphic")	*nivel* ("level")	*saga* ("saga")	*vida* ("life")
cualidad ("quality")	*gráfico* ("graphic")	*objetivo* ("goal")	*salto* ("jump")	*videojuego* ("videogame")
deber ("duty")	*habilidad* ("skill")	*objeto* ("object")	*sangre* ("blood")	*volante* ("steering wheel")

was later filtered to eliminate those lexical units that do not develop a new meaning relevant for the domain.

To this list of the key words resulting from the corpus analysis, we added the following: *cruceta* ("d-pad"), *disco* ("disc"), *expansion* ("expansion"), *jefe* ("boss"), *jugabilidad* ("playability"), *localización* ("localization"), and *mapa* ("map"). These words are also found in the corpus but with a lower frequency. However, from the point of view of their semantics, they are highly relevant for the domain as they refer either to essential concepts of videogames, as in the case of *expansión*, *jugabilidad*, and *localización*, or are an instance of a word with a new meaning in Spanish, such as *cruceta*, *disco*, *jefe*, and *mapa*.

The next phase of the study was the extraction of the concordances by means of the *Concordance* tool. These concordance lines were first analyzed in order to verify the meaning of each lexical unit and discard any usage that was not relevant to the domain of videogames. A second analysis allowed us to establish the combinations of the nouns with the verbs and adjectives. In some cases, concordances did not give enough contextual information, and the whole sentence was needed to establish the new uses and disambiguate several cases of polysemy.

In order to store the lexical units, their meaning, the collocations extracted from the concordances, and their context of use, a software tool, *Herramienta de Creación y Mantenimiento de Diccionario Electrónico Especializado en Videojuegos (HDE)* (Cortés and Hungría 2012), was specially developed.

3 Lexical units in context: Classifying and categorizing the domain of videogames in Spanish

A key step of the analysis was to study the lexis of videogames from a conceptual point of view. This analysis allowed us to identify the relationships between the lexical units within the domain of videogames from the point of view of the reality they refer to and to establish the hierarchical organization among them. According to our findings, the lexical units in the corpus can be classified into four areas: game, narrative, new technologies, and industrial production.

First, we found that much of the lexis related to the action of gaming, especially to its participants and development, such as *jugador* ("player"), *competición* ("competition"), *contrincante* ("opponent"), or *reto* ("challenge"). Secondly, many lexical items involve narrative elements, including formal features, such as *entrega* ("instalment") *escenario* ("scenery"), and *género* ("genre") as well as plot-related ones, such as *héroe*

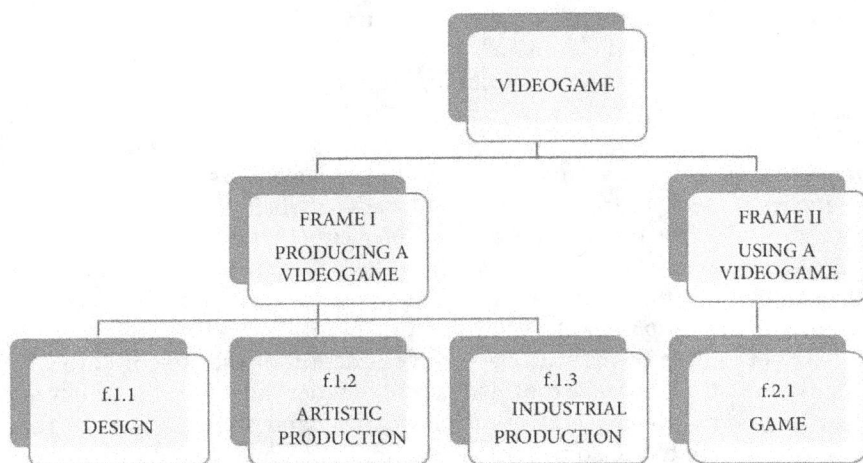

FIGURE 1.1 *Videogame frames.*

("hero"), *monstruo* ("monster"), and *princesa* ("princess"). This content is clearly related to the fact that videogames are usually fiction based. Thirdly, we found words referring to new technologies and the internet, such as *código* ("code"), *consola* ("console"), and *plataforma* ("platform"). Finally, and integrating all three areas—games, narration, and technology—we found lexis related to industrial manufacturing with terms such as *compañía* ("company"), *fabricante* ("manufacturer"), and *lanzamiento* ("launch"). The contextual analysis also revealed that these four frames could be further grouped into just two: the videogame as an industrial product (Frame 1. Creation of a Videogame) and the videogame as a vehicle for entertainment (Frame 2. Use of a Videogame), as shown in Figure 1.1.

For Frame 1, "Producing a Videogame," we established several subframes: Design, Artistic Production and Industrial Production. Frame 2, "Using a Videogame," refers to only one subframe, Game. Both frames and the words associated with them are shown in Table 1.3.

4 Analyzing the lexis of the videogame domain in Spanish: Some relevant findings

The analysis of the concordances and the meaning of the lexical units in the context of each frame revealed that some of these units developed new meanings resulting in semantic neologisms, that is, words already existing in a language that present new meanings. The lexical units classified as semantic neologisms are *avatar* ("avatar"), *aventura* ("adventure"), *combo*

TABLE 1.3 *Videogame frames and associated specific vocabulary*

Frame name	Words in Spanish
PRODUCING A VIDEOGAME: DESIGN	*arcade* ("arcade"), *aventura* ("adventure"), *carreras* ("racing games"), *consola* ("console"), *diseñador* ("designer"), *diseño* ("design"), *documentación* ("documentation"), *estrategia* ("strategy"), *género* ("genre"), *juego de disparos* ("shooting game"), *minijuego* ("minigame"), *móvil* ("mobile device"), *ordenador* ("computer"), *plataforma* ("platform"), *presupuesto* ("budget"), *programador* ("programmer"), *prototipo* ("prototype"), *rol* ("role-playing game"), *usuario* ("user"), *videojuego* ("videogame")
PRODUCING A VIDEOGAME: ARTISTIC PRODUCTION	*argumento* ("story"), *acción* ("action"), *carga poligonal* ("poligonal load"), *componente gráfico* ("graphical component"), *escenario* ("scenery"), *física* ("physics"), *guion* ("script"), *misión* ("misión"), *objeto* ("object"), *personaje* ("character"), *pruebas* ("tests"), *sombra* ("shadow"), *textura* ("texture")
PRODUCING A VIDEOGAME: INDUSTRIAL PRODUCTION	*actores de doblaje* ("dubbing actors"), *compañía desarrolladora* ("developing company"), *código* ("code"), *desarrollador* ("developer"), *disco* ("disk"), *diseñadores de niveles* ("level designers"), *doblaje* ("dubbing"), *equipo artístico* ("artistic team"), *entrega* ("instalment"), *expansión* ("expansion"), *guionistas* ("screenwriters"), *jugador* ("player"), *lanzamiento* ("launch"), *localización* ("localization"), *localizadores* ("localization experts"), *motor gráfico* ("graphics engine"), *mundos* ("worlds"), *músicos* ("musicians"), *niveles* ("levels"), *productor* ("producer"), *programa* ("program"), *pruebas* ("testing"), *sonorización* (sound), *título* (title), *traducción* ("translation"), *traductores* ("translators"), *tiempo de carga* ("loading time"), *universo* ("universo"), *usuarios múltiples* ("multiple users"), *versiones* ("versions")
USING A VIDEOGAME: GAME	*avatar* ("avatar"), *barra de vida* ("health bar"), *bonus* ("bonus"), *enemigo* ("enemy"), *energía* ("health"), *habilidades* ("skills"), *hechizo* ("spell"), *jefe* ("boss"), *juego* ("game"), mapa ("map"), *modo multijugador* ("multiplayer mode"), *motor de videojuegos* ("game engine"), *personajes tridimensionales* ("3D characters"), *puntos* ("points"), *truco* ("cheat")

("combo"), *cruceta* ("D-pad"), *disco* ("disc"), *escenario* ("scenario"), *expansión* ("expansion"), *golpe* ("blow"), *jefe* ("boss"), *localización* ("localization"), *mundo* ("world"), *nivel* ("level"), *objeto* ("object"), *personaje* ("character"), *plataforma* ("platform"), *rol* ("role"), *título* ("title"), *truco* ("trick"), and *vida* ("life"). A more detailed analysis of

TABLE 1.4 *Semantic neologisms in the CVC corpus*

Lexical unit	Collocations	Example from CVC	Meaning
AVATAR	*control de avatar*	*Una cámara situada inmediatamente detrás del **avatar** permite observar toda la zona circundante e interactuar con el entorno.* ("A camera located behind the avatar allows you to observe the entire surrounding area and interact with the environment")	Graphical representation or personification of the player in a game
AVENTURA	*aventura de acción aventura de amistad Aventura gráfica aventura de plataformas aventura de rol aventura de terror*	*La compañía Péndulo Studios lanza al mercado su última **aventura*** ("Péndulo Studios launches its latest adventure")	A videogame in which the player has to solve a series of riddles helped by some clues provided by other characters
COMBO	*combos básicos combos concatenados*	*A la vez, permite consultar y practicar una y otra vez las decenas de **combos*** ("At the same time, it allows the player to check and practice again the dozens of combos")	A combination of buttons that once pressed achieves a special move
CRUCETA	*cruceta de dirección*	*El personaje principal se controla con el mando analógico de la consola, mientras que con la **cruceta** se mueve la cámara)* ("The main character is controlled by the console analogic control while the d-Pad moves the camera")	Cross-shaped digital pad

Term	Forms	Example	Definition
DISCO	disco de expansión disco demo disco de misiones	De esta forma, ha conseguido *un disco que* presenta decenas de pequeños retos. ("In this way, they have created a disk that presents dozens of small challenges")	Optical disc. Videogame
ESCENARIO	escenarios de tres dimensiones escenarios en dos dimensiones escenarios en tres dimensiones escenarios tridimensionales	Ya en primera persona y pulsando el gatillo izquierdo se escanea el **escenario** y los Smiles se sintetizan ante el personaje. ("In first person and pressing the left trigger, the background is scanned and the Smiles are synthesized before the character")	Virtual setting, either bidimensional or three dimensional where a videogame takes place
EXPANSIÓN	expansión oficial	La **expansión** presenta nuevos enemigos, gráficos más trabajados y con un mayor número de texturas. ("The expansion presents new enemies, graphics and more textures")	Software product adding new content to a videogame
JEFE	jefe de final de fase jefe de final de nivel jefe de proyecto	Solamente en contadas ocasiones, cuando se lucha con los **jefes** de final de nivel se tiene una sensación de tres dimensiones ("Only rarely do we have a three-dimensional feeling when we fight with final bosses!")	An enemy, usually more powerful than the rest of enemies in the game
LOCALIZACIÓN	localización de software	La empresa ha abogado por profundizar en la **localización** de los productos. ("The developer has chosen to deepen the localization of the products")	Videogame preparation for sale in a specific region or country

(Continued)

TABLE 1.4 (Continued)

Lexical unit	Collocations	Example from CVC	Meaning
MUNDO	(no collocations in the corpus)	*En este Harry Potter Creator se presentan ante el chaval doce* **mundos** *que explorar* ("In this Harry Potter Creator, the boy is presented with twelve worlds to explore")	General setting where a videogame is played. Each of the thematic zones where part of the action takes place
NIVEL	*diseñador de niveles* *editor de niveles* *selector de nivel*	*El juego presentará quince* **niveles** *diferentes, por donde se moverá el* ("The game will present fifteen different levels where the user will move")	Scenery into which the game is subdivided due to programming or designing reasons
OBJETO	(no collocations in the corpus)	*para conseguirlos tendrán que utilizar una gran variedad de hechizos,* **objetos** *y ataques.* ("To get them, they will have to use a great variety of spells, objects and attacks")	An item that gives the character new abilities or powers
PERSONAJE	*personajes controlados por el PC* *personajes no jugadores*	*Si el jugador juega solo el propio ordenador generará* **personajes** *llamados bots* "(If the gamer is playing alone, the computer will generate characters known as bots")	Any entity in a game controlled by the player or by the machine
PLATAFORMA	*aventura de plataformas* *género de plataformas* *juego de plataformas* *título de plataformas*	*En resumen, una gran aventura tridimensional de la familia Bros para todos los aficionados a* **las plataformas** ("In short, a great three-dimensional adventure of the Bros family for all fans of platforms")	Videogame genre in which the character must jump from one setting to another

Term	Collocations	Example	Definition
ROL	género de rol, juego de rol, juego de rol en línea, juego de rol estratégico, juego de rol fantástico, juego de rol multijugador masivo, rol de aventura, título de rol, videojuego de rol multijugador masivo online	*Aventura y **rol** vuelven a cruzar sus destinos en Arcatera* ("Adventure and role playing cross their destinations again in Arcatera")	Videogame genre in which the player assumes the role of a character
TÍTULO	*título clásico, título comercial, título deportivo, título original, título superventas, título de acción, título de acción en primera persona, título de acción tridimensional, título de carreras, título de culto, título de estrategia táctica, título de fútbol, títulos del género, título de lucha, título de lucha de espadas, título de plataformas, título de referencia, título para Game Cube, título para PC, título para Xbox*	*lo que más llama la atención son los pequeños detalles que ambientan todo el **título**.* ("The most striking feature is the small details that set the whole title")	Electronic entertainment software which seeks to practice mental, musical, or sportive abilities
TRUCO	*menú de trucos, modo trucos*	*en Torrente depende de la habilidad del jugador o de teclear los siguientes **trucos**.* ("In Torrente it all depends on the player's skill or on typing the following cheat codes")	Specific sequence of letters or numbers that allows the player to alter the game or access some hidden content
VIDA	*barra de vida, fuerza de vida, nivel de vida, número de vida, pociones de vida, puntos de vida, vida virtual, vidas de complemento, vidas paralelas, vidas en línea, vidas infinitas, vida extra, vida restante*	*Si la dificultad es mantener excesiva se consigue **vida** infinita pulsando en el menú principal esta combinación:* ("If the difficulty is keeping excessive life, you get infinite life typing on the main menu the following combination.")	Gameplay duration given to the player according to his abilities or the number of challenges solved during the game

personaje, plataforma, título, aventura, and *rol* can be found in Álvarez de Mon and Álvarez-Bolado (2013: 63–84). Table 1.4 shows the information on these lexical units extracted from CVC. For each unit we provide the disambiguating collocations, that is, the noun accompanied by a modifier that is specific of the domain, an instance of use in a sentence, and, finally, the meaning in the videogame domain.

In CVC, we also found some instances of anglicisms such as *arcade* ("arcade"), as well as some calques. By *calque* (Aguado de Cea 1995: 9), we mean a process by which a foreign word is adapted to a language using translation, as in the case of *jugabilidad* ("playability"), *minijuego* ("minigame"), and *multijugador* ("multiplayer"). One possible explanation for the relatively low presence of English words and expressions in the lexis of videogames in this corpus could be the fact that it is made up of articles from a written source, *EL PAÍS*, with a higher level of formality than digital ones such as web reviews or fora. Furthermore, videogame reviewers are committed to following the stylistic rules of the newspaper.

To illustrate the results of our study, we present an analysis of some of the words whose meanings have evolved after analyzing their use in the corpus. To verify if the meaning of the word was a new one or was already recorded, we consulted the *Diccionario de la Lengua Española*, 22nd ed., 2001, and English dictionaries such as the *Oxford English Dictionary* (OED) online, 3rd ed.; the *Merriam-Webster Online Dictionary*, 11th ed.; and *The Videogame Style Guide and Reference Manual* by Thomas et al. (2007), the examples in the tables are extracted from the corpus, and the origin of each concordance appears in brackets.

ARCADE ("arcade"): This lexical unit shows a very interesting feature which was revealed after the analysis of the collocations. *Arcade* is used with two different meanings. On the one hand, it is one of the categories or genres of games, as evidenced by the expressions *género arcade* and *género de los arcades*. On the other hand, it refers to a simplified mode of playing another type of videogame genre, that of action. This is confirmed by Thomas et al. (2007: 45), who define arcade as "a simplified version of the game intended to give the player an immediate gratifying experience without requiring tutorials or significant practice." From a semantic point of view, *arcade* is a product belonging to the frame of Industrial Production, as examples 3 and 4 in Table 1.5 show. The fact that it is a manufactured product is revealed by its use accompanied by the verbs *crear* ("create") and *fabricar* ("manufacture").

From the point of view of its origin, *arcade* is one of those lexical units that are loans from the English language, and as such, it presents some problems of use. The fact that writers often have doubts when using this word is confirmed in our corpus. In CVC, *arcade* appears sometimes with a capital letter, between inverted commas, to signal its foreign origin; and even

TABLE 1.5 *Examples of use of* arcade

1. Según la opinión de algunos aficionados, está más cercano al género **arcade** o el de la acción trepidante (CVC/cnc/4684975/18)

   ("According to the opinion of some gamers, it is closer to the arcade genre or to that of action")

2. La opción **Arcade**, donde se ofrece un único combate contra otro jugador o la propia máquina, y Tournament, el torneo propiamente dicho (CVC/cnc/3156626/8)

   ("The Arcade option, which offers a single combat against another player or the machine itself, and Tournament, the tournament itself")

3. Desde hace años SNK se ha caracterizado por fabricar grandes "**arcades,**" el clásico género vuelve a la carga (CVC/cnc/5675979/37)

   ("For many years, SNK has been characterized by making great "arcades," the classical genre fights back")

4. Con los ingredientes básicos que les hicieron triunfar en el pasado han creado Metal Slug 3, un genuino **arcade** que propone acabar con todo enemigo que se cruce en el camino (CVC/cnc/5675979/32)

   ("With the basic ingredients that made them succeed in the past, they have created Metal Slug 3, a genuine arcade that proposes to kill all enemies that cross the path")

in the plural, *el género de los arcades*, as shown in Table 1.5. The metonymic relationship existing in English between the game and the place where it was played, the amusement arcade, is lost in Spanish, resulting in the formation of a false anglicism. In English, although the combination is mainly used in British English as the OED online states, it is defined as "an indoor area containing coin-operated game machines."

COMBO ("combo"): According to the uses found in our corpus, *combo* refers to a fixed combination of buttons that, when activated, produces a special movement for the characters in the game, thus allowing extra points to be obtained. This is clarified by the following example: *A los conocidos combos, combinaciones de botones que desembocan en un movimiento especial más poderoso, se añadieron los contracombos* (CVC/cnc/6113424/15) ("To the well-known combos, a combination of buttons that lead to a more powerful special movement"). This example also shows how by means of a prefix, *contra* ("against"), another neologism is created, *contracombo* ("countercombo").

In English, the word refers to a small group of musicians who play jazz, dance, or popular music; or, informally, to any combination, as can be seen in the *Merriam-Webster Online Dictionary of English*, although this use is considered to be informal and is mainly American. The word *combo* can be

TABLE 1.6 *Examples of use of* combo

1. Si utiliza las combinaciones adecuadas de teclas y ratón podrá conseguir increíbles movimientos o **combos** (CVC/cnc/3605137/2)

   ("If you use the right combinations of keys and mouse, you can get incredible moves or combos")

2. Si al terminar la pelea se han realizado suficientes movimientos espectaculares y **combos**-golpes especiales concatenados—la puntuación en estrellas será máxima (CVC/cnc/5065625 /12)

   ("If at the end of the fight there have been enough spectacular movements and special combos-concatenated strokes—the star rating will be maximum")

3. A la vez, permite consultar y practicar una y otra vez las decenas de **combos** de cada personaje, viendo en pantalla los botones pulsados (CVC/cnc/6113424 /13)

   ("At the same time, it allows you to consult and practice again and again the dozens of combos for each character while you can see the buttons pressed on the screen")

4. Se añadieron recientemente los contra-combos, cuatro movimientos que no sólo anulaban el efecto del **combo,** sino que aprovechaban la energía (CVC/cnc/6113424/16)

   ("Countercombos were recently added, four movements that not only canceled the effect of the combo, but took advantage of the power")

found in the *Diccionario de la Lengua Española*, 22nd ed., 2001, both as an adjective and a noun, but the meaning is not related to videogames. As an adjective, it is a variation of *combado*, an adjective meaning "warped," and as a noun, it refers to "*tronco o piedra grande*," that is, "a large trunk or stone." For this reason, *combo* is an example of a word coming from English, but, as the form is already existent in Spanish, it can be classified as a semantic neologism or semantic loan.

DISCO ("disk"): The analysis of the uses of this word in the CVC corpus reveals that it refers to the videogame as a product. In this regard, *disco* can be considered as an instance of a metonymy where the physical object stands for its content, as can be seen in the examples shown in Table 1.7. These examples illustrate the fact that the word can be used to refer either to the videogame as a whole, as in examples 3 or 4; or just to one aspect or component of the game as, in examples 1 and 2. In example 1, "the disc will not be dubbed," *disco* refers to the language content. In 2, *disco* denotes another part of a videogame, the training tools required to prepare for certain combat skills. However, in 3 and in 4, *disco* refers to the videogame in itself. The adjectives found in the corpus accompanying the noun *disco—favorito*

TABLE 1.7 *Examples of use of* disco

1. Pese a las buenas vibraciones del título, hay algo que no ha gustado a sus seguidores españoles, ya que por primera vez el **disco** no saldrá doblado al castellano (CVC/cnc/6409247/17)

   ("Despite the good vibrations of the title, there is something that Spanish fans didn't like, namely that for the first time the disk will not be dubbed into their language")

2. Prepararse para los combates requiere entrenamiento y observación. El **disco** posee varias herramientas para obtener esta formación (CVC/cnc/6113424/25)

   ("Preparing for combat requires training and observation. The disk has several tools to get training")

3. De esta forma, ha conseguido un **disco** que presenta decenas de pequeños retos (CVC/cnc/5428596/37)

   ("In this way, the result is a disk that presents dozens of small challenges")

4. El **disco** está enfocado como un título de acción en primera persona (CVC/cnc/6055380/21)

   ("The disk is designed as a first person action title")

("favorite") and *original* ("original")—confirm this use of the word for referring to the game. *Disco* is also found in the phrases *disco de expansión* and *disco demo*, which refer to some additional contents for a videogame and to the demonstration version of a videogame, respectively.

When comparing the use in both corpora, in the reference corpus, we found combinations such as *disco de arranque* ("boot disk"), *disco duro* ("hard disk"), and verbs like *conectar* ("connect"), *desconectar* ("disconnect"), or *compartimentar* ("partition") where *disco* is used as a storage device, revealing that the meaning in computing is different from the one in the videogame domain.

ESCENARIO ("scenery"): In the context of videogames, *escenario* refers to the virtual space in which the action of the game takes place. This meaning of a "physical" space for the action of the game is confirmed in the CVC corpus by the presence of the prepositional group *por el escenario* ("around the stage"), preceded by several verbs of action and movement such as *avanzar* ("move forward"), *caminar* ("walk"), or *desplazarse* ("move about") *por el escenario*. In the electronic game, the player moves the characters along the *escenario* so that they can accomplish their mission. The physical nature of this virtual setting is also reflected in the two most frequent combinations in CVC: *escenarios tridimensionales* ("three-dimensional scenery") and *escenarios en dos dimensiones* ("two-dimension scenery"). For this reason, its meaning seems to be related to its use in the theater ("stage"), where it refers to the place where the actors perform. But as the plural forms "*escenarios*" in the two examples show, in the videogame, there is more than one "stage." As in a film or in a theater play, *escenario* is the sequence of scenes one after the other (see examples 2 and 3).

Escenario also refers to the scenery or background, that is, the setting for the action taking place as in a theater play or a film. However, there is an important difference; in the electronic game, that visual environment where the player acts is a virtual, simulated environment. It is interesting to note that in CRC, the adjective *virtual* ("virtual") appears in combination with *escenario*, while in CVC, that adjective is never present since the *escenario* is always something virtual. Another important quality of *escenario* is its active role in providing the player and the characters with the necessary elements to continue with the game, as seen in Table 1.8, example 1. Therefore, in the domain of videogames, the *escenario* is a game component, which is controlled by the graphics engine and can be downloaded, as shown in Table 1.8, example 1. In Spanish, *escenario* covers the meaning of several English words: "stage," "scenery," "setting," "environment," and even "scenario."

EXPANSIÓN ("expansion"): In videogames, *expansión* acquires a new meaning different from the one in general language and also different from

TABLE 1.8 *Examples of use of* escenario

1. Ya en primera persona y pulsando el gatillo izquierdo se escanea el **escenario** y los Smiles se sintetizan ante el personaje … (CVC/5962846/254)

   (*"Already in first person and pressing the left trigger the background is scanned and the Smiles are synthesized before the character …"*)

2. A medida que el jugador avance por los distintos **escenarios**, irá adquiriendo distintas armas que irán completando su inventario (CVC/cnc/2770505/52)

   (*"As the player advances through the different backgrounds, he will acquire different weapons that will complete his inventory"*)

3. El juego se desarrolla en **escenarios** 3-D por los que el jugador, desde un punto de vista de primera persona, podrá moverse a su antojo (CVC/4307072/283)

   (*"The game is developed in 3D backgrounds, around which the player can move at will from a first-person point of view"*)

4. Deep Silver publica una aventura gráfica al más puro estilo tradicional con buen guión, buenos personajes y **escenarios** en dos dimensiones (CVC/6773747/349)

   (*"Deep Silver has published a graphical adventure in the purest traditional style with a good script, good characters and 2D backgrounds"*) the meaning in the domain of computing. According to the online edition of *Diccionario de la Lengua Española, expansión* refers to the action or the effect of expanding, which is the meaning found in the combinations in CRC: *expansión del universo, fase de expansión, grado de expansión,* or *plena expansión.* The combinations *ranura de expansión* and *tarjeta de expansión* show that, in computing, it is a physical object forming part of the hardware. However, in the domain of videogames, *expansión* is also a software product that provides the user with additional content for the original game, as shown in Table 1.9, examples 1 and 3. There is an interesting semantic difference between these two uses: the effect of expanding, and the object, both physical and digital. *Expansión* as an uncountable noun indicating an action turns into a countable noun within the specialized domain of computing, and then changes its semantic nature again, in the field of videogames.

The verbal actions associated with *expansión* in CVC, such as *comprar* ("buy"), *editar* ("edit"), *incluir* ("include"), *ofrecer* ("offer"), *presentar* ("introduce"), and *probar* ("try") confirm its physical nature as a product. This is especially obvious when compared with those verbs found in CRC related to the action of expanding rather than its result: *frenar* ("stop"), *impulsar* ("promote"), *incrementar* ("increase"), *perjudicar* ("harm"), and *preparar* ("prepare"). This difference between action and result can be further seen in the lexical combinations of *expansión* in the two

TABLE 1.9 *Examples of use of* expansión

1. Los chicos de Ensemble Studios presentan la **expansión** de Age of Mythology con el subtítulo de The Titans, con una cultura completamente nueva (CVC/cnc/5336979/17)

   ("The guys from Ensemble Studios present the expansion of Age of Mythology with the subtitle of The Titans, with a completely new culture")

2. Pero, a pesar de las mejoras, hay un factor, el de las partidas en línea, que no se ha incluido y que previsiblemente aparecerá en un disco de **expansión** (CVC/cnc/2020789/29)

   ("But, in spite of the improvements, there is an element, the online games, that has not been included and that will foreseeably appear in an expansion disk")

3. La **expansión** presenta nuevos enemigos, gráficos más trabajados y con un mayor número de texturas (CVC/cnc/2950546/39)

   ("The expansion presents new enemies, elaborate graphics and more textures")

4. La **expansión** incluye 15 nuevas misiones en las que el jugador deberá intentar vencer a los romanos (CVC/cnc/3288818/44)

   ("The expansion includes 15 new missions for the player to try and beat the Romans")

corpora. In CRC, *expansión del universo* ("universe expansion"), *fase de expansión* ("expansion phase"), *grado de expansión* ("expansion degree"), and *plena expansión* ("full expansion") signal a process of development or amplification. Conversely, lexical combinations in CVC stress both the material quality of the software product as in *disco de expansión* ("expansion disk") and the fact that this additional content is manufactured by the original developer, in *expansión oficial* ("official expansion").

JEFE ("boss"): *Jefe* is a semantic neologism introducing new connotations to the original meaning of the word in Spanish. In English, there are two words for translating the Spanish one, "chief" and "boss," which are almost synonyms in their meaning of "a person in charge of or employing others" (CCDEL 1987). However, according to the definition in the same dictionary, "boss" is used in American English with a negative connotation and is defined as "(mainly US) a professional politician who controls a party machine or political organization, often using devious or illegal methods." Example 2 in Table 1.10 illustrates this negative connotation in the use of *jefe* in the videogame domain, as *jefes* are the worst category of enemies. This negative meaning is confirmed by the verbs the word combines with in Table 1.10, examples 1 and 4, *luchar con un jefe* ("fight") and *enfrentarse a un jefe* ("confront") and the fact that the only adjective accompanying the word in CVC is *mafioso*. Thomas et al. (2007: 16) verify this negative meaning for the term *boss*, "A notable enemy, usually one possessing much greater power than other foes in the game."

TABLE 1.10 *Examples of use of* jefe

---

1. Tras cada episodio tocará enfrentarse con un **jefe** de final de fase contra el que tendrá más poder la inteligencia que la fuerza bruta (CVC/cnc/4667716/2)

   ("After each episode, the player will have to face a final boss against whom intelligence will be more powerful than brute strength")

2. En cuanto a los malos, serán de todo tipo, pero los peores serán los **jefes** a cargo de cada una de las misiones (CVC/cnc/3867538/8)

   ("As for villains, there are all kinds of them, but the worst will be the bosses in charge of each of the missions")

3. Spawn, del cómic de igual nombre, el de Xbox e Heihachi, **jefe** entre los jefes de los juegos de lucha Tekken (CVC/cnc/5206989/12)

   ("Spawn, from the comic book of the same name, Xbox and Heihachi, boss among the bosses of the Tekken fighting games")

4. Solamente en contadas ocasiones, cuando se lucha con los **jefes** de final de nivel se tiene una sensación de tres dimensiones (CVC/cnc/6074656/21)

   ("Only rarely do we have a three dimensional feeling when we fight with final bosses")

TABLE 1.11 *Examples of use of* jugabilidad

1. Parasite Eve 2 es un juego que roza el sobresaliente en todos los apartados, gráficos, musicales, sonoros y de **jugabilidad** (CVC/cnc/3030982/3)

   (*"Parasite Eve 2 is a game almost outstanding in all sections, graphics, music, sound and playability"*)

2. Pero pese a ofrecer la misma **jugabilidad** que el Wipeout original, éste Fusion no logra batir el impacto que supuso el origen de la saga (CVC/cnc/4296399/5)

   (*"But despite offering the same playability as the original Wipeout, Fusion fails to beat the impact that was the origin of the saga"*)

3. Esta rigidez persigue conseguir una fractura artística en cada fotograma, aunque vaya en detrimento de la **jugabilidad** (CVC/cnc/5962846/8)

   (*"This rigidity seeks to achieve an artistic fracture in each frame, although it is detrimental of playability"*)

4. Pese a ello, la **jugabilidad,** o lo que es lo mismo, la capacidad de entretener, se mantiene en cotas muy elevadas (CVC/cnc/3086491/4)

   (*"Despite this, playability, or the ability to entertain, remains at very high levels"*)

JUGABILIDAD ("playability"): Due to its lower frequency in CVC, *jugabilidad* was not included in the initial keyword list although it is a concept of special importance, and, for that reason, it was selected for this chapter. The meaning of *jugabilidad* shows two important features of a game. On the one hand, it refers to the quality of being playable, that is, a trait of the game, such as in Table 1.11, examples 1, 2 and 3. On the other hand, it refers to the pleasure the game offers to the player. In example 4, Table 1.11, the author of the review explains its meaning as the "ability to entertain"; *jugabilidad* implies that a videogame provides fun and entertainment. The fact that *jugabilidad* is a desired quality is revealed in CVC because it appears together with other nouns also expressing assessment criteria, such as originality, music, or graphics, as shown in Table 1.11, example 2.

Recently the word has been added to the *Diccionario de la Lengua Española*, where it is defined as *facilidad de uso que un juego, especialmente un videojuego, ofrece a sus usuarios* ("Ease of use that a game, especially a videogame, offers to its users"), signaling the relevance of this lexical unit.

5 Conclusions

This chapter has showcased a specific lexical approach to videogame discourse by presenting the results of a corpus-driven analysis of videogame reviews in a Spanish journal in order to discover the presence of specialized vocabulary. The purpose was to uncover the features of those Spanish words specific to videogames. One significant finding is that many videogame words are semantic neologisms, that is, words already used in Spanish that acquire a new meaning in the domain because of the influence of an English word. The cases of words coming directly from English to refer to a new concept are a minority in this corpus. One of those cases is *arcade*. The analysis of its uses in the corpus reveals it is also a case of polysemy within the domain, as it can be used to refer both to a genre and to a simplified way of playing some action games. The fact that it is an English word that has not yet been completely assimilated into the Spanish language is confirmed by the formal variation of its written form in the corpus, where both *Arcade* (with a capital letter) and *"arcade"* (between inverted commas) are present. *Disco* and *expansión* confirm that in the videogame domain, some words referring to technology develop a different meaning from the one present in the domain of computing. *Escenario* stands out for its complexity, as the Spanish word has a broader range of use and includes the meaning and uses of several English words: "stage," "scenery," "background," "setting," "environment," and even "scenario." A very interesting case is that of *jefe*, which is represented as meaning an enemy. Both *jefe* and *combo* reveal the influence of English usage on Spanish word formation and semantic shift, but according to the dictionaries consulted, it is mainly American English. *Jugabilidad*, finally, is an excellent example of a lexical unit specific of the videogame domain, which, because of its relevance, is now included in the *Diccionario de la Lengua Española*.

The results presented in this study of a corpus of Spanish contemporary press are highly dependent on its specific media context, a general broadsheet newspaper, *El PAÍS*, and the written videogame review in a specific time span. It will be interesting to carry out new analyses in different contexts and in different time spans to verify if the new meanings were used in previous sources or if they are still being used, and if so, if they might be included in a dictionary of the Spanish language.

The lexis of videogames as analyzed in this chapter contributes to the enrichment of standard Spanish, as many common words develop new uses in the domain. The domain of videogames can be classified as a specialized language, but it is really an intermediate domain between specialized knowledge and general knowledge as it involves different users with different levels of expertise.

Note

1 Videogames were declared a "cultural industry" by the Spanish Parliament in 2009 (*Cortes Generales. Diario de sesiones. Diario de Sesiones del Congreso de los Diputados. Comisión de Cultura. IX Legislatura* [March 25, 2009], 235: 2).

References

Aguado de Cea, G. (1995), *Diccionario comentado de terminología informática*, Madrid: Paraninfo.

Algeo, J. (1991), *Fifty Years among the New Words: A Dictionary of Neologisms*, New York: Cambridge University Press.

Álvarez de Mon, I. (2006), *Lenguaje y Comunicación Científica y Técnica. Los nuevos caminos de la lingüística aplicada en el siglo XXI*, Madrid: Acta.

Álvarez de Mon, I. and C. Álvarez-Bolado (2013), "Neology in the domain of videogames," *Ibérica*, 25: 61–82.

Bergenholtz, H. and S. Tarp (2010), "LSP lexicography or terminography? The lexicographer's point of view," in P. Fuertes-Olivera (ed.), *Specialized Dictionaries for Learners*, 27–37, Berlin/New York: De Gruyter.

Cabré, T. (1993), *La Terminología: Teoría, métodos, aplicaciones*, Barcelona: Antártida.

Cortés, E. and J. Hungría (2012), *Herramienta de Creación y Mantenimiento de Diccionario Electrónico Especializado en Videojuegos (HDE)*, Madrid: UPM.

Coseriu, E. (1973), *Teoría del lenguaje y lingüística general*, Madrid: Gredos.

Firth, J.R. ([1951] 1957), *Papers in Linguistics, 1934–1951*, Oxford: Oxford University Press.

Francis, J.G. (1993), "A corpus-driven approach to grammar: Principles, methods and examples," in M. Baker, G. Francis, E. Tognini-Bonelli, and J. Sinclair (eds.), *Text and Technology: In Honour of John Sinclair*, 137–56, Amsterdam: John Benjamins.

Furiassi, C. (2010), *False Anglicisms in Italian*, Monza: Polimétrica.

Görlach, M., ed. (2001), *A Dictionary of European Anglicisms*, Oxford: Oxford University Press.

Guerrero Ramos, G. (1995), *Neologismos en el español actual*, Madrid: Arco Libros.

Halliday, M.A.K. (1978), *Language as Social Semiotic: The Social Interpretation of Language and Meaning*, London: Edward Arnold.

L'Homme, M.C. (2009), "A methodology for describing collocations in a specialised dictionary," in S. Nielsen and S. Tarp (eds.), *Lexicography in the 21st Century*, Amsterdam: John Benjamins.

Miller, G.A. (1999), "On knowing a word," *Annual Review of Psychology*, 50: 1–19.

Onysko, A. (2007), *Anglicisms in German: Borrowing, Lexical Productivity and Written Codeswitching*, Berlin: Walter de Gruyter.

Rey, A. (2005), "The concepts of neologism and the evolution of terminologies in individual languages," *Terminology*, 11 (2): 311–31.

Scott, M. (2004), *Wordsmith Tools version 4.0*, Oxford: Oxford University Press

Sperber, D. and D. Wilson (1986), *Relevance: Communication and Cognition*, Oxford: Blackwell.

Van Dijk, T.A. (2009), "Discourse, context and cognition," *Discourse Studies*, 8 (1): 159–77, London, Thousand Oaks, CA and New Delhi: SAGE Publications.

Dictionaries

Hanks, P., (ed). (1987), *The Collins English Dictionary*, London and Glasgow: William Collins Sons & Co. Ltd.

Merriam-Webster Online Dictionary of English, 11th edn. Available online: https://www.merriam-webster.com (accessed June 26, 2018).

Oxford English Dictionary Online, 3rd edn. Available online: http://www.oed.com/ (accessed June 26, 2018).

Real Academia Española (2001), *Diccionario de la Lengua Española*, 22nd edn. Available online: https://dle.rae.es (accessed June 26, 2018).

Thomas, D., et al. (2007), *The Videogame Style Guide and Reference Manual*, *Power Play Publishing*. Available online: http://www.gamestyleguide.com/Video GameStyleGuideeBook.pdf (accessed June 26, 2018).

2

Lexical and Morphological Devices in Gamer Language in Fora

Isabel Balteiro

1 Introduction

Today's technology is developing very rapidly, not only providing people with new means of communication but also creating new forms of enjoyment, and with them new groups that need to share information, feelings, and so on about their common interests. This chapter deals with the intersection between these three components, and particularly the use of online fora by gamers to talk about videogames and gaming in general.

Videogames and gaming are, according to Jones (2008: 1), one of "the most influential form[s] of popular expression and entertainment in today's broader culture," which no longer occur in isolation but in groups: gamers interact, creating new (sub)groups or communities, which depend largely on the medium where the communicative acts occur. In this respect, as Ensslin (2012: 70) acknowledges, "[m]uch of the communication that takes place between gamers, game developers, journalists and other stakeholders happens online, in chat channels, on blogs and discussion fora." Accordingly, gamers construct their own identities and language by negotiating meanings and making lexical choices which are highly conditioned by the communicative channel—that is, people do not use or select the same words or do not use them in the same way in face-to-face interactions and in online ones, for example, *oh wait* (see Balteiro 2018b). Thus, gamer language is affected by online communication, in that this "require[s] users to engage in quick forms of turn-taking" which makes "words and phrases [...] be shortened" (Ensslin 2012: 70). Apart from that, the vocabulary used is also affected by general gamer jargon, slang, and techspeak (see Michael 2007). Although it may be considered as a combination of these three, I argue that this vocabulary may also present special and distinctive features, which places the lexis in gamer fora between jargon (for the technical words used) and slang (because of its informality and its colloquial and relaxed features).

Videogames and/or gaming have been generally associated with subcultural movements and, hence, linked to negative effects on people, especially younger audiences. This may partly explain why they have been disregarded for some time by academic studies. Nowadays, however, being aware of their impact and importance as an integral part of our society, researchers from different disciplines (psychology, education, language teaching/learning, discourse analysis, media and cultural studies, intercultural communication, sociolinguistics, etc.) are devoting careful attention to them (see, for example, Bobosh 2006; Keats 2011; Crystal 2001; Portnow 2011). Yet, not much consideration has been paid to gamer lexis, terminology, and/ or lexical creativity, especially in computer-mediated interactions among gamers when talking about videogames (see Ensslin 2012); or to the language of online fora in general (see Balteiro 2018) and also as regards gamer uses. This study attempts to fill this gap in the literature concerning language creativity, lexical choices and word-formation mechanisms and features used by gamers when they "talk" about videogames, videogame-related issues (including feelings, reactions, etc.), and gaming in online specialized fora.

2 Internet fora and techspeak

The emergence of the internet and its variety of situations or environments (see Crystal 2001), including fora, have contributed to the creation of new social and linguistic groups, which have largely influenced and changed the way people communicate. Consequently, language and languages in general have also changed and developed new, specific, dynamic, and flexible genres and registers, as is the case of internet English as well as the English used in other technology-related forms of communication. Online or internet forums/fora (also called *discussion groups*), in particular, may be defined as in-group websites that allow people, mostly registered members, to either synchronically or asynchronically talk about or discuss common specific topics or themes in the form of posts, which are organized by subject matter and thread, and which are, at least temporarily, archived.

Although synchronous online forums/fora and/or chatrooms have been largely analyzed in other disciplines (see, for example, Turkle 1995, 1999; Preece 2000; Miller and Slater 2000; Flinkfeldt 2011) or even in linguistics (see Reed 2001; Ooi 2008; Balteiro 2018), it seems that asynchronous fora (those that Crystal 2001: 22 calls "bulletin boards," where the discussion does not necessarily take place in real time) are still underexplored and underrepresented in the research literature in linguistics and, especially, in lexicology. For this reason, this study will contribute toward redressing this imbalance by analyzing how gamers make and adapt their lexical and morphological choices to their interactions, as well as to the topics addressed and, probably most importantly, to fora constraints and allowances.

The earliest studies on internet communication focused on whether online language was more like spoken or written discourse, or whether it was a completely new type of language. For example, Ferrara, Brunner, and Whittemore (1991) held that online written discourse comprises a new register of language, distinct from both spoken and written language (see also Baron 1998, 2003; Herring 1996, 2001; Herring et al. 2004, 2005). While Crystal (2001: 18) also argued that internet language or "Netspeak" is to be defined as "a type of language displaying features that are unique to the Internet" which has little in common with speech (see Crystal 2001: 41), other scholars have remarked that it has "strong oral qualities" (see, for example, Benwell and Stokoe 2006: 255; McDaniel et al. 1996). Nowadays, it seems, however, that academics agree on the fact that the language used in the internet, and hence forum language, is actually a combination of written and spoken language. Even though the medium is primarily written, the language used shares many characteristics of oral discourse, unlike written language, which tends to be more formal, elaborated, meticulous, precise, and explicit (see also Chafe and Tannen 1987; Halliday 1989). Furthermore, the label "Netspeak" given to online communications also quite transparently and clearly identifies online language with oral language, despite Crystal's opinion. In fact, online interaction, as in chats and synchronous fora, demands not only immediate responses, as messages or posts may be "lost" as the screen scrolls down, but also short, quick, and dynamic responses as well as proactive, emphatic, and efficient ones, as would be the case in face-to-face interactions.

Asynchronous fora (those analyzed in the present study), however, are closer to written language than synchronous online or technology-mediated interactions because they are less transient, they allow editing, and they give readers more time to read and respond to them. Hence, they are also more precise and meticulous. This does not mean that fora participants do not try to compensate for keyboard constraints and for the lack of linear progression, speed, and face-to-face interactions. As a result of all this, I argue that gamer language in asynchronous fora exhibits a mixture of formal and informal words, combining features of both written and spoken language.

Furthermore, apart from the use of special(ized) terms given gamers' topics, activities, feelings and so on, participants in online fora, as in any other community or sociolinguistic group, develop feelings of shared identity (see, for example, Collier and Thomas 1988; Preece 2000; Byram

2006; Wach 2014: 192), which is necessarily reflected in the language used. Accordingly, this determines and defines common, shared, and distinctive microlinguistic choices, innovations and even deviations, as Crystal (2001: 92) suggests, that make outsiders feel excluded from the group and their language and rather like uninitiated "newbies."

In general, it may be argued that online interactions allow participants, in this case gamers, to construct their own dynamic and evolving language more freely and not strictly in adherence with rules. Within these special codes, I shall focus on lexical and terminological choices as well as morphological and semantic innovations that gamers participating in fora use acquire and which are a sign of their belonging to a given social group or community.

3 Gamer language: Slang or jargon?

Gamers and videogames have often been regarded as related to marginal people and contexts, but more recently, they have come to be increasingly associated with young people (see, for example, Osgerby 2004) and, in general, with negative stereotypes (see, for example, Parkin 2013; Leigh 2014; Kowert 2014). For these reasons, especially, because of its association with marginal and young people, the language used by gamers has often and generally been assumed to be slang rather than *jargon*, a term usually restricted to more technical and formal discourse.

The *Oxford English Dictionary* (OED 2009; see slang n[3], 1c) defines *slang* as "[l]anguage of a highly colloquial type, considered as below the level of standard educated speech, and consisting either of new words or of current words employed in some special sense," which seems a rather old-fashioned or biased definition. In my opinion, slang certainly differs from standard language and is more colloquial and, therefore, inappropriate for formal contexts and hence prototypically found in rather informal ones, but this does not necessarily mean that it is "inferior." After all, even educated people may use slang, as shall be discussed here. As Mattiello (2008: 36) puts it, "slang differs from jargon in its lack of prestige and pretentiousness ... [and it] is much more familiar and spontaneous." Lighter (2001: 222) asserts that slang appears in "same-sex groups composed of peers of comparable age and social status,"[1] and this determines who belongs to a particular community and who does not. On the other hand, *jargon* may be defined as "the specialized vocabulary and phraseology of a set of people sharing a trade or profession" (Mattiello 2008: 36). However, in my view, the distinction is rather more complicated, as both types of language—slang and jargon—may appear, exist and be used by the same limited group or community. Medical nurses, for example, may refer to "a hypochondriac" or "repetitive strain injury (RSI)" in jargon, and the same condition may be called "a frequent flyer" or "Nintendo thumb or gamer's grip," respectively, in their slang. Thus, a given limited group may create slang when they speak or write informally and subjectively, even in technical or professional contexts where jargon is assumed.

Gamers use a highly specialized style of discourse, especially as regards vocabulary, terms[2] and/or lexical chunks. As suggested and as Ensslin (2012: 33) claims, "[gamers] need to learn highly specialised in-group codes, or sociolects, insofar as the language they use face-to-face with their gaming buddies is concerned," and this "tends to be even more specialised and paralinguistically and multimodally orchestrated than familiolects." Gamers are not only literate in games metadiscourse, which allows them to understand, use and refer to in-game phenomena, but they also have their in-group language, which they use in computer-mediated communication and which makes them part of a larger group of gamers. Thus, Ensslin (2012: 67) argues for the existence of a spectrum which ranges from highly specialized to highly accessible game-related discourses in such a way that "[g]eneral types of gaming-related discourse are enriched with specific lexis that taps into expert jargon and gamer slang for specific social and semiotic purposes." Similarly, I would contend that the language used by gamers in fora ranges from game-technical language and terminology ("techspeak," see McGrath 1998) or gamer jargon to more relaxed styles or games-related slang(s), such as those used in online communication and interactions. As a rule, gamers' lexical choices in computer-mediated communication may not differ much from those used in their discourse(s) about games and gaming in general. However, although I acknowledge this and the existence of "a diversity of ludolects, some of which are closer to expert jargon, and others closer to mainstream varieties, and [which] also differ according to which game and game genre is being referred to" (Ensslin 2012: 66), I maintain that, inevitably, gamers use slang words and expressions similar to oral speech ones in online communication, as this channel, to some extent, imposes or rather favors more relaxed forms of language.

The language and lexis of gamers in fora, whose description constitutes the main objective of this study, is expected to combine highly technical language proper to gamer jargon or techspeak with more colloquial language proper to computer-mediated communication. Accordingly, I hypothesize that not only short and colloquial forms of words, typically found in online communication, especially fora, but also technical and game-specific terms will be found. Furthermore, new and striking lexical, morphological and semantic creations may also be encountered as well as highly playful and emotional terms and emoticons (e.g., expressing anger or laughter), given the medium (fora) where gamers deliver their messages or posts. All of these— technical elements and more colloquial ones, new and striking but also old and well-known forms, standard and slang words, emoticons and so on— will also serve gamers to identify themselves and to mark their belonging to the gamer community and to keep nonmembers out. In other words, as already suggested and as will be argued, I expect the object of study—the lexis or vocabulary used by gamers in for a—to be mainly a combination of jargon and slang, although other standard and more general colloquial words may also be found.

4 The Study

4.1 Objectives and methodology

The study that follows is based on the analysis of a sample of over 130,000 words taken from around 90 threads of the popular gamer forum named "NeoGaf" (https://www.neogaf.com/). As of the time of writing, this forum has a total of 843,646 threads, 128,042,011 messages, and 126,458 members (see https://www.neogaf.com/; last accessed April 6, 2018)

The general aim of this study is to account for gamer language in fora; more particularly, by adopting a microlinguistic approach, the focus is on those word-formation mechanisms that seem to be most productive, or at least frequent, in gamers' interactions in fora. In addition, this chapter (1) focuses on how gamers make and adapt their lexical and morphological choices to their interactions, to the topics addressed and to fora constraints and allowances, on the one hand, as well as how all these condition and determine gamers' selections and uses, on the other; furthermore, it (2) also attends to the degree of technicality, colloquial or slang character of the elements used. Finally, considering these two main objectives, (3) it attempts to conclude on the proximity of gamers' language to written or spoken discourses.

The sample was compiled by gathering gamers' posts in an approximately two-month period—April and May 2017—from NeoGaf forum discussions. Not all the posts in a given thread were gathered, as I wished to explore as many threads as possible to ensure diversity in topics and, hence, in lexical choices. Gamers who participate in NeoGaf discuss different topics related to videogames and gaming, such as technical issues, news, strategies for playing videogames and so on. As the NeoGaf webpage reads, it deals with "[v]ideo game news, industry analysis, sales figures, deals, impressions, reviews, and discussions of everything in the medium, covering all platforms, genres, and territories." Apart from being open to any person in the world, this forum has three different membership types (depending on the time and number of participations, they may be "neophytes," "neo members," and "members"); however, for the purposes of the present study, these variables are not considered as factors affecting gamers' language use.

Once the data were compiled, they were analyzed, considering the main and specific objectives mentioned above. Many of the data repeated themselves, which helped me to discover specific linguistic (lexical and morphological) features and patterns that gamers use in their posts, as seen in the following section, where a qualitative discussion of the gamers' lexical resources in fora is provided.

4.2 Analysis and discussion of results

As explained, a sample of over 100,000 words from gamers' posts was analyzed for the purposes of this chapter. From those, 5,802 words were analyzed in more detail; a general account with selected examples of such an analysis is provided in this section, which, due to spatial limitations, cannot be as exhaustive as might be desirable.

I have identified not only new formations abbreviating or clipping old words but also those adding affixes or two bases together, alongside completely new words and semantic shifts, to account for new videogame products as well as for the developments in gaming and the technology involved. In general, acronyms and initialisms seem to be by far the most productive or at least more frequent word-formation mechanisms found in gamers' interactions in fora, followed by compounds, clippings, conversions and probably derivations. Blends, borrowings and semantic shifts are less numerous, while backformations and clipped compounds are almost nonexistent. Purely new creations, that is, words that do not come from any previous or existing word, like *Wii* (mentioned by Ensslin 2012: 70), have not been found.

Among acronyms and initialisms, that is, shortened or contracted forms of words or phrases created by using their initials, I found a variety of combinations as regards the inclusion of capital and lower-case letters, their combinations, numbers and so on. Examples of these words or terms refer to:

(1)   technical videogames issues, genres and so on(some common to computer science or photography) such as *fps* (frames per second), *HDR* (high dynamic range), *4k* (a collective term for digital video formats having a horizontal resolution of approximately 4,000 pixels), *DL* (download), *UI* (user interface), *OS* (operating system), *VR* (virtual reality), *VPN* (virtual private network), *SLI* (scalable link interface), *SSD* (solid state disk), *BIOS* (basic input output system), *HD* (high definition), *ND* (neutral density), *SKU* (stock keeping unit), *MP* (mega pixel), *SDK* (software development kit), *KPI* (a key performance indicator for online games), *PvP* (player vs. player), *PSVR* (PlayStation virtual reality), *RPG* (role-playing game), *Gif* (graphics interchange format), *RNG* (random number generator), and *MMO* (massive multiplayer online);

(2)   gaming, for example, *4v4* (four against, or versus, four), *NDA*
      (not down anymore), *H2K* (hard to kill), *SL* (*Second Life*), *ETA*
      (estimated time of arrival), *SFF* (small form factor), *OG* (operation
      group), *MAU* (monthly active users), *GwG* (game winning goal),
      FXAA (fast approximate anti-aliasing, an anti-aliasing algorithm
      created by Timothy Lottes under NVIDIA), *F2P* (free to play);

(3)   names of videogames, consoles or games-related companies/
      networks/fairs and so on, for example, *CTF* (*Capture the Flag*), *FW*
      (*Forsaken World*), *WoW* (*World Of Warcraft*), *MK8* (a racing game,
      Mario Kart number 8), *NES* (Nintendo Entertainment System), *IGN*
      (Internet Gaming Network), *E3* ([The] Electronic Entertainment
      Expo, commonly referred to as E3, an annual trade fair for gaming
      fans all around the world), *Konami* (acronym from the surnames
      Kozuki, Nakama, Matsuda, and Ishihara, the name of a videogame
      company), *PSN* (PlayStation Network), *N64* (Nintendo's third home
      videogame console for the international market), *TW3 / W3* (*The
      Witcher III*, game), *XB1* (Xbox One), *Civ6* (*Civilization VI*), *CCG*
      (collective card game), *TCG* (trading card game), *GOG* (good old
      games), *PAD* (puzzle and dragons), *ToS* (this acronym may refer to
      either *Town of Salem* or *Tree of Saviour*), *TGS* (*Tokyo Game Show*),
      TLOU (*The Last Of Us*), *H1Z1* (Humans 1 Zombies 1, A Zombies'
      virus), *MEA* (*Mass Effect Andromeda*, an action role-playing third-
      person shooter videogame), *DAI* (*Dragon Age Inquisition*), *KOTOR*
      (*Kings of the Old Republic*, a role-playing videogame set in the
      Star Wars universe), *Sega* (Service Games, a Japanese multinational
      videogame developer and publisher).

Another important but less frequent word-formation mechanism that shortens
words is clipping or, rather, words created by dropping one or more syllables
from a polysyllabic word (the eliminated syllable may be at the beginning
or at the end of the existing form). Gamers have manipulated some words
related to videogames, actions in them and gaming to obtain, among others,
the following: *ammo* (ammunition), *coop* (a game mode for playing with
other players), *dev* (developer), *demo* (demonstration, a trial of a videogame
before it is realized the full version), *improv* (improvisation), *mid* (among or
in the middle of), *hex* (hexadecimal), *sims* (simulations), *sync* (synchronize),
ppl (people), *dmg* (damage), *vid* (videogame), *rezzed* (resurrected, something
that was made to appear on the interface). Close to these and to compounds
are the so-called clipped compounds, that is, compounds whose shape
has been shortened, as in *joy-cons* (a kind of controllers made up by
Nintendo), *QooApp* (Qoo + App [clipping of application], an alternative
"market" of Asiatic videogames), *Ubisoft* (Union des Bretons Independants
[Ubi] + Software, a French multinational videogame publisher), *mobo*
(motherboard) and *comic-con* (comic convention).

Similarly to the preceding, some backformations, that is, the process of creating a new word by removing affix-like morphs, have been identified in my sample. However, only one of them seems worth mentioning for its actual relation to videogaming: *nitpick* (from *nitpicker*: of, relating to, or characteristic of a nitpicker or nitpicking).

Gamers in the fora selected for the present chapter have also made extensive use of *compounding*. This mechanism, which creates new words by combining and/or joining two bases or two words (sometimes even three), is apparently the second in importance as regards frequency and productivity. Some of the examples found in my sample are directly and exclusively used in videogaming, such as *videogame, gameboy, gameplay* (the features of a videogame, such as its plot and the way it is played, as distinct from the graphics and sound effects), *lootbox* (a box found in games that one can loot to gather objects), *Roguelike* (of a computer game sharing certain characteristics similar to the original game *Rogue*, especially involving a single character in a large dungeon/tower, with a key focus on inventory management and stat-building), *framerate* (the number of frames or images that are projected or displayed per second), *playthrough* (the act of playing a game from start to finish), *gamepad* (a handheld controller for videogames), *horde mode* (a typical feature from *World of Warcraft*, it refers to a great number of players fighting), *motherboard* (the mainboard of a computer), *downsample/downsampling* (to make a digital audio signal smaller by lowering its sampling rate or sample size [bits per sample], the process of reducing the sampling rate of a signal), *firmware* (type of software that provides control), *savescumming* (copying your saved game file to another directory to circumvent auto-deletion, so that you are required to start over every time you die), *bug out* (to be faulty and have problems in it), *blowback* (the backwards escape of unexploded gunpowder when firing a handgun), *cutscene* (a noninteractive scene that develops the storyline and is often shown on completion of a certain level, or when the player's character dies), *fanboy* (a person who is completely loyal to a particular game or company), *moveset* (the group of all attacks which can be used by a particular character or thing in a fighting game; these are generally considered to be masterpieces and/or works of art by those who create them, and utter gibberish by those who do not), and *Rocksteady* (British videogame developer based in Kentish Town, London). Besides these, many other compounds have also been detected, but they are not exclusively used in gamers' language, for example, *benchmark, godsend* (when a situation is tense and unexpectantly something or someone arrives that completely eases the situation), *upsell* (the "art" of tacking on high-priced options to an existing sale under the guise of customer service), *mockup* (a scale or full-size model of a design or device, used for teaching, demonstration, evaluating a design, promotion, and other purposes), *bottleneck, signups, barebones, outdated, rollback, out-lashes, low-keyed, one-hit, ward-off, showdowns*, *break-aways*, *backup*, *flashflight*, *forthcoming*, *upgrade*, or *manhunter*, among others.

Similar to compounds are *blends*, which are the result of combining or joining two parts or splinters of two separate words (rather than two words, as in compounds) to form a new one. However, the number of blends in my data is surprisingly scarce (cf., for example, Bryant 1976; Cannon 1985; Balteiro 2017) and indeed almost nonexistent. Accordingly, only three examples deserve detailed attention (*bullshot*, *Castlevania*, *Comicon*), one of them, *Comicon*, being a doubtful case which may even be regarded as a misspelling of a clipped compound, *Comic-con*, as mentioned here. *Bullshot*, a blend of *bullshit* and *screen shot*, refers to any videogame screen shot that is modified by a publisher to generate hype for its product, while *Castlevania*, a blend of *castle* and *Transylvania*, is a videogame series created by Konami where players battle undead creatures and demons and fight Count Dracula, who lives in Castlevania.

Derivation, or the process of creating new words by adding a prefix or a suffix to an already existing word, also provides a good number of lexical resources that gamers have created and/or use in fora to talk about their interests. Both prefixations and suffixations have been identified in the sample; however, some of the resulting forms are not new but already existed in English. Apart from other examples, some of which will be mentioned here, I would like to highlight the use of hyphens marking either the prefix or the suffix, as in *in-universe* (this refers to a perspective or view from the context of a fictional world, in contrast to a perspective from the real world), *de-list* (to take the IHS [IBM HTTP Server] off of the central processing unit [CPU]), *post-launch* (the action after the game is run), *Google-fu* (the ability to quickly answer any given question using internet resources, *-fu* being a new suffix), *invite-less*, or *e-sports-wise* (a special form because it presents prefixation *e-* and suffixation *-wise*). Other examples of videogame-related words that have been created by adding a prefix are *supernatural*, *revamp* (redo, change, update or upgrade); *uncracked*, *multiplay*, *remaster*, *debuff* (to debilitate a character in a videogame); *overclocking* (the configuration of computer hardware components to operate faster than certified by the original manufacturer); *outperform* (to do well in a particular job or activity compared to others of a similar type); *superhappy*, *replay*, *premade* (a pre-arranged group of people who will do an activity together); and *multiplayer* (a videogame designed for or involving several players). Examples of the use of suffixes include *doable* (word used to describe the possibility of an action taking place in the near future); *beefy* (something that's really nice); *cheapo* (cheap quality; the use of the suffix *-o* is usually associated with slang); *firmware* (a set of instructions that form part of an electronic device and allow it to communicate with a computer or with other electronic devices); *modders* (those who create modifications or "mods" to videogames); *bummer* (something that has happened that is bad or unpleasant); *platformers* (informal for platform games); *gaffer*; *standardize*; and *customization*.

Conversions, functional shifts or *zero-derivations*, that is, the derivation of a new lexeme from an existing one without a specific morphological marker indicating the change of word class and meaning, affect the two major categories in English: nouns and verbs. However, they are not very prolific if compared to abbreviations, compounds and derivations. Among the conversions from noun to verb, *mirror* (in relation to videogaming, to keep a copy of the contents of [a network site] at another site, typically in order to improve accessibility), *gif* (to create a gif image out of a video file), *pirate/pirating* (to illegally download software) and *frag* (to throw an object at another player in a videogame. The noun *frag*, which has undergone conversion to verb, is a clipping of *frag grenade*). Only one example of conversion from verb to noun has been reported, namely, *spawns* (places where the subject of a game reappears once they are killed).

In addition to the creation of new words by combining and/or reducing other base words or even lexical chunks, gamers also use *borrowings* or *loanwords*, that is, lexical elements from other languages, either with similar meanings or uses to those in other contexts or registers or with new ones. Among the former, instances like the following have been detected: first, from Spanish, *playing solo* (from Spanish *solo*; the meaning is maintained, "doing something by yourself"), *basura* (used as an exclamation to refer to something which is useless or not worth at all, literally "rubbish"), *por favor* (used as an exclamation to express negation or disagreement with something, similar to English "oh, please, no!"); second, from French, *prótegé* (a person who is guided and supported by an older and more experienced or influential person), *coup-de-grace* (a death blow, especially one delivered mercifully to end suffering); third, from Latin, ad nauseam (used to refer to the fact that something has been done or repeated excessively so that it has become annoying or boring); and, finally, from Japanese, *Sensei* (a master, a teacher) and *gacha* (a monetization technique used in many successful Japanese games, onomatopoeic for the hand-cracking action of a toy-vending machine). Unlike these, in the example Closed Beta/alpha, the meaning and uses of the Greek terms *beta* and *alpha* has changed. Thus, the expression refers to two groups of players, probably by influence of the military where different groups use this code to name themselves, for example, *alpha squad*.

Apart from combining word forms (or word/lexical chunks) and reducing them or borrowing lexical elements from other languages, gamers also "recycle" or reuse words by changing their meanings, providing words with new meaning or extending them. Some verbs and nouns in the data appear to have undergone some kind of semantic shift, as follows: *boss* (a significant computer-controlled enemy), *spawns* (places where the subject of a game reappear once they are killed), *frag* (to through the specific object at another player in a videogame), *clip* (to kill or to hit), *patch a game* (to "fix" a part of a game or make it better because it has a bug), *strafe jump* (to dodge or move sideways for the purpose of dodging), *mirror* (to keep a copy of the contents of [a network site] at another site, typically in order to improve accessibility), *windowed* (having or using framed areas on a display screen for viewing information), *streamer* (a member, participant or onlooker, sometimes called a *friend* or *follower*, in an online user's social media community), *boot* (to kick a dying computer), *ace* (a single player that kills all enemy players by himself; often used in first-person shooters [FPS] such as *Call of Duty* and *Halo*), *snipers* (a highly skilled marksman who uses his skills in fieldcraft, marksmanship, and will to survive to take the vital shot which could decide a battle; sniper games celebrate the marksmen with the most deadly skills), *port* (to create a new version of [an application program] to run on a different hardware platform) or *homebrew* (see above).

In addition to the preceding account of word-formation mechanisms used by gamers in their creation or production of new words, I also attend to the degree of technicality and the colloquial or slang character of gamers' lexical selections, as it is my aim to prove whether, as hypothesized in the present work, the language of gamers in fora is highly conditioned by the technicalities surrounding and involving gaming and videogames, computers, for a, and so on and, hence, is a mixture of jargon and slang. As expected, the data analyzed suggests that gamers' fora language emerges as a result of combining elements from different technical, standard, and colloquial registers which are implicitly related in different ways to gamers' interactions in fora. Accordingly, among many others, some of which have already been mentioned, are

(1) technical terms (with different degrees of technicity) from videogames and/or gaming language or jargon related to types of consoles, systems, problems with videogames and gaming, and so on, namely *PS4, Pro with Boost Mode on, home videogame consoles, speed zone v2, triple grenade launcher upgrade, quake clone, framerate, playthrough, homebrew, gamepad* (a handheld controller for videogames), *horde mode, lootbox* (see above), *vouch* (to agree with someone on a point), *rig* (a computer; word used by computer geeks or hardcore computer gamers), *VRAM* (video RAM or VRAM, a dual-ported variant of dynamic RAM [DRAM], which was once commonly used to store the frame buffer in some graphics adapters), or *NES* (Nintendo Entertainment System); but also names of videogames such as *Tower of Guns, Quake 3* and so on;

(2) technical terms related to computer language, such as *USB* (universal serial bus), *PC* (a personal computer), *DLC* (downloadable content), *PSU* (power supply unit), *HDD* (hard disk drive), *SDD* (software design and development), *GPU* (graphics processing unit), *BIOS* (basic input output system), *CPU* (central processing unit), *GB*

(gigabyte), *RAM* (random access memory), *GUI* (graphic user interface), *MSAA* (multisample anti-aliasing, a type of spatial anti-aliasing, a technique used in computer graphics to improve image quality), *Vcore* (bore voltage of a computer system), *hardware, software, post, wireless, reboot* (to restart a computer), *keyboard, input, framerate, motherboard, bottleneck* (when a part of your computer internals is slowing your entire system down), *app, touchscreen, backups, overclock* (to run a computer processor at a speed higher than that intended by the manufacturers), *bitrate* (the number of bits that are conveyed or processed per unit of time), *pixel* (the smallest element of an image that can be individually processed in a video display system), *devkit* (a software development kit [SDK], typically a set of software development tools that allows the creation of applications for a certain software package, software framework, hardware platform, computer system, videogame console, operating system or similar development platform), *Capcom* (a blend of "capsule" and "computers"; name given to a developer and publisher of videogames founded in 1979, and one of the largest software companies in the world. The term may also refer to a capsule communicator);

(3) words proper to fora such as *OT* or *off-topic* (a message board, thread or newsgroup that does not deal with the main topic), or *trolling* (to troll: to post inflammatory or inappropriate messages or comments on the internet in order to upset other users and provoke a response from them);

(4) lexical elements from standard general language like *update, improvement, hater, remake, deductible, disheartened, brightness, cultureless, homeowner, all-new, worrisome, long-lost, the move, to review, research, IMHO* (in my humble opinion), *ATM/atm* (at the moment), *FWIW* (for what it's worth), *FYI* (for your information), *asap* (as soon as possible), *specs* (spectacles, eyeglasses), *pro* (professional), *ppl* (people) and *demo* (demonstration);

(5) colloquial and slang words, determined by online communication features, which are widely used; they range from colloquial general words to swearwords, as shown here and below. First, suffixes *–y* and *–o*, which are often added to words in slang, have been distinguished, as in *matey* (a word often used to address friends, which may sometimes sound aggressive), *beefy* (thick, not at all slim), *bulky, goofy, buddy, catchy, punchy* (dazed), *clacky* (extremely awesome), *flimsy* (very thin, or easily broken or destroyed), *no biggy* (no biggie, no big deal), *whammy* (an adversary; also a hex), *cheapo* and so on. Second, pronunciation spellings feature, like *gonna* (going to), *dunno* (don't know), *sup* (what's up?), *it ain't* (It isn't), *nah* (slang

for "No"), *nop* (nope) and so on. Third, general and gamers' specific slang and colloquial words are used, mainly nouns and adjectives, for example *Cool? superb, quirk* (a special, unusual characteristic), *thang* (thing), *gooners, mishmash, swap* (exchange), *chunk*, but also verbs like *revamp* (redo; change; update; upgrade), or *wait* (as in *wait, you can upgrade nonstarter weapons?*). Fourth, abbreviations are also very common, as already seen. Some slang and colloquial examples identified in gamers' fora are *tomm* (tomorrow morning), *imo* (in my opinion), *LMO* (let's move on), *BTW* (by the way), *IIRC* (if I recall [or remember] correctly), *LOL* (laughing out loud), *LOTF* (laughing on the floor), *ROFL* (rolling on floor laughing). Fifth, rhyming slang expressions feature, like *easy peasy* (very easy; short form of "easy-peasy-lemon-squeezy"). Sixth, taboo and euphemistic expressions are used such as *piss, kick off somebody's bum, to miss the boat* (to miss an opportunity, a chance) and *darnit* (damn-it).

As mentioned earlier and as expected (due to the fact that fora language constitutes a written genre where participants "chat"), the data exhibit linguistic features of both written and oral language. Among the former, technical and formal words like *maneuverability* in *Double jumps made maneuverability amazing* as well as long and elaborated sentences like *Generally speaking, I think the game is fun for a certain type of player, but I feel their biggest barrier will be that there's nothing for players who can't aim super well to contribute to their team, or to achieve in general* may be highlighted. However, these are not as frequent and numerous as colloquial and slang expressions which clearly outnumber the former. Accordingly, many words and expressions from colloquial and slang oral language have been identified in my data:

(1) Personal pronouns deleted, as in *Haven't been able to get into the Closed Beta? Must've gained, can't wait to frag the rest of yous*;

(2) Contractions and pronunciation spellings, for example, *gonna, kinda* (as in *it's kinda hectic*), *dunno* (don't know), *it ain't* (it isn't), *ain't, yall* (you all), *nah* (no), *s'pose*;

(3) Transcriptions of laughter and emoticons, like *hehe, lol/Lol, LOTF, ROFL, :D*;

(4) Exclamations, marks of reflection, hesitation, assertion, approval, negation, and so on, as in *yuck, hmm / hm, erm, yipes, yea ok* (as in *as a quake game yea ok*), *yeah, uhh yea, yep, yay, nah, zzZzzZzz*;

(5) Nonconcordance of personal pronoun and verb and other violations of grammatical rules, like *this maybe has issues that needs ironing out*;

(6) Reduplication of elements to reproduce (oral) length of phonemes, as in *I reallllllllllly don't like this game*;

(7) Adjectives expressing feelings and emotions, such as *devestated* [*sic*], *frustrated, off-putting, dazed, eye-bleeding* (very good/outstanding), *mindblowing*;

(8) Use of taboo and euphemistic expressions as well as swearwords and expressions containing them, like *Piss, [...] and them typing a load of shit afterwards, holy shit, holy crap, shitty gametypes, a damn shame, ugly-ass, bullshit, shit, Dumb as shit, dumbshit, a ton of shit, smartass, dang/damn/goddamn/yo damn/darnit, fuck you, go fuck yourself, FFS (for fuck's sake), utter horseshit, this poo sandwich, it's fuckin radical, DGAF* (don't give a fuck), *FML!* (fuck my life!), *GAF* (give a fuck), *POS* (piece of shit), and so on;

(9) Other oral-like expressions/discourse: *that's really dumb, gonna pass. How unfortunate, wait, you can upgrade non-starter weapons? also like, there is we don't even know, No wondering if you're going to get a code—you'll get it right away! Haven't been able to get into the Closed Beta? You're in luck! Must've gained, can't wait to frag the rest of yous,* amongst others.

Although gamers' interactions in fora take place in the written medium, and despite the presence of specialized or technical videogames and computer-related vocabulary, the general syntactic simplicity, shortness, incompleteness, and spontaneity of the expressions in fora (see above and also Greenfield 1972; Chafe 1982; Ochs 1979; O'Donnell 1974; Tannen 1982; Jonsson 2015), as well as the use of a large number of colloquial and slang words, lead me to tentatively conclude that gamers' language features in fora are closer to oral or spoken language than to written language ones. Still, it may be argued that gamer fora belong to a written register which presents many stylistic and lexical features proper to oral or conversational discourses, where, for example, emotional and taboo expressions such as *dumb as shit, go fuck yourself,* or *this poo sandwich,* as aforementioned, have an important semantic load and contribute to the flow of their discourse. Accordingly, as Merchant (2001) suggests, this constitutes a new linguistic genre best described as *rapid written conversation.*

5 Conclusions

In this chapter, I have not only acknowledged the role of internet fora for gamers as they provide them with a new medium of communication that allows the gamer community to interact but I have also, and mainly, analyzed the language used in those interactions. More specifically, I argued that, as expected, the language used is highly homogeneous in the sense that the expressions and lexis employed do not differ much among gamers, as it is the language itself (and/or a common lexical core) that contributes to their identification within their community or group. In fact, it may be suggested that gamers as a community have their own register, which presents an equilibrium between jargon and slang, but also specific lexical, semantic and morphological features which may be global or universal for all gamers. Furthermore, I have identified specific creative characteristics of gamers' lexical stock in fora, concluding that gamers' lexis and terminology are highly prolific in abbreviations (mainly, acronyms and initialisms), and that they contain not only technical videogames and gaming-related words but also many colloquial and slang words. Such words are probably conditioned by the online written medium where the interactions take place, and they make this register lexically and stylistically closer to spoken discourse.

This study does not, however, attempt to extinguish the possibility of carrying out other research works on the language, lexis, and word-formation mechanisms used by gamers in general and, more specifically, in online communication or fora. Rather, this chapter is only a preliminary approach to these. Larger and more detailed analyses of data should be performed in order to describe gamers' uses of lexis and morphology both qualitatively (as in the present paper) and quantitatively. Moreover, supralexical data and tendencies would deserve further attention as it is in a discourse approach where more and highly emotive and creative data may be found, rather than in a purely lexico-morphological approach, where these are generally overlooked. In the present study, however, though a lexico-morphological approach has been primarily adopted, data may be contextualized within the general discourse approach of the book. Thus, considering this broader perspective, it may be argued that the analysis and the data themselves allowed the identification of highly expressive, emotional, colloquial, and slang expressions which break the standard rules of language and which make gamers discourse linguistically distinctive.

Notes

1 Note, however, that I do not agree with the restrictive nature of Lighter's definition, as I believe that groups may be composed of people of either sex.
2 The difference between vocabulary and terminology depends on the degree of technicality of the lexical units.

References

Balteiro, I. (2018a), "Emerging hybrid Spanish–English blend structures: '*Summergete con socketines*,'" *Lingua*, 205: 1–14.

Balteiro, I. (2018b), "Oh wait: English pragmatic markers in Spanish football chatspeak," *Journal of Pragmatics*, 133: 123–33.

Baron, N.S. (1998), "Letters by phone or speech by other means: The linguistics of email," *Language and Communication*, 18: 133–70.

Baron, N.S. (2003), "Language of the Internet," in A. Farghali (ed.), *The Stanford Handbook for Language Engineers*, 59–127, Stanford: CSLI Publications.

Benwell, B. and E. Stokoe (2006), *Discourse and Identity*, Edinburgh: Edinburgh University Press.

Bobosh, J. (2008), "O Rly? Slang and communication development in the World of Warcraft gaming community." Available online: https://digitalclaxon.wordpres s.com/2008/11/05/o-rly-slang-and-communication-development-in-the-worl d-of-warcraft-gaming-community (accessed May 25, 2018).

Bryant, M.M. (1976), "Blends are increasing," *American Speech*, 39: 3–4.

Byram, M. (2006), "Languages and identities." Available online: https://www.coe.int/ t/dg4/linguistic/Source/Byram_Identities_final_EN.doc (accessed May 25, 2018).

Cannon, G. (1985), "Blends in English word formation," *Linguistics*, 24 (4): 725–53.

Chafe, W. and D. Tannen (1987), "'The relation between written and spoken language," *Annual Review of Anthropology*, 16: 383–407.

Chafe, W.L. (1982), "Integration and involvement in speaking, writing, and oral literature," in D. Tannen (ed.), *Spoken and Written Language: Exploring Orality and Literacy*, 35–54, Norwood, NJ: Ablex.

Collier, M.J. and M. Thomas (1988), "Cultural identity: An interpretive perspective," in Y. Y. Kim and W.B. Gudykunst (eds.), *Theories in Intercultural Communication*, 99–120, Newbury Park, CA: Sage.

Crystal, D. (2001), *Language and the Internet*, Cambridge: Cambridge University Press.

Ensslin, A. (2012), *The Language of Gaming*, New York: Palgrave Macmillan.

Ferrara, K., H. Brunner, and G. Whittemore (1991), "Interactive written discourse as an emergent register," *Written Communication*, 8 (1): 8–34.

Flinkfeldt, M. (2011), "Filling one's days': Managing sick leave legitimacy in an online forum," *Sociology of Health & Illness*, 33 (5): 761–76.

Greenfield, P.M. (1972), "Oral or written language: The consequences for cognitive developed in Africa, the United States, and England," *Language and Speech*, 15: 169–77.

Halliday, M.A.K. (1989), *Spoken and Written Language*, Oxford: Oxford University Press.

Herring, S.C., ed. (1996), *Computer-Mediated Communication: Linguistic, Social and Cross-Cultural Perspectives*, Amsterdam: John Benjamins.

Herring, S.C. (2001), "Computer-mediated discourse," in D. Schiffrin, D. Tannen, and H.E. Hamilton (eds.), *The Handbook of Discourse Analysis*, 612–34, Malden, MA: Blackwell.

Herring, S.C., L.A. Scheidt, S. Bonus, and E. Wright (2004), "Bridging the gap: A genre analysis of weblogs," in *Proceedings of the 37th Hawaii International Conference on System Sciences* (HICSS-37), Los Alamitos, CA: IEEE.

Herring, S.C., L.A. Scheidt, E. Wright, and S. Bonus (2005), "Weblogs as a bridging genre," *Information Technology & People*, 18 (2): 142–71.

Jones, S.E. (2008), *The Meaning of Video Games: Gaming and Textual Strategies*, New York: Routledge.

Jonsson, E. (2015), *Conversational Writing*, Frankfurt am Main: Peter Lang.

Keats, J. (2011), *Virtual Worlds: Language on the Edge of Science and Technology*, Oxford: Oxford University Press.

Kowert, R. (2014), "The gamer identity crisis: Towards a reclamation." Available online: http://www.firstpersonscholar.com/the-gamer-identity-crisis/ (accessed May 25, 2018).

Leigh, A. (2014), "'Gamers' don't have to be your audience: 'Gamers' are over." Available online: http://www.gamasutra.com/view/news/224400/Gamers_d ont_have_to_be_your_audience_Gamers_are_over.php?utm_source=dlvr.it&ut m_medium=tumblr (accessed May 25, 2018).

Lighter, J.E. (2001), "Slang," in. J. Algeo (ed.), *The Cambridge History of the English Language, Vol. VI: English in North America*, 219–52, Cambridge: Cambridge University Press.

Mattiello, E. (2008), *An Introduction to English Slang: A Description of Its Morphology, Semantics and Sociology*, Milan: Polimetrica.

McDaniel, S.E., G.M. Olson, and J.S. Magee (1996), "Identifying and analyzing multiple threads in computer-mediated and face-to-face conversations," in *Proceedings of the ACM Conference on Computer Supported Cooperative Work* (CSCW'96), 39–47, Cambridge, MA: ACM Press.

McGrath, M. (1998), *Hard, Soft and Wet*, London: Flamingo.

Merchant, G. (2001), "Teenagers in cyberspace: Language use and language change in internet chatrooms," *Journal of Research in Reading*, 24 (3): 293–306.

Michael, R. (2007), "Gamer lingo: How to speak like an elite gamer," *Durhamregion.com*. Available online: https://www.durhamregion.com/commun ity-story/3485785-gamer-lingo-how-to-speak-like-an-elite-gamer/ (accessed May 25, 2018).

Miller, D. and D. Slater (2000), *The Internet: An Ethnographic Approach*, Oxford: Berg.

O'Donnell, R. (1974), "Syntactic differences between speech and writing," *American Speech*, 49: 102–10.

Ochs, E. (1979), "Planned and unplanned discourse," in T. Givón (ed.), *Discourse and Syntax*, 51–88, New York: Academic Press.

Ooi, V.B.Y. (2008), "The lexis of electronic gaming on the Web: A Sinclairian approach," *International Journal of Lexicography*, 21 (3): 311–23.

Osgerby, B. (2004), *Youth Media*, London/New York: Routledge.

Parkin, S. (2013), "If you love games, you should refuse to be called a gamer." Available online: http://www.newstatesman.com/if-you-love-games-you-are-n ot-a-gamer (accessed May 25, 2018).

Portnow, J. (2011), "Gaming languages and language games," in M. Adams (ed.), *From Elvish to Klingon*, 135–60, Oxford: Oxford University Press.

Preece, J. (2000), *Online Communities: Designing Usability, Supporting Sociability*, Chichester, UK: John Wiley & Sons.

Reed, D. (2001), "Making conversation: Sequential integrity and the local management of interaction on Internet newsgroups," *Proceedings of the 34th Hawaii International Conference on System Sciences 2001*. Available online:

https://www.computer.org/web/csdl/index/-/csdl/proceedings/hicss/2001/0981/
04/09814035.pdf (accessed May 25, 2018).

Tannen, D. (1982), "Oral and literate strategies in spoken and written narrative,"
Language, 58: 1–21.

Turkle, S. (1995), *Life on Screen: Identity in the Age of the Internet*, New York:
Simon & Schuster.

Turkle, S. (1999), "What are we thinking about when we are thinking about
computers?" in M. Biagioli (ed.), *The Science Studies Reader*, 543–52, London:
Routledge.

Wach, S. (2014), "Language practices on Internet game fora," *Styles of
Communication*, 6 (1): 191–204.

3

Phraseology and Lexico-grammatical Patterns in Two Emergent Paragame Genres

Videogame Tutorials and Walkthroughs

Christopher Gledhill

1 Introduction

When looking at the discourse of videogames, it is possible to focus on highly visible features of language such as novel terminology (*combo*, *to plink*, *whiffed*, etc., see Álvarez-Bolado and Álvarez de Mon, this volume), original combinations of existing forms (*cutscene*, *sidequest*, etc., see Balteiro, this volume), or markers of oral interaction and emotion (*ouch!* see Ensslin and Finnegan, this volume). However, in this chapter, I concentrate on "phraseology," which I define here very informally as "the preferred way of expressing meaning in a particular discourse." Whereas many linguists consider phraseology in terms of idiomatic expressions, proverbs, fixed phrases, and the like, in this chapter I adopt the "contextualist" approach, first proposed by J.R. Firth and then developed during the early days of corpus linguistics by J. Sinclair and others (Firth 1957; Sinclair 1991; Stubbs 1993; Hunston and Francis 2000; Hoey 2005; Sinclair and Mauranen

2006). Following this approach, I suggest that videogame tutorials and videogame walkthroughs are not only well-defined varieties of language, but they also have their own particular phraseology, that is to say, their own particular configuration of lexico-grammatical patterns (as defined by Hunston and Francis 2000; Gledhill 2000a, b). In the following sections, I set out a methodology for establishing phraseological patterns, which starts off by examining "key" (statistically significant) grammatical items, and then proceeds to analyze the most typical ways that these items are used in characteristic phrases on the basis of corpus evidence (an approach first set out in Gledhill 1995 and developed in other studies, e.g., Groom 2007, 2010).

The advent of corpus linguistics has changed the ways that analysts think and talk about routine patterns of expression. Corpus linguists have developed a variety of terms to talk about the units of phraseology, such as *collocational frameworks*, *lexical patterns*, *collostructions*, *discourse routines*, and so on (Renouf and Sinclair 1991; Hunston and Francis 2000; Stefanowitsch and Gries 2003; Tran et al. 2016). While phraseology is often discussed in terms of the general language, analysts have more recently explored how routine patterns are used in different registers, in particular in specialized areas such as scientific writing (Gledhill 2000a, b), academic discourse (Groom 2007, 2010), technical instructions (Coutherut 2016), business communication (Née et al. 2017), and so on. However, while there is now a sizable literature on the phraseology of specific genres in English and other languages, only a few studies (Ensslin 2011) have examined routine phrases in videogame discourse.

In this contribution, I claim that the basic unit of phraseology is the "lexico-grammatical pattern" (LG pattern, for short). A typical LG pattern can be defined as a recurrent sequence of lexical items which extends beyond the syntactic group (i.e., it can be longer than a nominal group, verbal group, etc.). In addition, each LG pattern has an identifiable semantic or rhetorical function, which is specific to the particular discourse in which it is observed. In the following sections, I set out a replicable methodology (as set out below) for identifying the most typical LG patterns in videogame tutorials (henceforth VGT) and videogame walkthroughs (VGW). While I suggest that this methodology is systematic, I do not claim that it is very sophisticated. Indeed, corpus-based methods such as multifactorial analysis (Biber et al. 2004, 2010), the analysis of n-grams and tag-grams (Née et al. 2017), and more recent approaches (such as textometrics) are now widely used by corpus linguists in order to identify regularities of expression. However, I find that such advanced methods pose problems for those analysts who want to look at the behavior of a specific type of discourse without prior training in statistics or programing.

For this reason, I have previously suggested (Gledhill 2000a, b, 2015) that the analysis of grammatical keywords provides an efficient way of conducting a preliminary analysis of the main LG patterns in a particular corpus. For example, in a corpus of VGTs, the pronoun *it* is found to be a statistically significant key word when this specific corpus is compared with a corpus of general English. This observation is not significant on its own, but in the VGT corpus, it can be seen that the word *it* is associated with longer patterns of expression such as <*it is good for VV+ing NN*>[1] in which the embedded verb (here marked *VV*) refers regularly to a specific type of attack (*dodging fireballs, punishing whiffed air attempts, starting combos*, etc.). In other words, we have moved from the observation of a highly frequent grammatical pronoun to the observation of game-related terminology (and the way it is evaluated, in the phrase <*it is good for VV+ing*>). It can be shown that micropatterns such as these may vary according to context but often express an abstract meaning as a whole which goes beyond the local meaning of its constituent units and their frame of reference (Adam 2011 calls these "macro-propositions" [131]; in Gledhill 2000a, b, I call these "discourse functions"). Thus, in the context of VGTs, the pattern <*it is good for VV+ing*> has a rhetorical function which can be paraphrased as "summarizing the main advantage of a previously mentioned fighting ability." It is in this respect that lexico-grammatical patterns resemble more traditional phraseological phenomena, such as idioms, proverbs, routine formulae, and similar multiple-word units.

I would argue that by identifying lexico-grammatical patterns and associating them with rhetorical functions in this way, it is possible to arrive at a description of the most characteristic features of a particular type of text. In addition, I suggest that if it is possible to show that a particular type of text or discourse has a predictable and productive repertoire of lexico-grammatical patterns, then it is likely that this discourse has evolved into a mature LSP, that is to say, a variety of language that serves the purposes of a self-defining group or community, which participates in the adaptation of its own conventionalized lexico-grammatical patterns of language (i.e., phraseology), as well as its own codified channels of communication (i.e., genres, as described by Swales 1990; Gee 2005; and others). In the concluding section of this chapter, I return to this notion in the light of my analysis of the VGT and VGW corpora.

2 Data selection: General characteristics of the VGT and VGW corpora

Before looking in detail at corpus data, it is worth pointing out some of the general features of VGTs and VGWs. The two corpora analyzed in this chapter were collected by my students as part of a course titled "Technical Discourse Analysis" (TDA).[2] This course is part of a two-year master's course at the Université Paris Diderot, France (*Master ILTS—Industries de la langue et traduction spécialisée*). The aim of the master's is to train technical communicators and translators, with particular emphasis on the acquisition of terminology and phraseology in different technical and specialized domains. The specific aim of the TDA course is to raise awareness about the different types of technical genres in English, as well as to promote the systematic use of corpus analysis as a transferable research skill.

The TDA course requires students to build and then analyze a corpus which is representative of a particular genre of written English. The students can explore any genre, as long as it belongs to a technical register.[3] The exercise therefore excludes literature, fiction, and journalism but includes expert-to-expert genres (such as dissenting opinions, oil refinery operating manuals, scientific research articles, etc.) as well as expert-to-nonexpert genres (organ donation brochures, political manifestos, social network privacy policies, etc.). Over the years, my students have worked on all of these genres (the characteristics of some of these are set out here for the purposes of comparison). However, more recently, some students have also asked to study genres which do not easily correspond to the canonical notion of a technical text, and they have shown increasing interest in texts related to gaming and videogames. In 2016, two groups of students chose to study VGTs and VGWs. Although these texts present a number of contextual differences in relation to more traditional technical genres, my students argued convincingly that VGTs and VGWs constitute legitimate topics for the TDA project.

The VGT corpus (111,695 tokens) is described by my students as "character guides, general tutorials and glossaries from various sources, such as videogame websites (*IGN, Gamefaqs, Supercheats*), websites [specializing] in competitive fighting games, and online versions of paperback guides." The two students working on this project decided to concentrate on the tutorials available for one game, *Streetfighter 4*,[4] published by two well-known websites on fighting games: *Eventhubs* and *ShoRyuKen*. Each of these sites was mined to obtain 97 texts of around 1000 words each. Each text describes in detail the fighting abilities of the characters encountered in *Streetfighter 4*. Although these texts are essentially instructional, they can also be seen as recreational: the authors are assumed to be experts in the game, and they take pleasure in exploring the different abilities and tactics employed by the player characters and their different adversaries, as well as passing on their experience and overt evaluation about these specific fighting styles.

The VGW corpus (568,998 tokens) is made up of eight "Japanese" RPG walkthroughs. These texts are available from various sources (the one used by my students was *Gamefaqs*). As the wordlist data suggest, these texts are extremely long: their average length is comparable to a short novel or PhD thesis (over 70,000 words in this corpus). Walkthroughs originated as strategy guides in videogame magazines (they are related to, but not the same as video "longplay" and "playthrough"), although the term *walkthrough* itself is related to the development of computer software (*software walkthrough, software technical review*). While tutorials focus on combat techniques, RPG walkthroughs present a comprehensive description of an entire game world, thus allowing players to explore every location and to succeed (or make informed choices) in every encounter in the game. Yet while the settings and the types of activities are very different, walkthroughs seem to share a number of rhetorical aims that are similar to fighting game tutorials: the author talks the reader through a world that is mutually recognized and enjoys recounting the hazards and solutions to specific tasks that have to be overcome before further exploration.

As mentioned above, VGTs and VGWs are different but related text types. Both are examples of what Ensslin (2011: 8) calls "paratextual genres," since they represent fan literature about the game rather than discourse which emerges within the game. Yet they also appear to belong simultaneously to two subcategories ("language about games by gamers," and "language used in instruction manuals," Ensslin 2011: 6). I would suggest that many of the linguistic differences between these texts and the general language corpus (in this case, my reference is the British National Corpus [BNC]) can be explained in terms of *technicity* (the extent to which these texts engage with the specificities of the fictional game world, or the mechanics of the game) and *interactivity* (the extent to which these texts provide a space for gamers to interact). The following two features (labeled F1 and F2 for "Feature 1" and "Feature 2") are evidence of technicity:

F1) *Impersonal expression.* Both VGTs and VGWs use a variety of ergative, passive, and other impersonal structures. Such structures focus on inanimate objects (by topicalizing or "thematicizing" entities in sentence-initial position), thus allowing the omission of animate participants. As has often been observed, these are typical of scientific or technical registers:

(1)   **Crafts work** in the same fashion as arts, some are **target-fixed...** [*Trails in the Sky*, walkthrough][5]

(2)   To use talent arts, the talent gauge must be full—**gauge fills up** every time **an auto-attack hit connects**. [*Xenoblade*, walkthrough][6]

F2) *Terminological networks.* VGTs and VGWs use a wide range of technical uses of general language items (*juggle, meaty, poke*), as well as specialized collocations (*lag tactics, negative edge, tick throw*) and abbreviations (*c.mk* "crouching medium kick," *SJC* "super jump cancel") that are often mentioned without comment or definition (it is assumed that users have access to online glossaries such as "Terminology and glossary guide for fighting games"[7]). In addition, many terms involve novel forms of amalgamation such as nominal-verbal conversion: *cutscene, rushdown, sidequesting, super cancel*, and so on (in the VGT corpus, many of these examples involve particles, as discussed in Section 4.1 below).

In the following discussion, I suggest the addition of two further features (F3 and F4) on the basis of corpus analysis:

F3) *Multiple embedding*. As shown in sections 4.1 and 4.2, below, some particles and prepositions are highly statistically salient in both the VGT and VGW corpora. Observation in the corpus suggests that these items are involved in technical nouns and verbs which are made up of embedded sequences of particles and prepositions (often associated with particle verbs/prepositional verbs), as in:

(3) Spin Drive Smasher—Fairly easy to combo [short for "combination"] **into off of** a HK Spiral Arrow or any version of the Quick Spin Knuckle. [*StreetFighter 4*: Cammy, tutorial][8]

(4) During the middle of the game, Junpei won't be as available, so you can miss **out of** maxing **out** this SL. [*Persona 3*, walkthrough][9]

A further example of embedding can be seen in premodification and postmodification of nominal groups, as in the following examples (here the complex nominal is enclosed in single brackets [...]):

(5) Starting your punish combos with [**the first hit of close Heavy Kick canceled into Heavy Punch Whip of Love**] can be a little easier to land than attempting to start the same combo with Heavy Punch… [*Xenoblade*, tutorial]

(6) The Heavy Kick version of Lynx Tail is active almost twice as long, creating a bit of an unsafe mix-up for foes expecting [**the less active Medium Kick Lynx Tail**]. [*StreetFighter 4*: Elena, tutorial]

(7) It's a 3-frame start up and is probably [**the furthest teaching 3 frame normal**] in the cast. [*StreetFighter 4*: Cody, tutorial]

F4) *Complex subordination*. As discussed in Section 4.3, both corpora (but especially VGWs) use a variety of complex conditional clauses, as well as other forms of subordination, a characteristic which they share with other technical genres such as procedural manuals (Coutherut 2016). This can be seen in the following examples (following the conventions of Systemic Functional Grammar, embedded clauses are signaled by square brackets [[...]] and bound/subordinate clauses by slash symbols: // A = main clause // B = bound clause):

(8) (A) [[Starting your punish combos with the first hit of close Heavy Kick [[canceled into Heavy Punch Whip of Love]]]] can be a little easier [[to land]] // (B) than attempting to start the same combo with Heavy Punch, // (C) though this easier combo does lose a bit of damage./// [*Xenoblade*, walkthrough]

(9)   (A) The Heavy Kick version of Lynx Tail is active almost twice as
       long, // (B) creating a bit of an unsafe mix-up for foes [[expecting
       the less active Medium Kick Lynx Tail.]] /// [*StreetFighter* 4: Elena,
       tutorial]

Thus far, we have seen that features F(1–4) are typical of formal, technical
registers. We now turn to a further set of features (labeled F5–8 here), which
show that both VGTs and VGWs also make use of resources that are thought
of as typically interactive and oral (many of these characteristics have also
examined elsewhere, for example in Balteiro, this volume):

F5) *Marked appraisal and evaluation*. VGTs and VGWs both involve
numerous asides, comments, judgments, and other markers of authorial
stance. These functions can be realized by various structures ranging from
vague quantifiers to idiomatic expressions:

(10)   This boss is a **bit tricky** to hit, for he is located at the very edge of
       the battlefield [*Trails in the Sky*, walkthrough]

(11)   Battle Basics 1. **Piece of cake**, just gang up on the ghosts. [*Trails in
       the Sky*, walkthrough]

F6) *Lexical reduction*. Although VGTs and VGWs involve a number of
technical terms, they also make use of highly informal terms, including
vague nominals (*pantonyms*):

(12)   Don't buy any weapons, though, you can get better **stuff** soon.
       [*Tales of Symphonia*, walkthrough][10]

(13)   Hit and run won't work, this **thing** has too great a range. [*Tales of
       Symphonia*, walkthrough]

F7) *Lexical expansion*. Since both VGTs and VGWs often deal with combat,
they expand the lexical repertoire for this area, thus introducing a rich
set of quasi-synonyms. While some items are euphemisms or attenuating
expressions (*clean up, deal with, get rid of, finish off, make quick work out
of, pick off*, etc.), others refer to death and destruction more directly, often
using slang or taboo language to express these meanings with more force
(*destroy, hack away, kill, pound the crap out of, do serious damage, happy
slaughtering! take out, whack, wipe out*, etc.).

F8) *Interaction markers*. These items also contribute to the high degree of
engagement (authorial stance, subjectivity, oral style) in VGTs and VGWs.
This category includes oral discourse markers and typographic and other
features below the level of the word (speech marks, exclamation marks,
contractions, etc.):

(14)   Fight, then you'll be at the exit. **Ouch**. [*Radiant Historia*,
       walkthrough][11]

(15) A lone knight. **Easy, right? Wrong,** this knight will wipe the floor with you if you're not careful. [*Disgaea*, walkthrough]

Again, on the basis of corpus analysis (below), I suggest two further features of interactivity:

F9) *Directed imperatives.* As discussed in sections 4.1, 4.2, and 4.4, a particularly characteristic feature of the VGW corpus is the widespread use of imperative instructions (including widespread ellipsis) to direct to the player/reader through their adventure. Here are some typical examples of this:

(16) Ask Murray about "Sound," then [ELLIPSIS] about "Bell toll." [*Trails in the Sky*, walkthrough]

(17) Also **note that you can use** a LK Scarlet Terror (Vf5 downleft. gif charge Vf5 right.gif + LK) kick instead of the HK **and then juggle** with an EX Scarlet Terror afterwards. [*StreetFighter 4*: Vega, tutorial]

(18) **Then** [ELLIPSIS] back to the first room and [ELLIPSIS] west. [*Devil Survivor*, walkthrough][12]

F10) *Deontic modality.* As discussed briefly in Section 4.4, VGTs and VGWs use a rich variety of modal forms (which are more typically encountered in oral registers) to express advice, directions, and evaluation. Here are just two examples from a wide variety of potential forms in the corpus:

(19) If you kill one enemy with a crit/weakness and get "1 More," you **gotta** kill another enemy and get ANOTHER "1 More"... [*Persona 3*, walkthrough]

(20) He'll start by using Vampire's Mist, which is a Mystic-type skill that targets all groups in the field and heals him up. WTF [NB what the fuck] **are we supposed to** do?! Mmm... [*Devil Survivor*, walkthrough]

It should be clear that not all of the characteristics listed here can be identified using the methods I set out below. The corpus-based analysis of LG patterns I set out below only reveals indirect evidence for evaluation, terminology, lexical expansion (or reduction), and markers of interaction. On the other hand, other features emerge from the corpus analysis: there is more emphasis on clause structure (complex subordination and embedding) as well as the construction of verbal groups (transitivity, direct imperatives, deontic modality). In the following sections, I examine these features in more detail and gradually make the case for identifying extended patterns of expression (LG patterns) as an important first step in the analysis of these texts.

3 Data collection: The identification of grammatical keywords

In this section, I describe the statistical methods used to identify the first ten[13] salient grammatical keywords in the VGT and VGW corpora. As a first step in the characterization of a particular corpus, it is useful to identify the key lexical and grammatical items of that corpus using a tool such as *Keywords* (Anthony 2014). The AntConc Keywords program compares the lexical frequency lists from two comparable corpora; when a word has a significantly higher than average frequency in one corpus compared to the other corpus, that word is placed toward the top of the Keywords list. The position of each keyword depends on its "keyness" score. The keyness score is based on a comparison between the probability of encountering a particular word in the corpus under study and its probability of occurrence in the reference corpus. Thus, for example, the item *opponent* has a frequency of 878 per 111,695 words in the VGT corpus (or a relative probability of occurrence of 7.8 per 1000 words). In contrast, the same item has a frequency of 1,428 per 100,000,000 in the BNC (and thus a probability of occurrence of 0.01428 per 1000 words in that corpus). In this instance, the difference in probability is so great that the item *opponent* achieves a very high keyness score, and as can be seen in Appendix 1, AntConc consequently places *opponent* at rank 5 in the table of Keywords for the VGT corpus.

To give an idea of the kind of data that the Keywords tool reveals, Appendices 1 and 2 show the first ten items which emerge as the highest scoring Keywords for the VGT and VGW corpora.[14] In a project on technical terminology, many of the lexical items on these lists (such as *combo, enemy, opponent, quest*) could be considered as candidate terms for a glossary in this domain. However, in this study, I am interested in the key use of particular grammatical items. I contend that if function words such as *you* or *up* obtain a relatively high score as Keywords in a particular corpus (even if they are not at the top of the Keywords list), then these items nevertheless have greater significance in that corpus, because—when one compares different registers of English—grammatical items are not usually expected to have as high a degree of variability as their lexical counterparts. In other words, most observers would agree that lexical items (such as *opponent* or *combo*) are expected to vary quite markedly in their frequency of occurrence. By contrast, it is less obvious that the relative frequency of grammatical words such as *you* and *up* can also vary, according to different registers (e.g., *you* will occur much more in spoken conversations than in scientific research articles, a fact that is behind the multifactorial analysis of linguists such as Biber et al. 2004, 2006). However, it is also the case that grammatical items are so ubiquitous and frequent in the English language as a whole that their degree of relative frequency does not vary as much as lexical items. Thus, even if we observe a relatively small movement in the overall frequency of an item such as *you* (say an increase by 5 percent), it is likely to be more significant than an equivalent movement in the frequency of a typically less frequent lexical item such as *opponent*.

In Appendices 3 and 4, I have set out the first ten grammatical keywords found in the VGT and VGW corpora. These items are identified as "grammatical" because they belong to the closed lexical classes of

(1)   Adverbials and particle-like items such as *up* (AV),

(2)   Coordinating and subordinating conjunctions such as *if* (CJ),

(3)   Pronouns and deictic items such as *here* (PN),

(4)   Prepositions and grammaticalized items such as *right* (PR),

(5)   Auxiliaries, modals, and grammaticalized verbs items such as *get* or *let* (VX).

Of course, it is not useful to analyze grammatical items in isolation. It is, however, important to have some idea of the extent to which different types of grammatical items are salient in different types of texts. I have therefore set out in Table 3.1 the distribution of the first ten key grammatical items in the VGT and VGW corpora in comparison with the first ten key grammatical items in three other major registers: legal, informative, and specialized (these being the main registers that are studied in the TDA course, as mentioned in Section 2).

Table 3.1 gives a general idea of some the broad similarities and differences that can be observed in a sample of major technical registers in English. For example, it can be seen that coordinating conjunctions (*and, or*) are salient in legal, informative, and specialized registers, while subordinators (*that, until, when*) are salient in legal and informative texts. In the data analysis below, I explore some of the reasons why a different set of conjunctions is preferred to these in the VGT and VGW corpora (namely *if* in the VGT corpus, and *if/after/once* in the VGW corpus).

Explanatory Note 1. Some items are in parentheses because they belong to more than one part of speech.

Explanatory Note 2. All the corpora mentioned here were compiled by students on the TDA course, with the exception of scientific research articles, which are reported in Gledhill (2000, 2015).

A similar observation can be made with regard to prepositions (PR). Table 3.1 shows that legal, informative, and specialized registers share a preference for items which typically introduce circumstantials (*for, in, with*), or the item *of*, which is used in complex nominals. In contrast, VGTs and VGWs have a marked preference for AVs of direction (*down, up*), AVs of time

TABLE 3.1 *Key grammatical items in a sample of four technical registers*

Register	Genre	Texts	Words	key AV	key CJ	key PN	key PR	key VX
Legal	Social network privacy policies	30	105,321		or (that)	(any), (that), (to), you, we	For, (to), with	May, is
	U.S. dissenting opinions	489	1,881,336	not	or, (that)	it, (that)	At, by, for, in, of	is
Informative	Organ donation leaflets	58	112,526		and, or (until) when	(this), you,	For, of, (to),	will
	U.K. political manifestos	53	313,180		and	you, we	for, of, on, (to)	Are, have, will
Specialized	Oil refinery operating manuals	39	162,366		and		by, from, in, of, (to), with	are, be, is
	Scientific research articles	150	515,073	not	and (that)	(that)	For, in, of	Did, have, is, was
Instructional	Videogame tutorials (VGTs)	97	111,695	(down), forward, (into), (off) right, (up)	if	it, you	(down), (Into), (off), (up)	Can, lets
	Videogame walkthroughs (VGW)	8	568,998	(down) now then, (Up)	once	here, you	(down), (up)	Can, get

(*now*, *then*) and PRs and other particles expressing a direction (*forward*, *into*, *off*). None of these items is significant on its own; rather, the fact that these items vary from one register to another points to significant differences in the phraseological patterns that are typical of these genres.

Finally, looking at Table 3.1, it is also interesting to note that that there are certain affinities between genres or registers. For example, privacy policies, donation leaflets. and political manifestos all share a preference for the PN *you*, an item which is also a key word in VGTs and VGWs. One explanation for this is that all these genres are "directive"; they aim to make their readership respond to (or be responsible for) the content of the text. This is not a rhetorical function found in dissenting opinions or research articles (although it is perhaps surprising to find that *you* is not a key item in operating manuals). Having said this, as discussed below, I suggest that it is longer stretches of expression such as <*you can get VV/AJ*>[15] which account for the particular distribution of items such as *you* in VGTs and VGWs.

4 Data description

In this section, I explore the first ten grammatical items that rank as salient keywords in the VGT and VGW corpora, with a particular focus on the phraseological patterns that are associated with these items. Rather than looking at each corpus in turn, in the following sections I divide the analysis into five "semantic zones," each corresponding to a different set of grammatical items which are salient in VGTs and VGWs. These categories are set out in Table 3.2.

Although Table 3.2 sets out five categories of analysis, there is only space in Section 4 to discuss categories 1–4. Category 5, "causation and transitivity," is partially discussed in the other sections (the attributive use of *get*, which is its most frequent use, is discussed in Section 4.3 in relation to the pattern <*if your opponent gets AJ*>, while the modal *can* is discussed in Section 4.4 in relation to the pronoun *you*).

Finally, the following conventions are observed in the data analysis:

(1)   A simplified version of the BNC tagset is used to signal abbreviated parts of speech: AT (article, determiner), AJ (adjective), AV (adverb), CJ (conjunction), NN (Noun), PN (pronoun), PR (preposition), VX (verb auxiliary or modal), VV (verb).

(2)   Authentic corpus examples are presented in a different font (or in italics if quoted in the text). Each bullet point represents the start of a new example.

(3)   Lexico-grammatical patterns are presented in triangular brackets, with generic items labeled by part of speech, as in <*CJ (and, but, so) it's AJ (good, great) for VV (catch, dodge, knock)+ing*>.

TABLE 3.2 *Semantic zones and grammatical keywords in the VGT and VGW corpora*

Semantic zone	Grammatical keyword
1. AV/PR direction and spatial extent	down, forward, into, off, right, up
2. AV/PR deictics and transitional location	here, now, then
3. CJ conditional advice and choice	if, once, after
4. PN reformulation and evaluation	it, this, you
5. VX causation and transitivity	can, does, get/gets, lets

4.1 AV/PR directions and spatial extent

In grammatical terms (Halliday and Matthiessen 2014), the semantics of adverbial and prepositional phrases can be analyzed in terms of either *location* (expressing a static point in space or time) or *extent* (expressing dynamic movement across space or time). Generally speaking, the items found in the VGT and VGW corpora tend to have a usage that is closer to extent. This notion is expressed in terms of relative space (*forward, into*), cardinal directions (*north*), and somatic relations (i.e., body-oriented items such as *right*).

In the general language, each of these items has many different potential contexts of use. But in technical corpora such as VGTs and VGWs, the relative frequency of each of these items can often be explained in terms of one or two very regular, recurrent patterns of use. For example, a number of these words, notably *forward, left, right, up, and down*, refer specifically to inputs on a joystick or remote control, and this usage accounts for the majority of their occurrences. Other items, especially *off, up,* and *down*, are predominantly used with particle verbs, such as *hold up* and *knock down*. Another very frequent use of *up* in both the VGTs and VGWs is as a post-modifying particle in a variety of converted (deverbal) nouns such as *cross-up, follow-up, jump-in, mix-up, start-up, wake-up* (these being the most frequently encountered occurrences). This usage is particularly prevalent in the VGT corpus. As can be seen in the following sample, the particle *up* provides a productive way of creating neologisms in this domain, with each verb + participle compound premodifying another noun which expresses a specific type of attack or maneuver (or more generally a fight, as in *mix-up*):

(1)   Best use is the EX version which will knock the opponent into the air for a **follow-up** juggle.

(2) She's got solid pokes, a decent **mix-up** game, and fairly simple combos into her super and ultras.

(3) Guacamole Leg Throw—In EX form, this is Fuerte's best anti-air attack. Use it especially as a **wake-up** counter to meaty **jump-ins**.

These examples are significant, but they have more to do with the formation of new terminology than the phraseology of this particular register. Perhaps the clearest example of an extended LG pattern which I can identify in this category involves *into*. In general English, *into* is often used with transitive verbs to form a causative construction of the type *<VV NN into NN>*. In contrast, in the VGT corpus, *into* typically occurs in combination with an intransitive verb, as in the following structure: *<VV (buffer, cancel, combo, land, tick) into AT NN (combo, move, [specific attack]>*. Semantically, this pattern refers to a transitional maneuver from one type of attack to another:

(4) Pressure with low attacks and **cancel into the Hazanshu** to both maintain pressure and go for the mix-up.

(5) Gouken can also **combo into his Shin Shoryuken** off of a backward throw.

(6) As a meaty cross-up, if it gets blocked you can **tick into a grab** attempt.

Although the notion of extent is typically expressed by adverbial and prepositional phrases, many other constructions convey direction and movement (including verbal ellipsis—signaled below as [Ø], cohesive markers, and temporal AVs such as *again, then...*). This is especially the case in the VGW corpus, which employs an impressive repertoire of constructions expressing directionality. The following example shows how directed movement can be expressed by the same extended pattern within the same stretch of text (I suggest that the pattern occurs several times in the following extract, and has the form: *<CJ (and, but)/AV (again, then) VV (go, [Ø]) AV (back, east, left, north, right)/PR (down, to) NN>*):

(7) **<Go north>** for a Panacea Plus. **<Then [Ø] back to the first room>** **<and [Ø] west>**. On the north side is a hidden Freikugel Mercy. **<Again [Ø] to the first room>**, **<but [Ø] east>** this time. **<[Ø] Down the vine>**, grab the Nirvana Plate, **<and go east>** for another vine."

Versions of this "directed movement" pattern will also be discussed in the following sections.

4.2 AV/PR deictics and transitional location

The adverbials (AV) *then*, *here*, and *now* all occur within the first ten key grammatical items in the VGW corpus. In grammatical terms, these items express *location*, that is, a static point of reference, whether spatial or temporal, or a point in the development of the text itself. In functional terms, as we see below, each of these items is regularly associated with a distinct set of constructions which express the management of space and time, or distinct stages within the text itself (all of these being functions that are more typical of VGWs).

Looking at *then*, one common pattern with this item takes the form <*then VV [imperative]*>. This pattern introduces the next step in a sequence of instructions and generally involves a verb of action or movement (as mentioned above, with a potential ellipsis of the VV) or, less frequently, a verb expressing a communicative or mental process:

(8)   Turn right immediately after exiting, then [Ø] **right and left**.

(9)   Head up to his office on 2F. To the item shop! **Then** [Ø] **to the bar**.

(10) Have 100 Kills by the time you finish the stage. **Then choose** the option "to kill"

The adverbial *now* has a similar directional function to *then*. However, while *then* instructs the player to select a direction or to engage in the next action in a sequence, *now* appears to operate at a higher level of in-game activity: the player is instructed to travel to the next scene or to a separate location. As with *then*, sentence-initial *now* is sometimes accompanied by ellipsis of the following verb, hence the commonly encountered pattern <*now VV to NN [location]*>:

(11) **Now return to 5F to Bridge to Apocrypha** by retracing your steps.

(12) **Now** [Ø] **south to the next screen**. Across the bridge, then north.

(13) **Now zoink over to Central Seal Island**, SAVE YOUR GAME, ascend the stairs and...

Sentence-initial *now* also has a function to play in expressions which refer to the metagame; in this pattern, the AV introduces a summative comment on the current state of play:

(14) **Now comes the fun bit,** pick up the Defense penalty symbol, but don't throw it just yet,

(15) **Now comes the hard part** in dealing with the Crusaders.

(16) **NOW, this is important**. If you try to rush it, it will just run away...

We have seen that the AV *then* appears to have a "staging" function in terms of giving immediate directions, while *now* appears to signal a transition between scenes or a commentary on the gameplay. In contrast, *here* (used as a pronominal) is typically found in contexts in which it presents new items (these are metagame items about the game world, such as *lists, locations, rewards*, etc.):

(17) **Here's a list** of those tags.

(18) **Here are the locations** of all of the frogs in the city:

(19) **Here are your rewards** for winning: ...

Finally, another significant use of *here* is as a deictic adverbial to briefly describe an encounter, especially in the recurrent pattern *<here you VX ('ll, will) find NN>* (to some extent, this usage brings us back to the sequential use of *then*):

(20) 1F: **Here you find** 3rd Lift Engine Room,

(21) **Here you'll find** the final pedastal [*sic*], and the final change for your ring.

(22) **Here you will find** some houses with strange computers that offer information on the final dungeon.

4.3 CJ conditional advice and choice

In this section, I examine three key subordinating conjunctions (CJ). It is not surprising that items such as *if* and *once* (and related items such as *after*) are salient in the VGT and VGW corpora, as they are associated with the expression of conditional instructions (which as pointed out by Hawreliak [this volume], are part of the underlying computational "source code" of this type of discourse). It is also interesting to note that this pattern is also found in other procedural genres such as boardgame rules, laboratory protocols, recipes, and so on (Coutherut 2016).

The subordinator *if* has a high Keyness score not only in the VGW corpus (rank 8) but also in VGTs (albeit slightly lower down, at rank 15). Although *if* is used in many different contexts, I suggest that it is associated with three general patterns in these corpora. In pattern IF-1, *<if you VV (cancel, cause, connect, land, hit) a NN (attack, poke, knockdown,* etc.)>, the author offers advice to the player in the main clause, although this is conditional upon a move or attack described in the subordinate clause with a technical VV or NN:

(23) Lightning Kicks are very safe when blocked and so give you some free chip damage **if you can cancel a poke into them**.

(24) And **if you cause a knockdown**, try using Zangief's s.HK to purposely whiff over the opponent...

(25) **If you land a Tenshin throw**, this is a good way to capitalize and use up your super meter.

In contrast, pattern IF-2 (<*if an opponent get/s AJ / AV [evaluative]*>) tends to occur in the VGT corpus and is semantically more restricted. This construction involves a main clause offering advice modified by a subordinate clause, in which an *opponent* is described in terms of his/her behavior, movement, or some other quality expressed by an attributive use of the VV *get*. In many cases, the main clause offering advice involves a serial VV such as *try*, as we can see in these examples:

(26) **If an opponent gets comfy** behind a low guard, *try* jumping backward and tagging the opponent with an instant overheadj.HP[16]

(27) **If your opponent gets wise** to your cross-up attempts, *try* mixing it up with this diving kick.

(28) **If your opponent somehow get** [*sic*] **out of the corner**, don't panic and *try* to switch…

Finally, pattern IF-3, <*if you VV (expect, know, sense)*>, formulates advice to the player in terms of a choice expressed in the main clause, with the conditional clause qualifying this choice as an aspect of the player's affect or mental state. This rather more subtle pattern of advice is typical of the VGW corpus:

(29) This is an optional battle, so do it only **if you feel like it.**

(30) It also hits twice, so throw it out **if you sense** a Focus Attack coming.

(31) **If you want** your arts to wear down all that HP faster, you may consider…

The other key conjunctions to be found in the VGW list are *once* (rank 6 in the VGW) and *after* (rank 12 in the VGW, although this item may be an adverb, preposition, or conjunction). While *if* expresses a clause relation which affects whether the propositional content of the main clause is realized or not, items such as *once* and *after* express circumstantial meanings which affect the manner or means by which the main clause is realized (Halliday and Matthiessen 2014). In general, clauses introduced by *once* refer to in-game situations that have to be accomplished before the next action or event can occur:

(32) **Once you successfully make the ramp jump**, go NW someways and find Zain […]

(33) Have Hero and Atsuro do the same and attack him. **Once he's gone**, gang up on the one at E07 (white dude).

(34) You have to do the Priestess Door first. **Once that's done**, you can do whichever door you want.

In contrast, *after* as a subordinator is typically used with meta-comments, that is to say, instructions that relate to game controls, a particular stage of the game, or more general gameplay:

(35) **After this stage is cleared** you cannot go back to any of the maps in Episode 13

(36) Asgard, **after you've completed disc 1.** This will be your final encounter.

(37) Her second unique cancel is triggered by tapping PP immediately **after inputting the command** for her Seismic Hammer

Generally speaking, the patterns discussed above suggest that VGWs exploit a much wider range of clause expressions than VGTs, an observation that is also confirmed when looking at the PN *you* and auxiliary verbs, as will be described here.

4.4 PN reformulation and evaluation

There has been much previous research on the key role of pronouns such as *you* in gaming discourse and in other procedural genres (Lassen 2003; Ensslin and Bell 2012; Coutherut 2016, among others). It is therefore not surprising to find that there are several pronouns (PN) toward the top of the Keywords list of salient grammatical items in both the VGT and VGW corpora. For space reasons, I cannot here provide a full analysis of these items, but instead I will focus on a small handful of patterns that are associated with *you, it,* and *this* (the latter two both ranking highly in the VGT).

Broadly speaking, it is possible to identify three main patterns for *you* in the VGT and VGW corpora: YOU-1: definitions of the player's specific abilities using *can*, as in *<you can VV NN>*); YOU-2: advice expressed by deontic modals, such as *<you VX (better, had better, should,) VV>*; and YOU-3: conditions and comments expressed in subordinate clauses, such as *<CJ (if, when, whenever) you are ready>*.

I suggest that pattern YOU-1 accounts for many uses of both *you* and *can* in both corpora. This usage corresponds to the "radical" (enabling) use of the modal VX *can*. In many contexts, *you can* defines a player's special ability, generally in terms of a technical verb. The following examples come from the VGT:

(38) Here's a quick list of effective moves during which **you can buffer** the inputs: ...

(39) **You can easily juggle** the ultra for most hits off of a High Step Kick, an FADC'd Tiger Uppercut, ...

(40) **You can plink** lk and mp together (which the system will use the mp) and tap out EX legs for a quick confirm into a juggle.

Pattern YOU-2 is similar, but here *you* is the subject of a variety of modal expressions expressing advice, such as <*you VX (be better off, gotta, got to, have better, wanna, want to, will need to) VV (learn, rely on, try, use)*>. Here, the lexical verbs are not as specific as in pattern YOU-1; rather they tend to be "conatives" (such as *to learn to, to try to*), that is to say, items that express the relative success or failure of a process rather than the process itself, or that express the subject's attempts to accomplish a process:

(41) Really, **you're better off** *using* the meter for EX attacks and FADCs.

(42) A good general combo into Gen's solid super, which **you should learn** well.

(43) As a Honda player, **you want to try and make** the fight an up-close battle...

Pattern YOU-3 involves a variety of subordinate clauses, in which the player's state of readiness is presented as a condition on which the next piece of advice or the rest of the adventure is dependent. One extended version of this pattern that is typical of VGWs has the following wording: <*CJ (once, when, whenever) you are AJ (ready, set)*>:

(44) **Once you are set**, focus on taking them down by one.

(45) **When you're ready** to proceed, go to the lower level.

(46) **Whenever you are ready** to leave, speak with Mitsuru.

We now turn to the pronouns *it* and *this*. Both are especially salient in VGTs. Unsurprisingly, *it* has a wide range of uses, most notably as an anaphoric item referring to a specific object or skill. In the VGT corpus, *it* as subject is often used to introduce an evaluating clause. One regular example of this involves the extended LG pattern <*CJ (and, but, so) it VV AJ (good, great) for VV (catch, dodge, knock, pressure, throw+ing)*>. This pattern provides a positive judgment of a skill or attack that has been defined in the antecedent context, while at the same time specifying what type of attack *it* can be used for:

(47) Dash Low Straight—Hits low, **so it's good for catching** opponents as they try to back away from Balrog's mean face.

(48) The invincibility on the teleport starts instantly, **so it's good for dodging** full screen fireballs...

(49) It'll pass over the heads of crouching opponents... **But it's good for knocking** opponents out of the air.

As complement, *it* is also used in another clear example of an extended LG pattern *<use it as a NN (counter, poke)>*, in which the VGT author describes an alternative way of using a previously defined attack:

(50) The EX version will even go through projectiles so **use it as a counter** to fireballs from mid-range.

(51) Instead, Gouken's c.MP—go figure—acts a lot like most characters' c.MK, hitting low with decent range and cancelability, so **use it as a poke.**

(52) If you get knocked down, **use it as a wake-up counter.**

The general grammatical difference between *it* and *this* is that *it* typically reiterates a specific antecedent referent, while *this* potentially introduces an element of reformulation or evaluation, with the possibility of broadening or narrowing the frame of reference (Huddleston and Pullum 2002). One "narrow" use of *this* in the VGT corpus is to introduce the definition of an immediately adjacent term (a type of attack) while introducing an evaluation (clearly a fundamental function of this kind of instructional discourse, as pointed out in Ensslin 2012). As these contexts are typically definitions, the subsequent reformulation renames the specific term for an attack using a hypernym: *<this is AT AJ (good, solid) NN (attack, move, overhead, super, way, etc.)>*:

(53) Shienkyaku—**This is a good anti-air attack** with good priority.

(54) Fuhajin—**This is a very interesting projectile move.**

(55) **This is a pretty good super** that's very easy to combo—for example.

In the VGW corpus, *this* is typically used in a pattern that sums up an entire situation or reformulates an in-game item or event in order to evaluate it. Since this usage involves various different patterns, the following sample gives an idea of the different kinds of structure which share this "summarizing" function (as mentioned above, a similar set of expressions are introduced by the AV *now*):[17]

(56) However, they're very powerful, and not easily beaten **at this point in the game.**

(57) **The key to this battle** is to exploit the knock back effect of the enemies' normal attack. How? Simple.

(58) If you want to polish off an opponent **this isn't a bad way to do it.**

(59) **For this reason,** it's best to use the teleport defensively or to cover distance after knocking down an opponent.

5 Data summary and discussion

In this chapter, I have shown that certain particles (such as *up*, *off*), adverbs (*now*, *here*), conjunctions (*if*, *once*), and pronouns (*you*, *it*) are statistically more likely to occur in VGTs and VGWs in comparison with the general language (and in comparison with other major technical genres). Rather than analyzing these items in isolation, I have argued that function words provide the framework for longer stretches of expression—LG patterns, which I claim are a useful focus of analysis when we are looking initially for the main linguistic characteristics of a particular register or genre.

The analysis of LG patterns shows a number of interesting similarities and divergencies between VGTs and VGWs. Generally speaking, there are three overall sets of patterns which emerge. First, a dominant phraseology[18] that emerges in both corpora involves the *management of the player's moves within the imaginary space of the game*. In VGTs, many LG patterns (formed around particles and prepositions) deal with the manipulation of the joystick or provide a precise definition of fighting moves as in examples such as *Fairly easy to **combo into off of** a HK Spiral Arrow*. Similarly, in VGWs, many LG patterns (often built around adverbials and textual adjuncts) serve to stage events or tell the player to transition from one scene to the next (Now, *you can do two things.*). A second set of phraseological patterns involves the *framing of advice*: in VGTs (as in many other instructional genres), conjunctions such as *if* set out the conditions for concrete actions, especially types of maneuver (*If you land the cross-up, combo into a LK Tatsumaki*), while in VGWs, we see a rich diversity of clause relations, which serve to summarize necessary actions or set the scene for events (***Once** the first group of Demons is defeated, an enemy raid group will appear at A01*). A related set of phrases involves "directed imperatives," often structured around sequences of verbs, ellipses, and adverbials such as *then* (*Down back charge **then** half circle forward **then** dash*). This usage is also related to the verbal construction of instructions which involves a variety of deontic modals (*If this still doesn't work, you **might want** to try over compensation.*). Finally, a third set of phraseological patterns is concerned with *tracing the relationships between discourse referents*. Thus, in VGTs, pronouns such as *it* are used to introduce definitional and evaluative constructions (*it's a great anti-air and easily juggles*), while in VGWs, items such as *this* are used to comment on the ongoing state of play (***This** is the final event in the 7th day*). Other patterns exist, but the above-mentioned examples represent some of the most recurrent and regular sequences of expression that can be observed in these corpora. It is notable that in each case, we are not only concerned with the behavior of a single grammatical item, but also with its role within an extended pattern of use.

It is important to note that the LG patterns found in VGTs and VGWs are not unique. Even if some patterns have the appearance of specific turns of phrase, such as (*<if an opponent get/s AJ/AV [evaluative]>*) (a pattern more typical of VGTs), or an even longer stretch such as *<CJ (and, but) / AV (again, then) VV (go, [Ø]) AV (back, east, left, north, right)/PR (down, to) NN>* (a prototypical pattern in VGWs), it is likely that these constructions can be found in other, similar, contexts elsewhere in the language, especially in related types of texts, especially procedural genres such as (written) games rules, (oral) street directions, and so on. In addition, it is worth noting that the patterns observed in this study are only the most routine expressions to be found in these corpora. Thus the LG patterns identified above should really only be seen as a core linguistic "background," against which the more specific features of these texts can be brought into focus.

This point leads me back to the notion of LSP. As I mentioned in the first part of this chapter, the view of language I have adopted here is the systemic functional model (Bloor and Bloor 1986; Stubbs 1993; Halliday and Matthiessen 2014, among others). From this perspective, it is considered that all discourse is necessarily adapted to a specific context of situation (i.e., a *genre*), and as such involves distinct rhetorical functions (such as reporting, recommending, exploring, expounding, etc.). According to this model, specific discourse functions are realized by specific lexical and grammatical phrases, and these are in turn are derived from the repertoire of potential LG resources (thus, a functional genre is realized by a *register*, a specific set of linguistic choices taken from the overall system). On the basis of this approach, I would suggest that it is not useful to characterize VGTs or VGWs as genres which can exist somewhere on a continuum between LSP and LGP. Rather, I suggest that there is no such thing as "LGP"; all instances of language can be seen as belonging to one specific register or another, and thus are all forms of LSP. If this hypothesis can be entertained (following Bloor and Bloor 1986), then the important question is not whether VGTs and VGWs are closer to LSP or to LGP, but rather the whether these genres represent highly recognizable, codified, conventionalized genres or whether they are hybrid, emergent, indeterminate text types. On the basis of the above study, and having looked at the prototypical forms of expression to be found in these texts, I am tempted to say that VGTs and VGWs belong to the rather more codified end of the spectrum. Indeed, I would suggest that the phraseological patterns that can be observed in VGTs and VGWs display a very high degree of regularity as well as expressing functions that have been adapted very specifically for the particular purposes of the gamers who have produced them. This is perhaps surprising, because—as we have seen—these texts are produced by amateur fans who are not necessarily professional writers or even proficient speakers of English. It is this degree of stability or "maturity" that I would claim is the most important factor in determining whether these texts represent a recognizable genre or register, since it can be shown that they have developed not only a series of conventions relating to terminology and other local features of language, but also much broader features such as phraseology.

However, although I claim here to have identified some of the core regularities of expression in VGTs and VGWs, this is only a partial picture. The analysis of LG patterns I have set out above still does not tell us much about patterns of regularity at higher levels of analysis, such as "rhetorical moves" (Swales 1990). In addition, the analysis of phraseology has little to say about the social functions of these texts within a broader "ecology of genres" (Spinuzzi and Zachry 2000), although some corpus analysts, such as Groom (2010), have claimed that the analysis of "semantic sequences" shows the underlying ideology of certain forms of academic writing. In this regard, it would be interesting to explore the relationship between VGTs, VGWs, and other paratexts such as the "post-match report" (a genre associated with sports such as chess, cricket, and tennis in English, etc.), as well as related "alternative" genres such as games-related webseries, game reviews, let's-play commentaries, and so on. In addition, it would be interesting to see to what extent the language forms appropriated in VGTs and VGWs are "fed back" into other genres of the language, so that texts such as VGTs and VGWs may be seen as nexus points in the development of new forms of English. As Gee (2005) has pointed out, both in-game discourse and paragame genres provide an "interaction space" in which authors and players enact (or re-enact) the social contract of gameplay in verbal form. Thus rather than seeing VGTs and VGWs as overly codified discursive "cul-de-sacs," it may be possible instead to see them rather as "discourse sandboxes," in which a significant subset of language users (including presumably many young gamers and speakers of English as a lingua franca, as defined by Seidlhofer 2001) enjoy a relatively safe communicative environment, and thus a high degree of freedom to expand their language use, at the same time as exposing themselves and others to very advanced forms of the English language.

Notes

1 In this paper, I use a simplified version of the BNC tagset, where NN = noun, VV = verb, PN = pronoun, PR = preposition, and so on. See Section 4 for the full list.

2 I am grateful to my past students Camille Croz, Walter Goguillon, Barbara Paul, and Arnold Savary for their part in designing and building the VGT and VGW corpora.

3 Here I assume Halliday's distinction between "genre" as a social/functional label and "register" as a linguistic/formal label (Halliday and Matthiessen 2014).

4 *Streetfighter 4* (2008), fighting videogame, Capcom. Available at: https://www.playstation.com/en-us/games/street-fighter-iv-ps3/.

5 *The Legend of Heroes: Trails in the Sky* (2004), Japanese role-playing videogame, Nihon Falcom. Available at: http://trailsinthesky.com/fc/.

6 *Xenoblade Chronicles* (2012), Japanese role-playing videogame, Monolith Soft. Available at: http://xenobladechroniclesx.nintendo.com/.

7 Available at: https://www.eventhubs.com/guides/2007/oct/21/street-fighter-terminology-acronyms-lexicon-and-glossary-guide/.

8 *Streetfighter 4* (2008), fighting videogame, Capcom. Available at: https://www.playstation.com/en-us/games/street-fighter-iv-ps3/.

9 *Persona 3* (2006), Japanese role-playing videogame, Atlus. Available at: https://www.atlus.com/persona3/.

10 *Tales of Symphonia* (2004), Japanese role-playing videogame, Namco Tales Studio. Available at: <http://www.bn-ent.net/cs/list/talesofsymphonia_ps2/>.

11 *Radiant Historia* (2011), Japanese role-playing videogame, Atlus. Available at: http://www.atlus.com/radianthistoria/.

12 *Devil Survivor* (2009), tactical role-playing videogame, Atlus. Available at: https://www.atlus.com/devilsurvivor/.

13 This methodology was first discussed in Gledhill (2000a, b). The reason why ten grammatical items are selected as a cut-off point is that it is difficult to find more than a handful of grammatical items within the first few hundred salient Keywords, as discussed in Section 3 (and as demonstrated in Appendices 3 and 4). If the corpus is large enough and representative enough of a specific genre, then I find that the analysis of up to ten grammatical items gives a good overall picture of the main *n-grams* (repeated expressions) that are particular to this type of text.

14 There are still many items which constitute "noise" in Appendices 1 and 2. For example, the keywords list for VGW includes symbols such as *o* and *x* which are not lexical items but are used to decorate the text.

15 Here the symbols *VV/AJ* refer to a past participle verb used as a predicative adjective.

16 This is one of many abbreviations in the VGT corpus (here *j.HP* = "jump High Punch").

17 As reformulating items, some of these examples of *this* are not PNs but rather adjectival determiners (AT).

18 As mentioned in the introduction, ever since the development of large-scale corpus analysis, lexicographers and descriptive grammarians have been interested in examining the routine expressions of not only the general language (Renouf and Sinclair 1991; Stubbs 1993; Hunston and Francis 2000) but also in specific types of discourse (Biber et al. 2004, 2010; Groom 2010, inter alia). Regardless of methodological differences (which are many), many of these linguists refer to such regularities as *phraseology*. It is worth stating here that such a broad use of the word *phraseology* happens to coincide with the way the term is used in the general language, as noted in the Collins COBUILD dictionary: "Phraseology: A set of phrases used by a particular group of people" (Cobuild 2018).

References

Primary sources

Collins COBUILD (2018), *Collins Online English Dictionary*. Available online: https://www.collinsdictionary.com/dictionary/english/phraseology (accessed March 3, 2018).

Demon King (2005), "Fighting videogame," *Nintendo*. Available online: https://www.nintendo.com/games/detail/demon-king-box-3ds (accessed March 3, 2018).

Devil Survivor (2009), "Tactical role-playing videogame," *Atlus*. Available online: https://www.atlus.com/devilsurvivor/ (accessed March 3, 2018).

Eventhubs, "Videogame tutorial archive." Available online: https://www.eventhubs.com/moves/sf4/ (accessed December 5, 2018).

Gamefaqs, "Videogame walkthrough archive." Available online: https://gamefaqs.gamespot.com/psp/933329-the-legend-of-heroes-trails-in-the-sky/faqs (accessed March 3, 2015).

Persona 3 (2006), "Japanese role-playing videogame," *Atlus*. Available online: https://www.atlus.com/persona3/ (accessed March 3, 2018).

Radiant Historia (2011), "Japanese role-playing videogame," *Atlus*. Available online: http://www.atlus.com/radianthistoria/ (accessed March 3, 2018).

Shoryuken (2011), "Tutorials." Available online: http://shoryuken.com/category/tutorials-2/ (accessed March 3, 2018).

Streetfighter 4 (2008), "Fighting videogame," *Capcom*. Available online: https://www.playstation.com/en-us/games/street-fighter-iv-ps3/ (accessed March 3, 2018).

Tales of Symphonia (2004), "Japanese role-playing videogame," Namco Tales Studio. Available online: http://www.bn-ent.net/cs/list/talesofsymphonia_ps2/ (accessed March 3, 2018).

The Legend of Heroes: Trails in the Sky (2004), "Japanese role-playing videogame," Nihon Falcom. Available online: http://trailsinthesky.com/fc/ (accessed March 3, 2018).

Secondary sources

Adam, J.M. (2011), "Les consécutives intensives: Un schéma syntaxique commun à plusieurs genres de discours," *Linx*, 64–65: 115–31.

Anthony, L. (2014), "AntConc (Version 3.4.3)" [Computer Software], Tokyo, Japan: Waseda University. Available online: http://www.laurenceanthony.net/ (accessed April 3, 2018).

Biber, D., S. Conrad, and V. Cortes (2004), "If you look at...: Lexical bundles in university teaching and textbooks," *Applied Linguistics*, 25 (3): 371–405.

Biber, D., R. Reppen, and E. Friginal (2010), "Research in corpus linguistics," in R. Kaplan (ed.), *The Oxford Handbook of Applied Linguistics*, 548–67, Oxford: Oxford University Press.

Bloor, M. and T. Bloor (1986), "Languages for specific purposes: Practice and theory," *Centre for Language and Communication Studies Occasional Papers 19*, Dublin: Trinity College CELTS.

Coutherut, M. (2016), "Les textes procéduraux en Anglais: Création d'une échelle de prototypicalité," PhD thesis, Université Paris Diderot.

Ensslin, A. (2011), *The Language of Gaming*, Basingstoke: Palgrave Macmillan.

Ensslin, A. and A. Bell (2012) "'Click = Kill': Textual you in ludic digital fiction," *Storyworlds* 4: 49–74.

Firth, J.R. (1957), *Modes of Meaning: Papers in Linguistics, 1934–1951*, Oxford: Oxford University Press.

Gee, J.P. (2005), "Semiotic social spaces and affinity spaces: From The Age of Mythology to today's schools," in D. Barton and K. Tusting (eds.), *Beyond Communities of Practice: Language, Power, and Social Context*, 214–33, Cambridge: Cambridge University Press.

Gledhill, C. (1995), "Collocation and genre analysis: The phraseology of grammatical items in cancer research abstracts and articles," *Zeitschrift für Anglistik und Amerikanistik*, 43 (1/1): 11–36.

Gledhill, C. (2000a), *Collocations in Science Writing*, Tübingen: Gunter Narr Verlag.

Gledhill, C. (2000b), "The discourse function of collocation in research article introductions," *English for Specific Purposes*, 19 (2): 115–35.

Gledhill, C. (2015), "On the phraseology of grammatical items in lexico-grammatical patterns and science writing," in P. Thompson and G. Diani (eds.), *English for Academic Purposes: Approaches and Implications*, 11–42, Newcastle upon Tyne: Cambridge Scholars Publishing.

Groom, N. (2007), "Phraseology and epistemology in humanities writing," PhD thesis, University of Birmingham.

Groom, N. (2010), "Closed-class keywords and corpus-driven discourse analysis," in M. Bondi and M. Scott (eds.), *Keyness in Texts*, 59–78, Amsterdam: John Benjamins.

Halliday, M.A.K. and C.M.M. Matthiessen (2014), *Halliday's Introduction to Functional Analysis*, 4th edn., London: Arnold.

Hoey, M. (2005), *Lexical Priming: A New Theory of Words and Language*, London: Routledge.

Huddleston, R. and G. Pullum (2002), *The Cambridge Grammar of the English Language*, Cambridge: Cambridge University Press.

Hunston, S. and G. Francis (2000), *Pattern Grammar: A Corpus-Driven Approach to the Lexical Grammar of English*, Amsterdam: John Benjamins.

Lassen, E. (2003), *Accessibility and Acceptability in Technical Manuals*, Amsterdam: John Benjamins.

Née, E., F. Sitri, M. Veniard, and S. Fleury (2017), "Routines discursives et séquentialité dans des écrits professionnels: La mise au jour d'une séquence évaluative," *Corpus*, 17: 145–68.

Renouf, A. and J. Sinclair (1991), "Collocational frameworks in English," in K. Aijmer and B. Altenberg (eds.), *English Corpus Linguistics*, 128–43, London: Longman.

Seidlhofer, B. (2001), "Closing a conceptual gap: The case for a description of English as a lingua franca," *International Journal of Applied Linguistics*, 11: 133–58.

Sinclair, J. McH. (1991), *Corpus, Concordance, Collocation*, Oxford: Oxford University Press.

Sinclair, J. McH. and A. Mauranen (2006), *Linear Unit Grammar: Integrating Speech and Writing*, Amsterdam: John Benjamins.

Spinuzzi, C. and M. Zachry (2000), "Genre ecologies: An open-system approach to understanding and constructing documentation," *Journal of Computer Documentation*, 24 (3): 169–81.

Stefanowitsch, A. and S. Gries (2003), "Collostructions: Investigating the interaction between words and constructions," *International Journal of Corpus linguistics*, 8 (2): 209–43.

Stubbs, M. (1993), "British traditions in text analysis. From Firth to Sinclair," in M. Baker, G. Francis, and E. Tognini-Bonelli (eds.), *Text and Technology: In Honour of John Sinclair*, 1–33, Amsterdam: John Benjamins.

Swales, J. (1990), *Genre Analysis: English in Academic and Research Settings*, Cambridge: Cambridge University Press.

Tran, T., A. Tutin, and C. Cavalla (2016), "Typologie des séquences lexicalisées à fonction discursive dans la perspective de la rédaction scientifique," *Cahiers de Lexicologie*, 108: 161–79.

Appendices

APPENDIX 1 *The first ten keywords from the VGT corpus (568,998 tokens)*

Rank	Freq.	Keyness	Item
1	4247	57788.557	Vf
2	4249	57594.019	gif
3	946	11684.395	combo
4	900	10988.616	downright
5	878	8885.698	opponent
6	438	5934.267	jpg
7	551	5678.751	ultra
8	451	5671.799	MK
9	444	5473.516	EX
10	362	4925.702	streetf

APPENDIX 2 *The first ten keywords from the VGW corpus (111,695 tokens)*

Rank	Freq.	Keyness	Item
1	54588	606116.703	
2	2305	25593.519	Lv
3	1346	14400.420	xx
4	1060	11069.711	HL
5	1076	10531.019	location
6	1272	8291.952	x
7	1384	7750.775	o
8	758	7618.154	enemy
9	1491	7077.486	E
10	681	6878.193	quest

APPENDIX 3 *The first ten grammatical keywords from the VGT corpus*

Rank	Freq.	Keyness	Item
13	1300	4193.935	down
15	1200	4105.332	right
39	1157	1675.753	can
45	1866	1542.512	you
58	927	1113.553	up
102	210	681.547	forward
103	611	656.297	into
164	1555	373.658	it
192	52	311.637	lets
211	272	273.407	off

APPENDIX 4 *The first ten grammatical keywords from the VGW corpus*

Rank	Freq.	Keyness	Item
21	6351	4983.754	you
30	608	3755.339	up
158	1120	1010.787	get
257	1230	712.373	then
308	1741	608.446	can
394	220	501.225	once
690	91	272.549	down
821	563	219.931	if
1049	445	162.058	here
1109	245	149.288	now

4

Playing with the Language of the Future

The Localization of Science-fiction Terms in Videogames

Alice Ray

1 Introduction

At the beginning of the twenty-first century, videogames represent a great part of the global entertainment industry with billions of dollars in sales and thousands of jobs created in the United States alone (O'Hagan 2007), and "[t]he computer and video game industry continues to grow each year" (Chandler and Deming 2012: 1). Videogame publishers want to export their products to other countries so that their audience can expand and their profits increase, but if the quality of videogames has increased along with the industry weight in the global economy, so have the demands of players across the world. Indeed, immersion into virtual worlds can only be maximized if the videogame uses the player's language. International versions of games have to be created, and their quality needs to be equivalent to the original game. Therefore, one of the main features of videogame internationalization is *localization*: the "actual process of translating the language assets in a game into other languages" (Chandler and Deming 2012: 4).

Localization is now an essential part of the videogame industry, and "[e]ven though the international versions of the game have the same functionality and features, the gamer can be easily pulled out of the gameplaying experience if the quality of the localization is not good" (Chandler and Deming 2012: 3–4); it "requires the skill and art of translation" (Bernal-Merino 2015: 1). As the localization needs of videogame publishers are growing, so are the needs of a research field on the subject. There are a great number of books and articles that focus on localization in videogames—*The Game Localisation Handbook* is a perfect example (Chandler and Deming 2012)—and many of these works concentrate their analysis on the technical constraints and cultural aspects of videogames localization (see e.g., Carlson and Corliss 2010; O'Hagan and Mangiron 2004, 2006; Bernal-Merino 2015).

This chapter aims to analyze one particular linguistic aspect of videogame localization: the translation of invented words in science-fiction videogames. Science fiction and videogames have always been inherently interconnected—even if the nature of this link is not clear[1]—and one of the specificities of the genre is to create new words in order to provide credibility to the world imagined by the authors.

For that purpose, I shall focus on two American science-fiction videogames and their localization into French:[2] *Starcraft II, Wings of Liberty* (Blizzard Entertainment 2010); and *Alien: Isolation* (The Creative Assembly 2014). *Starcraft* is a real-time strategy game in which the player embodies Jim Raynor, a marshal who becomes a rebel and fights against both the Dominion and its leader, Arcturus Mengsk; and the invasion of Zergs, an alien species. It is clearly a wargame: the player has to manage her/his army, think about new strategies, and improve new technological military items in order to thwart the enemy and win battles. *Alien: Isolation* is considered a survival horror game in which the player is immersed into an atmosphere of constant fear. The player embodies (with a complete internal point of view) the main character, Amanda Ripley, daughter of Ellen Ripley (the main protagonist of the *Alien* franchise), who decides to investigate the disappearance of the Nostromo, her mother's spaceship. She is sent on Sevastopol, a space station in orbit, to bring back the flight recorder of the Nostromo. However, the flight recorder is not the only thing the team brought back from the ghost spaceship, and Amanda will soon meet the creature that made audiences jump in their seats back in 1979. *Alien* is a specific kind of survival horror videogame as the player does not have to kill the monsters[3] because xenomorphs (the aliens) cannot be killed with human weapons. The player only needs to survive and avoid the creature (and other enemies).

These games have been chosen because, first, they do not involve the same gameplay that allows this analysis to determine if the gameplay has any influence on the lexical creativity and the translation in science-fiction videogames. Second, the environments of these videogames are distinctive. *Starcraft* offers the vision of a far future with advanced technologies, while *Alien* offers a future world in which technology and scientific innovations are not specifically a part of the narrative. Third, the games were successful and well acclaimed both by videogame critics and players.[4] Finally, the environments and narratives of these games introduce a great number of new lexical creations.

Using contrastive linguistic analysis (involving morphology, lexical semantics, etc.), I shall determine if some translation patterns are used to (re) create those invented words, the reasons why their localization is important as a way of immersing the player in a new alien environment, and how the gameplay can also be a constraint for the localization of the invented terms. After contextualizing videogame localization and the science-fiction genre, summarizing the creative strategies for producing new words in English and French, and introducing the methodology used in this study (see also Álvarez-Bolado and Álvarez de Mon, this volume), I shall analyze how invented words[5] are created in the original and French versions of the videogames.

2 Videogames: Localization and science fiction

In videogames—as well as in the translation of software—translation is referred to as *localization*. This process is the transfer of a videogame's language assets (i.e., connotation and denotation, cultural references, plays on words, lexical creativity, language register) from one language into another. None of the other aspects of the game will be changed during localization, as the previous step is the internationalization of the product, which means that it has already been adapted to receive the new language.

There are different kinds of videogame localization:

(1)   Minimal "Box and Docs" localization: only the packaging and documentation are translated (Bernal-Merino 2011: 14).

(2)   Partial localization: the user interface and interactive menus are translated and subtitles "for pre-rendered cut scenes, and in-game animations" are provided (Bernal-Merino 2011: 15).

(3)   Full localization: the entire game is translated (Bernal-Merino 2011).

Some games need to be internationalized (i.e., the design is altered) as the users have to be "convinced that the international versions were planned for them from the beginning" (Chandler and Deming 2012: 7). Examples can be found in which some of the game's elements are slightly different depending on the target culture, as "publishers and localizers have to guarantee that game elements that could be misinterpreted or lost in translation are deleted or adapted" (Bernal-Merino 2011: 17). For example, in a Japanese videogame entitled *Fatal Frame*, Francesca Di Marco (2007) noticed some differences in the female protagonist between the original version and the localized version (United States). In the original videogame, the female character looks young, she is wearing a school uniform, and her hair is black; in the localized version, she looks older, she's wearing casual clothes, and her hair has been lightened. The internationalization process takes into account the geopolitical and cultural forces at play to adapt the videogame design in each country where the videogame will be released. I contend that this process is beyond the language assets of the game and hence beyond the translator's role—except, of course, when the cultural references are in the text itself.

The decision to implement a full localization depends on the economic reliability of the publishers and the degree of internationalization desired for the game. In spite of being the most expensive option, full localization is "slowly becoming the standard" (Bernal-Merino 2011: 17). However, videogame localization is a specific kind of translation: not only does it include all the audiovisual translation features (subtitling, space constraints, voiceover, cultural and historical references,[6] etc.), but it also has to consider the gameplay and the players' experience of the game, as well as interactivity; the translated game "does not have to be loyal to the original text, but rather to the overall gaming experience" (Mangiron and O'Hagan 2006). Players are not only immersed in another world, but they are also a part of it as they are active and recreate the game as long as they are playing it: "the game is nothing else but what the player does when he plays" (Triclot 2011: 23). To be completely immersed in another universe and forget the surrounding reality, players need to understand the game and its environment without making any effort, and to enjoy "the same gaming experience as someone who plays the source version of the game" (Chandler and Deming 2012: 7). Furthermore, if "[g]ame localization is unique in the sense that it may require both the skills of a technical and literary translator" (Dietz 2007), the narrative aspect of the videogame cannot be translated as a novel. After all, the "stories in videogames are non-linear because they depend on players' decisions" (Bernal-Merino 2007: 5). The translation needs to take these factors into account. Videogames, more than any other audiovisual medium, have to be immersive as reality cannot interfere with the gameplay experience, and thus the translation has to retain "'the look and feel' of the original" (Mangiron and O'Hagan 2006: 14).

However, as videogames are both entertainment media and works of art, designers continuously have to come up with new ideas, new gameplay, and new stories. And even if their localization produces a great number of constraints induced by the content as well as the form, translators also need to be creative, not only to respond to the player's demands but also because videogame writing is a mixture of artistic and specialized writing. In other words, a great deal of specialized terms can be found in every single videogame (e.g., *load*, *save the game*, *cheat code*, *gameplay*). Yet, in each case, players are also dealing with a narrative with a specific environment, specific characters, and specific objects. Thus, videogame designers use both game-specific lexis already known by the players (*game over, loading, save the game*, etc.) and lexical innovations to polish their games and offer players a new and original product.

Science fiction is a very popular genre, especially in visual arts,[7] and videogames are not an exception. Although videogames are categorized in terms of gameplay,[8] unlike films or books, which are classified by genre, a great number of videogames display an obvious science-fiction setting or narrative: "the majority of new story games and many (perhaps most) examples of strategic planning games and massively multiplayer online games are still sf or fantasy" (Tringham 2015: 13). Unlike movies, but exactly like boardgames and tabletop RPGs, videogames give the player the opportunity to create an emergent story, or at least to be a part of it, and imaginary genres offer perfect settings for such personalized experiences.

The imaginary worlds of science fiction are an experimental field: the player knows that it is not real, but the whole narrative is so intense and so detailed that, during the game, she/he believes it could be true, in a future or faraway world. The genre deals with experimentation, and in videogames, the experiment becomes an experience: the player can be a part of a possible future (or past) and test it. Science-fiction videogames offer a real opportunity to be a part of a science-fiction experiment about humanity:

> One of the premier values of science fiction as literature is that it enables us to look at ourselves through alien eyes. It enables us [...] to see not only what is, but, submerged in it, what has been, and what will be: to perceive the linkages, the connections, the web of cause-and-effect that holds the world together. (Dozois 1976: 115)

3 Let's play with the language of the future

In this section, I will focus on a comparative study and the way lexical creations are translated from English into French in both *Starcraft* and *Alien*. First and foremost, I will briefly define some of the creative strategies used by both languages to produce new words, focusing on those most used in the videogames:

(1) Compounding: creates a lexical item "by combining existing words or lexical elements, leading to a new form" (Valeontis and Mantzari 2006: 5); for example, *chatroom* or *science fiction*.

(2) Derivation: creates a lexical item "by adding one or more affixes to a *root* or to a *word*" (Valeontis and Mantzari 2006: 5); for example *anti-aircraft* or *activistic*.

(3) Semantic shift:[9] describes a process during which "an existing term
     [...] is used in order to designate a different concept" (Valeontis
     and Mantzari 2006: 6); for example, the term *mouse*, which now
     designates the device used to move the cursor of a computer.

Some studies have been conducted in science-fiction neologism and language,
but there is no space here to discuss them.[10] Lexical creations of science
fiction are particularly significant in videogames as they not only describe
new items and concepts but also allow players to use them in the videogame:
if the items have a name, a purpose, and a specific way of functioning, it
makes them more real, more tangible, and more useful for winning the
videogame. Thus, the translation of these lexical creations is challenging, as
they are a key part of the gaming experience.

3.1 Corpus-based methodology

Both games were played in full in both languages, and each lexical creation
was manually noted in a table with its French equivalent. The neologisms
were determined according to: a) the terms do not appear in any dictionary[11]
or in any other text with the meaning they convey inside the videogame; and
b) the terms only appear in other science-fiction contexts. As the format of
this chapter does not allow me to include all the terms of our corpus (239
terms for *Starcraft* and 205 for *Alien*),[12] I decided to analyze a sample: the
chosen terms were selected according to their frequency, their construction
(equivalent in both the videogames), and the distinct challenges they offer
in regard to translation. All the terms were analyzed according to basic
principles of word formation (Valeontis and Mantzari 2006; Spencer and
Zwicky 2001; Tournier 1985; Humbley 2006). The aim of this study is to
offer an introduction to the subject based on a lexical analysis of the various
phenomena which occur during the creation and localization of science-
fiction invented words.

The comparative analysis will be divided in five parts. The first section
is an introduction of the lexical creations in the videogames; then, each
section deals with a specific word-formation pattern: N + N and Adj
(+ Adj) + N compounding; the items created by derivation; and, finally, the
ones created by semantic transfer. The results will be compared, and I will
analyze whether the videogames' gameplay and narrative differences have
an influence on the translation of their lexical creations.

3.2 *Starcraft II: Wings of Liberty* and *Alien: Isolation*: Digital space adventures

Starcraft contains a lot of lexical creations, whether they are formal or
semantic neologisms (or fictive words, as introduced by Angenot [1979]),

as the videogame creators have imagined three whole new species and societies—the Dominion (Terran), the Protoss, and the Zerg—with their own technological tree and military units (see Figure 4.1). The lexical creations enter the lexicon—or *xenoencyclopedia* (Saint-Gelais 1999)—of the player so that she/he can suspend her/his disbelief and immerse her/himself into the fictive and ludic world. Even though the players need to believe that all those things could be real someday/somewhere, there is no, or very little, link with our present technology.

Alien contains a great number of lexical creations due to a semantic shift, unlike *Starcraft*, which tends to create new morphological terms. *Alien* does not involve new societies, as the player is only immersed in the future of humanity and the videogame setting is an enclosed space: an orbital space station where discoveries are limited.[13] However, as the story is set in the future, it includes some new objects and technologies. Unlike *Starcraft*, *Alien* claims to be more technologically and scientifically realistic. Most of the lexical creations refer to technological items which could be within our reach someday.[14] None of the lexical creations in *Alien* refer to completely new items with no scientific or technical roots (unlike Protoss technology in *Starcraft*, for example).

In both videogames, I noticed some recurrent patterns in word formation: N + N; Adj (+Adj) + N, derivational words, and semantic transfer. To compare both the creation and translation of the new lexical items introduced in their gameplay, a sample of each pattern will be analyzed and compared.

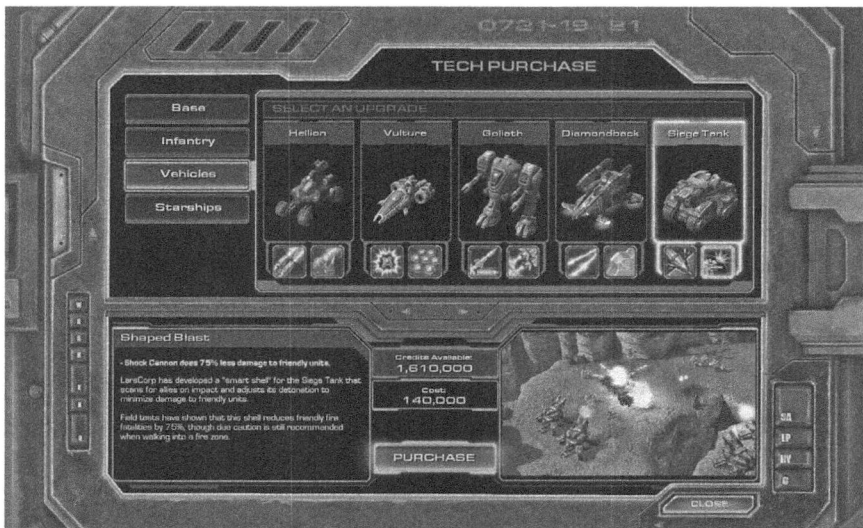

FIGURE 4.1 *Tech purchase menu for Terran units. Screenshot by Alice Ray,* © *Blizzard Entertainment.*

3.2.1 N + N compounds

Each example of N + N compounds was analyzed according to the relation between the modifier and the modified (head noun)—this relation creates a brand new object or concept in the science-fictional world. All these semantic relations between the two lexical items are relevant concerning translation issues, as the semantic relations[15] between them need to be clarified during the translation process in French: "The difficulty of translating the noun + noun sequences into French naturally stems from the fact that this structural pattern is not characteristic of French" (Bagge and Manning 2007: 567).

Some translation patterns can already be observed for N + N compounds (Table 4.1). However, it is obvious that the semantic relation between the head noun (N_2) and the modifier (N_1) is not the only element to consider during the translation process:

(1)   In the first example, the translated compound is agglutinated to form a noun using a V + suffix sequence. The verb *disloquer* in French means "to separate with more or less violence the parts that compose an object," ("disloquer," CNRTL),[16] but there is no specific word for describing something that does the action of *disloquer*. The suffix -*eur* allows the creation of the agent of the activity: *disloqueur*[17] (DSF n.d.: xx). The fact that the "void ray" is actually a combat unit is important for the translation, as the term *disloqueur* in French evokes the image of an item in the player's mind. The context of the game is vital here for disambiguation.

(2)   The semantic relation between nouns is more explicit in the translation through the preposition *de*. However, the preposition can be ambiguous, and it is the specific context of the videogame which helps the players to understand what *prison du vide* exactly means: a psionic power which disables the enemy by freezing it. The targeted unit is isolated and remains in a void matter for a while, unable to attack or use any special abilities. The preposition *de* can be misunderstood by the player as a prison located in a void instead of being a prison made of a nonmatter material, the "void."

(3)   The noun *terrazine* refers to a specific kind of gas. The translation deletes the noun *gas*, since the suffix -*ine* in French already refers to a chemical substance ("-ine," CNRTL). Moreover, space is one of the main technical constraints in a videogame: the translation cannot exceed a certain amount of signs which are inscribed in the videogame code. French translated texts are generally expanded when the source language is English because of a more important expansion factor.[18] In this example, the translator is able to reduce the amount of text in the French translation while keeping the semantic context of the term.

TABLE 4.1 *Analysis of N + N compounds*

N°	Videogame	Original term	Semantic relation*	French translation	Translation patterns[†]
1.	*Starcraft*	void ray	cause N_2 is responsible for N_1	disloqueur	V + suffix "-eur"
2.	*Starcraft*	void prison	material N_2 is made of N_1	prison du vide	N_b + preposition "de" + N_a
3.	*Starcraft*	terrazine gas	material N_2 is made of N_1	terrazine	N_a
4.	*Starcraft*	ghost academy	possession N_2 belongs to N_1	académie fantôme	$N_a + N_b$
5.	*Starcraft*	punisher grenades	type N_1 defines the kind of N_2	grenades punition	$N_b + N_a$
6.	*Starcraft*	psi disrupter	type N_1 defines the kind of N_2	disrupteur psi	N_b + Adj
7.	*Alien*	orbital outposts	location N_1 is the space location of N_2	comptoirs orbitaux	$N_b + Adj_a$
8.	*Alien*	ambulance shuttle	Type N_1 is the quality of N_2	navette-ambulance	N_b–N_a
9.	*Alien*	priming mechanisms	purpose N_1 defines the purpose of N_2	mécanisme de détachement	N_b + preposition "de" + N_a

(*Continued*)

TABLE 4.1 (*Continued*)

N°	Videogame	Original term	Semantic relation*	French translation	Translation patterns[†]
10.	*Alien*	supply ship	Purpose N_1 defines the purpose of N_2	vaisseau d'approvisionnement	N_b + preposition "de" + N_a
11.	*Alien*	hypersleep chambers	purpose N_1 defines the purpose of N_2	chambres d'hypersommeil	N_b + preposition "de" + N_a
12.	*Alien*	ion torch	Use N_2 uses N_1 to function	soudeur ionique	N_b + Adj_a

* I have used the semantic relations categories given in Moldovan et al. (2004) and Balyan and Chatterjee (2015) as they offer clear categories which can correspond to every nominal compound (i.e., all compounds with a noun as head). In order to be absolutely clear about the categorization of semantic relations, I also used seed verbs (i.e., verbs used to paraphrase semantic relations), as proposed by Balyan and Chatterjee (2015: 96–97). The semantic relations between the lexical items are mainly determined by subjectivity and interpretation. Some relations are ambiguous and can easily be stated otherwise.

† To compare the original structure with the chosen translation patterns: $N_a = N_1$ and $N_b = N_2$.

(4) The translation maintains the N + N sequence in this example. Both nouns are inverted in the French version, but the first component, *ghost*, is not turned into an adjective or explained through a preposition. Indeed, the term *fantôme* refers to a specific Terran unit:[19] the term describing the structure in which the players can invoke the ghost units needs to incorporate the name of the unit. The term *académie fantôme* is perfectly clear within the context of the videogame and respectful of the grammar rules of the target language: N_b functions as the label of N_a.

(5) The French translation maintains a N + N sequence but a modulation[20] is made on the modifier *punisher*. Indeed, the translation does not depict the agent ("punisher") but the concept in itself (*punition*, i.e., "punishment"). There is an adjective in French which describes the agent of punishment: *punisseur* ("punisseur," CNRTL). However, the use of this adjective as modifier would have diminished the produced effect: *grenades punisseuses* has a weaker impact on the player's mind, while *grenade punition* gives the same importance to both the head noun and the classifier and highlights the violence of the weapon. The change of point of view (from the agent to the concept) allows the translator to keep the meaning of the term and the terrific image it implies (powerful explosives) without disturbing the gaming experience.

(6) The word *disrupter* is adapted to French with the transformation of the suffix *-er* into the suffix *-eur*. The term *disrupteur* is used as an anglicism in some domains;[21] therefore, the technical connotation it implies transfers perfectly the idea of a new technological item. Moreover, the term *psi* is used to describe paranormal or extrasensorial faculties, as it is the case here ("psi," CNRTL).

(7) Obviously, the semantic relation between the two components is a relation of space location. The French could have kept two nouns linked with a preposition of location as *comptoirs en orbite*. However, the adjective *orbital* also refers to the orbit place around a planet (or a spaceship) and, in line with space constraints, using the adjective is shorter. Here, a transposition[22] keeps exactly the same semantic feature as the original as well as almost the same number of signs.

(8) The compound word is also translated with a N + N pattern. Indeed, both *navette* and *ambulance* refer to the same kind of item—a vehicle—except that the latter also refers to its type or function. In French, some vehicle denominations use this N + N structure in order to specify the vehicle (e.g., its function or a

physical feature): *bateau-mouche* or *camion-citerne*, for example. The new term *navette-ambulance* is formed on the same pattern and anchored in a linguistic reality familiar to francophone players. Therefore, the videogame context seems to be closer to the player's reality, and the gaming experience is improved.

(9) The French translation uses a preposition to highlight the semantic relation between the two nouns. However, *priming* and *détachement* do not have exactly the same meaning: the English term suggests that the mechanism is used to get the spacecraft ready for departure (ODE 2010: 1410); the French term is a hyponym as *détachement* gives the function of the mechanisms: it unfastens the ship so that it can move away from the station.

(10) The first noun clearly defines the purpose of the second one. A *supply ship* is a ship with a particular goal. In French, the translation uses the pattern N + preposition + N to explain the semantic relation. The construction *vaisseau-approvisionnement* could have been imagined, but the length of the two components are too different (*vaisseau* contains two syllables and *approvisionnement* six syllables).

(11) The term *hypersleep* is in itself a lexical creation. The translator has two choices: whether to translate this new noun as an adjective in order to obtain a translation pattern N + Adj, or to translate it as another noun and explain the semantic relation of function between the two components with the preposition *de*. The adjective derived from *sommeil* is *somnogène*, but this adjective would change the meaning of the fictive term as it means "which makes someone or something feel sleepy" ("sommeil," CNRTL); *chambres hypersomnogènes* would mean that the chambers themselves make the astronauts go into hypersleep (like a sleeping pill). However, *chambres d'hypersommeil* clearly indicates that the chambers are rooms used to get the astronauts into hypersleep: they are not the machine which causes the astronauts to sleep. The structure pattern N + preposition + N prevents any ambiguity.

(12) This example is a technological item translated with the N + Adj pattern, *torch* being replaced by the hyponym *soudeur*. The term *torch* can be ambiguous as it has many definitions (ODE 2010: 1876), and the translator needs to know exactly what kind of item the *ion torch* is in the context of the videogame (see Figure 4.2). The look and the use of the item clearly indicate that it is a kind of *soudeur* (i.e., blowlamp). The modifier term *ion* has an adjectival equivalent in French *ionique* ("ionique," CNRTL), and using this adjective makes the term shorter and more respectful of the space constraint.

FIGURE 4.2 *Ion torch. Screenshot by Alice Ray, © The Creative Assembly Limited and Sega Corporation.*

If some translation patterns are used several times in the translation of N + N compounds, it is obvious that semantic relations are not the only factors at play. Fictive words must be convincingly translated as if they were potential words, and the pronunciation also needs to correspond to something the language users are used to hearing. The localization of the terms is crucial because it must retain the same playability and interactivity in the translated game as in the original version. Moreover, the technical constraint may also be crucial in translation patterns.

Through the examples of N + N pattern localization treatment, it is obvious that, whereas *Starcraft* lexical creations tend to be translated with the same compositional patterns as in the original version to highlight the strangeness of the terms and create a real cognitive estrangement with the players, the lexical creations of *Alien* tend to be clarified (Berman 2000: 289); the semantic relations between the components are clearly expressed. The phenomenon can be explained by the difference of gameplay between the videogames: *Starcraft* is a wargame in which the player is consciously immersed in a completely new world; *Alien* is a survival game in which the player needs to be alienated in order to feel terrified. The lexical creations tend to reflect these differences and so do their localizations.

3.2.2 *Adj (+ Adj) + N pattern*

Some lexical creations are based on an Adj + N sequence; the sequence forms describe a new item or a new reality. The adjective in these structures is not only used as a modifier, but it cannot be detached from the noun it classifies as it identifies the entire new concept described. The semantic relation of the components can also be extracted from a paraphrase in order to help understand the term and its French translation.

(13) The first example in Table 4.2 is an Adj–Adj + N nominal compound. The two adjectives describe a feature of the nominal head: the flamethrower is composed of two parts bound together (it is vital for the translator here to have access to an image of the item in order to visualize the true semantic relation between the components). The word *jumelé* in French is used because it is strongly associated with military items, describing the assembly of "(two or more automatic weapons) for the purpose of a simultaneous shot" ("jumelé," CNRTL). Using an adjective strongly connected to the military world is a way to anchor the lexical creation in what the player already knows and to make her/him believe that the item could be used in an authentic military context. The translation adds another noun, *version*, and substitutes the first adjective *twin* with a prefix *bi-*, which has the same meaning (DEF n.d.: IV) but adds a scientific dimension to the word because of its Latin roots. The noun *version* highlights the fact that the referred item is a more powerful kind of flamethrower (deployed on the Hellion unit, Figure 4.3).

(14) The English version of the term uses a comparative element as modifier: *greater*. French does not possess affixes to express comparison; instead, a phrase has to be used: *la plus grande*. However, in order to respect space constraints and because of the nature of the referent in itself, the comparison can be deleted; indeed, *greater spire* is the name of a Zerg unit; the suffix *-er* has been added only to emphasize the difference between the simple *spire* and its evolution, the *greater spire* (see Figure 4.4). In the translation, this emphasis can be rendered by the adjective *grande*.

(15) In the translation of *short-range ambulance*, the semantic relation between the two components has been highlighted by the preposition *de*: one of the features of this kind of space ambulance is its short-range movement. In French, this feature is often expressed with the lexicalized prepositional phrases *de courte portée* or *à courte portée*: *force nucléaire à courte portée* and *appareil de courte portée*. The localized version displays an already known pattern to immerse the player into the novelties set by the videogame while using familiar patterns.

(16) The lexical creation *working Joe* describes a kind of android used to perform unpleasant tasks. As they are only used to work on the spaceship, they are all similar and look like giant dolls (see Figure 4.5). In English, the use of the proper noun *Joe* is a cultural reference: *Joe* refers to "an ordinary man" (ODE 2010: 943), thus reflecting the monotony of representation. In French, there is no

TABLE 4.2 *Analysis of Adj + N structure*

N°	Videogame	Original terms	Semantic relation	French translations	Translation patterns
13.	*Starcraft*	twin-linked flamethrower	property Adj$_{1-2}$ defines the kind of N$_1$	version bi-jumelée du lance-flammes*	N$_b$ + prefix$_a$ – Adj$_b$ + preposition "de" + N$_a$
14.	*Starcraft*	greater spire	property Adj$_1$ defines the kind of N$_1$	grande aiguille†	Adj$_a$ + N$_a$
15.	*Alien*	short-range ambulance	Property Adj$_1$ (Adj$_1$–N$_1$) is one of the N$_2$'s features	ambulance de courte portée	N$_2$ + preposition + Adj$_1$ + N$_1$
16.	*Alien*	working Joe	purpose Adj$_1$ defines the purpose of N$_1$	androïde lambda	N$_1$ + Adj$_1$

* Here, I haven't considered the term *lance-flammes* as a V + N nominal compound because the term is already a lexicalized compound in the general French dictionary ("lance-flammes," CNRTL).

† The term *greater spire* cannot be considered just as a nominal group in the context of the videogame as it describes a whole new entity; the adjective is a part of this new item.

FIGURE 4.3 *Hellion unit. Screenshot by Alice Ray, © Blizzard Entertainment.*

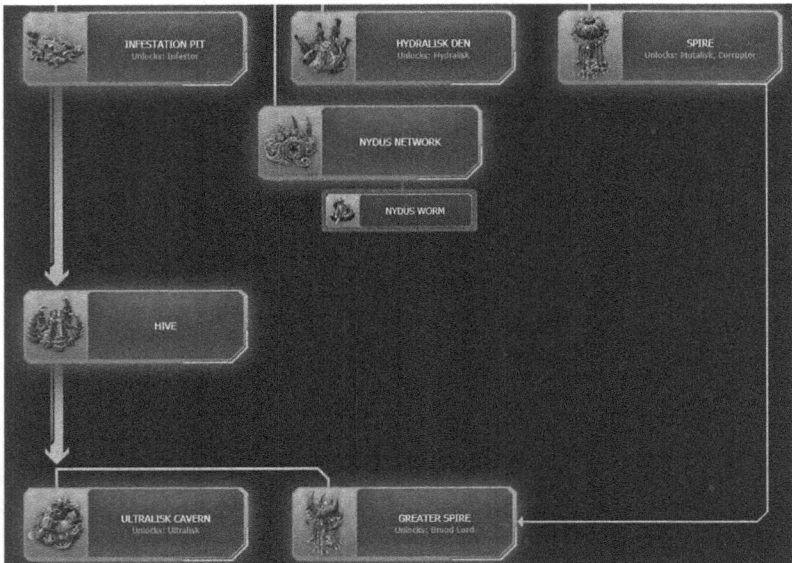

FIGURE 4.4 *Zerg tech tree. Screenshot by Alice Ray, © Blizzard Entertainment.*

FIGURE 4.5 *Working Joe. Screenshot by Alice Ray, © The Creative Assembly Limited and Sega Corporation.*

proper noun which bears the same idea. In order to maintain a short term,[23] the translator used a Greek term which refers exactly to the same idea except that it can be used with anything (whereas *Joe* is only used with human beings): *lambda*. The adjective *working* has been replaced in the translation by the noun *androïde*[24] as the adjective *lambda* needs a modified word. Perhaps, using the noun *travailleur* would have been too heavily connoted in French as it can relate to political and social discourses.[25]

The analysis of Adj + N nominal compounds has emphasized the fact that the context in which the lexical creations are made is extremely important. The semantic relations that are implicit in complex nominals can be deduced not only thanks to the definition of the terms but also through their visual representations. Through these examples, it becomes obvious that even though it was legitimate to think that a structure in Adj + N would be translated by a structure in N + Adj in French, the translation patterns used are more diversified and rely mainly on the semantic relation between the noun and the adjective, on the subjectivity of the translator, and on the context of the videogame. The nature of the imaginary referents and the context of their use determine the way the lexical creations will be translated.

3.2.3 Derivation[26]

Derivation (or affixation) is one of the most productive ways of creating new words (Lieber and Stekauer 2014), especially in science and technology (Table 4.3).

(17) and (20) All the terms created from Greek roots refer to new items which exist thanks to technological progress. As the words seem to be real scientific words, players will likely suspend their disbelief and immerse themselves completely in the videogame; the terms are not *ex nihilo* creations. Both in English and in French, Latin and Greek roots are traditionally used in scientific and scholarly terminology. Since the affixes have the same role in both languages, translators often keep the Greek and Latin affixes.[27]

(18) An English affix is used to create a new term, *over-*; this prefix means "above; upper; superior; eminent" (Sheehan 2008: 89). Its use in this lexical creation is significant as an *overlord* is a "semi-intelligent spacefaring leviathan" which was "inducted into the Swarm so that their heightened sense could benefit the Zerg in battle" (*Overlord*, Blizzard). An overlord is an aerial zerg unit which provides air supply and carries other units. The prefix *over* describes how and where the unit moves. In fact, the French translation of *overlord* only keeps the semantic feature of *over* as the word *lord* is not translated. The semantic feature of *dominant* conveys the idea of spatial superiority (the overlords float above the ground), but its connotation[28] also conveys the meaning of *lord* as the verb *dominer* can also mean "to exert a sovereign power" ("dominer," CNRTL). In order to create a noun (as the referent is a being), the French translation uses a suffix which creates nouns from verbs: *-ant* (DSF n.d.: XVII).

(19) The last example from *Starcraft* uses a Latin suffix *-ium* and refers to a specific resource: a blue crystal which amplifies the psi powers of some units. In English, the suffix *-ium* is used to form "names of metallic elements" (ODE 2010: 930); in French, it is used to

TABLE 4.3 *Analysis of derived nouns*

	Videogame	Original terms	Original affixes	French translations	Translation affixes
17.	*Starcraft*	holoboards*	Greek prefix	holo-tribune	Greek prefix
18.	*Starcraft*	overlord	English prefix	dominant	French suffix
19.	*Starcraft*	jorium	Latin suffix	jorium	Latin suffix
20.	*Alien*	hypersleep	Greek prefix	hypersommeil	Greek prefix
21.	*Alien*	vidicom	Latin prefix	vidicom	Latin prefix

* A *holoboard* is a Terran structure used to project a hologram in order to spread Dominion propaganda.

construct "terms belonging to the lexicon of chemistry of physics" ("-ium," CNRTL). The element *jor* has no particular meaning, but it can be derived from the words *jar* or *jorum*, both referring to a container. The meaningful semantic feature in the term *jorium* is the suffix, which allows the player to know it is a chemical material.

(21)   The lexical creation *vidicom* refers to a technologically advanced phone booth (see Figure 4.6), in which the caller can visualize her/his interlocutor. In French, the complete form *video-* is preferred rather than the shorter version *vid-*. However, the prefix remains the same in the translation, and the reason may be the fact that *vidicom* is inscribed on the phone booth in the videogame. Perhaps the translators could not change the linguistic elements inside the videogame décor.[29] The lexical creation has therefore been kept exactly as it is in English. Nonetheless, the term is perfectly understandable by francophone players, as *comm* is also the shorter version of *communication*; the prefix *vid-* is close enough to the prefix *video-*, and the image of the item itself is of great use in understanding the term.

The common Latin and Greek roots used both in English and in French in their scientific and technological lexicons are useful in the creation of new science-fiction words; they are used in both videogames in order to anchor the new inventions in a scientifically and technologically possible extrapolation. However, *Alien* contains fewer derived terms than *Starcraft*,

FIGURE 4.6 *Vidicom. Screenshot by Alice Ray, © The Creative Assembly Limited and Sega Corporation.*

and this can be explained by the developers' intent to anchor the game's lexical creations, and therefore its environment, in a reality that seems to be within players' conceptual reach.

3.2.4 Fictive meanings

Another way of creating new words is to give "new meanings to a form which already exists" (Béciri 2003: 60)—in other words, a shift in meaning. The science-fiction environment of the videogame gives a new meaning to already existing terms; the link between the official meanings of the terms can be either narrow or loose (Table 4.4).

(22) The first example of a fictive meaning is *reaper*, which becomes *faucheur* in French. This is a formal equivalent as both nouns refer to "a person who harvests a crop" (ODE 2010: 1479 and "faucheur," CNRTL). Moreover, in both languages, *reaper* is also a metaphor for death (*the Grim Reaper, la grande faucheuse*). So, the connotation of the term is kept in the translation: this combat unit is dreadful and murderous.

(23) The term *carrier* is expanded in the localized version with a V + N sequence: *porte-nefs*. The formal equivalent of *carrier* would be *porteur*. However, this term does not have exactly the same meaning as it mainly refers to a person who is carrying something ("porteur," CNRTL), whereas *carrier* refers to "a person or thing that carries [...] something" (ODE 2010: 267) with many examples of *carrier* used with an item rather than a human being. The term needs to be adapted to convey the same idea in the player's mind: a vehicle which carries other vehicles. The verb *porter* in French is often used in compounds such as *porte-monnaie, porte-feuille, porte-voix*, and so on, and the V + N pattern is a highly productive pattern: *prie-Dieu, ouvre-boîte, taille-crayon*, and so on. Using the verb *porter* allows the translator to save space and convey the same meaning as the original term. However, the verb needs a head noun to be understandable, and the French translation uses an element which is not explicit in the original term: what the carrier carries.

(24) The term *hellion* is used because its meaning in the real world describes one of the features of a new item in *Starcraft*: *hellion* is "a rowdy or mischievous person, [...]" (ODE 2010: 816), while in the videogame, it is "a high-speed raider [...] capable of causing damage disproportionate to their size" ("hellion," Blizzard). In the term *hellion*, the concept of "hell" can be found; in French, the verb which is mainly used with the idea of hell is *tourmenter*, from which the noun *tourmenteur* is derived: "a person who brutalizes

TABLE 4.4 *Analysis of fictive meanings*

N°	Videogame	Original terms	New meaning	French translations	Translation strategies*
22.	*Starcraft*	reaper	Infantry unit with a jet-pack	faucheur	Formal equivalent
23.	*Starcraft*	carrier	Protoss air unit	porte-nefs	expansion
24.	*Starcraft*	Hellion	Terran vehicle unit	tourmenteur	modulation
25.	*Alien*	synthetics	synthétique	N	N
26.	*Alien*	lifeboat	navette de sauvetage	N + N	N + preposition "de" + N with "de" > goal
27.	*Alien*	navigation officer	navigatrice	N + N	N

* The definition of our translation strategies are mainly extracted from Delisle et al. (1999) and Shuttleworth and Cowie (2014). The formal equivalent refers to "a Target Language item which represents the closest decontextualized counterpart to a word or phrase in Source Language" (Shuttleworth and Cowie 2014: 62).

another" ("tourmenteur," CNRTL). The connotation of the term is kept as well as the main meaning of the original term.

(25) In the real lexicon, the term *synthetics* means a material "made by chemical synthesis, especially to imitate a natural product" (ODE 2010: 1805). In the videogame, it refers to the androids aboard the orbital station. The French translation of the term is *synthétique* which, in the videogame, is used as a noun, whereas in a common dictionary, it is an adjective ("synthétique," CNRTL). This transposition (i.e., a grammatical shift, the replacement of one word class with another) can be explained by the fact that the noun *synthèse* refers to a "method of reasoning" ("synthèse," CNRTL) while the adjective *synthétique* is mostly understood as the opposite of being natural, which is the main meaning of the noun *synthetics* in the context of the videogame.

(26) and (27) These examples are taken from maritime terminology as they respectively refer to a "boat launched from land to rescue people in distress at sea" (ODE 2010: 1020) and a navigation officer "is [...] responsible for making decisions on steering and maneuvering the ship, controlling navigation and communications" ("Navigation Deck Officer," careers at sea). Their translations include the same idea by preserving the tenuous link between space

travel and sea travel. In example 26, *lifeboat* becomes *navette de sauvetage*; while the term remains attached to the marine metaphor in English (as *boat* never refers to spaceship, unlike *vessel* or *ship*), the translation uses the term *navette*, which can be applied to both fields ("navette," CNRTL). The second term, *navigation officer*, is translated by *navigatrice*. The French term used is not the official translation of *navigation officer* ("navigation officer," BTB Termium), and the jobs are different.[30] However, players are not supposed to be specialists in marine terms, and this distinction will not disrupt the gaming experience.

The fictive meanings always use one of the features or connotations of the original term. The translation mostly uses the same word-formation strategies, as is the case in our examples, only one is not a fictive meaning but a compound (*porte-nefs*) in French. The creative strategies of the science-fiction neologisms are kept in the translation whenever it is possible in compliance with the technical and cultural constraints.

Like *Starcraft*, *Alien* uses semantic shift to create lexical innovations. However, the fictive meanings in *Alien* are closer to their original definitions: only a feature of the referent is changed (location for example), or the degree of precision of the word. The videogame gives the player an immersive experience in which everything could be real;[31] the definition of the words must correspond to the science-fictional environment, but they also have to give the player a sense of familiarity.

4 Conclusions

Whether it is in *Starcraft* or *Alien*, the same semantic relations can be found in nominal sequences, and some translation patterns are recurrent. However, it is also obvious that while both videogames are science fictional, they do not display the same types of lexical creations: *Starcraft* tends to be more creative and invent completely new items the player cannot guess without the game context, whereas *Alien* tends to use terms that are more familiar. Even though the environments are based on the science-fiction genre, *Alien* and *Starcraft* do not offer the same translation challenges: by comparing translation patterns, it is clear that the creativity of the translator has more space in *Starcraft*, in which new items have to be plausible but also "entertaining"; in *Alien*, lexical creations have to be more anchored in players' scientific and technological reality while still giving them a vision of the future.

Science-fiction videogames need lexical creations to give players the feeling that they are playing in a possible future (or an alternative world), exactly like films and TV shows that show the audience new items and concepts and name them in order to anchor them in a linguistic reality (if the signifier seems to be plausible, so does the referent). Unlike audiovisual media, however, videogames have another dimension: interactivity. The player is not only the witness of new technologies and objects, but she/ he also has to use them in order to win (or finish) the videogame. Fictive words are an important part of this cognitive alienation, as the player knows these objects do not really exist but also voluntarily decides to believe in them and to use them as if they were real things.[32] In this context, the translation of these specific lexical creations is an important feature of the international impact of videogames. The translation will determine whether or not "game immersion can be achieved and maintained successfully by taking the suspension of disbelief a step further and creating a convincingly personal experience for players each time they enter the game world" (Bernal-Merino 2015: 40). For international audiences, the playability of the videogame relies on its localization,[33] and in a science-fiction videogame, lexical creations are an important part of the challenges.

By analyzing the creation and the translation of some fictive words from two science-fiction videogames, *Alien: Isolation* and *Starcraft II, Wings of Liberty*, this chapter has provided some evidence for the fact that creativity is an important feature of the videogame localization process—the subjectivity of the translator is obvious, as the choices made are sometimes very different for the same morphological patterns. But translation is also determined by technical constraints and by the differences in the ability of languages to create new words—especially compound words. The nature of fictive words depends heavily on the videogame narrative and aim: *Starcraft* aims to create an entire new universe with new technologies, new military units, and new species, and it is completely outside the present, while *Alien* tends to be more realistic in order to have a stronger impact on the players' feelings. The lexical creations reflect the atmosphere of the videogame and so do their translations.

Notes

1   "[T]here may be some connection between sf's affinity for logical extrapolation and [...] 'structural fabulation,' and the complexly simulative rule systems that underlie many recently developed types of game" (Tringham 2015: 1–2). Not to mention that the first videogame that includes an original gameplay is *Spacewar* (1962), a science-fiction videogame (Tringham 2015: 2).

2   There is no space here to discuss the localization market in France, but there are some figures: in 2012, "foreign videogames total 93.2% of the sales" (Forsans 2013).

3 In many survival horror videogames, the player needs to kill a variety of monsters in order to survive throughout the game: *Resident Evil* (1996), *Dead Space* (2008), and *Silent Hill* (1999), for example.

4 *Starcraft II* has been very successful, with 31 awards won ("Awards," Blizzard) and nearly 4.5 million copies sold throughout the world by the end of 2010 (Activision and Blizzard 2011: 13). Most of the critics are positive, even laudatory (see Onyett 2010; Welsh 2010). *Alien* has been acclaimed by both critics and players (see Kelly 2014; O'Brien 2014; RPS 2014), and more than 2.1 million copies have been sold (Makuch 2015).

5 There have been some interesting works written on language in videogames (Mon Rego and Sánchez 2013; Ensslin 2012) and especially about language learning through videogames (Berns et al. 2011; Aghlara and Tamjid 2011), but this paper only aims to give an introductory analysis of invented word localization in science-fiction videogames.

6 In the case of science fiction, the translation has to be coherent with the translations of previous science-fiction works. For example, *Alien: Isolation* has to be translated in accordance with the *Alien* franchise. Some invented words have to be localized according to their previous translation as some of them are re-used in different works and media (*blaster, terraform, ansible*, etc.).

7 This is especially true for cinema, with a great number of science-fiction movies in the top blockbuster lists (see "Top 100 des meilleurs blockbusters," *SensCritique* and "Top 100 Blockbusters," *AMC*).

8 "While the concept of genre usually applies to a choice of subject within a form, in the case of videogames, the term has typically been employed to describe a type of gameplay," (Tringham 2015: 4); some examples are first-person shooter or role-playing games.

9 The term *fictive meaning* is employed to describe science-fiction terms created through semantic shift.

10 See Vermes (2003), Wozniak (2014), and Guttfeld (2008), for example.

11 Both the *Oxford Dictionary of English* (2010) and *Merriam-Webster Online* have been used, as well as some science-fiction databases and glossaries such as *Technovelgy* and Prucher (2009).

12 As the extraction of the neologisms was made manually, our original corpus is not exhaustive.

13 Many words linked to maritime and aeronautic domains are used: *navigation officer, ship, flight record unit, shuttle, lifeboat*, and so on.

14 Spacecrafts and technologies inside the game are not possible yet (not as it is presented in the videogame), but the idea of creating a space station is not so far in the future, as we have already made tremendous technological advances in that regard.

15 The semantic relation discussed here is the "relation between the head noun, and the other nouns of the noun compound" (Balyan and Chatterjee 2015: 93).

16 The CNRTL is an online lexical database for French language.

17 Here is an example of the interreferentiality of videogames and science fiction, as the term can be seen in other games: *World of Warcraft* (Blizzard Entertainment 2004) and *Warhammer* (Games Workshop 1983). Some lexical creations are not new, as they have been used for decades in science-fiction universes and already have an "official" translation that needs to be used in order to keep coherence.

18 In translation terminology, expansion is "an increase in the amount of text that is used in the target language to express the same semantic content as compared to the parallel segment in the source text" (Delisle et al. 1999: 138).

19 The term *ghost* (and *fantôme*) is a semantic neologism as it refers to a unit able to hide from the eye of the enemy thanks to its psionic powers.

20 Modulation is one of the seven categories described by Vinay and Darbelnet (1958), and it "involves a manipulation of mental rather than grammatical categories" (Shuttleworth and Cowie 2014: 108); in other words, the translation slightly shifts the meaning of the terms in order to be semantically equivalent to the source text and to be idiomatic in the target language (e.g., *boire un verre* [to drink a glass] can become "to grab a drink").

21 It is used in electricity, for example.

22 The transposition is a translation strategy "where equivalence in meaning or sense is established by changing the word class or part of speech of a word of phrase" (Delisle et al. 1999: 171).

23 In the context of the videogame, people on the spaceship will not use a complicated and long term to address the androids.

24 However, in the videogame, the androids are mostly referred to as "Joe" and "Lambda."

25 The term *travailleur* is employed by left-wing political parties such as Lutte ouvrière and is very famous as the opening of every speech by Arlette Laguiller (Laguiller 1974).

26 See Spencer and Zwicky (1998) for a full analysis of the derivational process.

27 Examples can be found in science-fiction literature, but the use of pseudo-technical or scientific terminology can also be seen in nonfictive domains such as advertising (cosmetics, drugs, household products, etc.).

28 The noun *lord* describes "a man of noble rank or high office" (ODE 2010: 1045).

29 This theory is confirmed by the fact that the paintings on the walls are not translated either.

30 A *navigateur* is the person in charge of the ship route, whereas the navigation officer "direct[s] operations" ("navigation officer," BTB Termium).

31 Everything is scarier when the environment is realistic.

32 See Ermi and Mäyrä (2007), Wolf (2003), and Przybylski et al. (2010), among others, for studies on videogame immersion and engagement.

33 Some gamers play in the original language, but unless they are bilingual, their gaming experience is not as immersive as if it was their native language.

References

Abeillé, A. and D. Godard (1999), "La position de l'adjectif épithète en français: Le poids des mots," *Recherches linguistiques de Vincennes*, 28: 9–32.

Activision and Blizzard (2011), "Fourth Quarter and CY 2010 Results." Available online: http://files.shareholder.com/downloads/ACTI/4913798968x0x440263/2 a37de98-400f-4916-9bb3-ae5ddf1b86b8/ATVI%20C4Q10%20slides%20 FINAL.pdf (accessed April 5, 2018).

Aghlara, L. and N.H. Tamjid (2011), "The effect of digital games on Iranian children's vocabulary retention in foreign language acquisition," *Procedia— Social and Behavioral Sciences*, The 2nd International Conference on Education and Educational Psychology, 29: 552–60.

Álvarez-Bolado Sánchez, C. and I. Álvarez de Mon Rego (2013), "Semantic neology in the domain of videogames in Spanish," *Ibérica*, 25: 61–82.

Angenot, M. (1979), "The absent paradigm: An introduction to the semiotics of science fiction," *Science Fiction Studies*, 6 (17). Available online: https://www. depauw.edu/sfs/backissues/17/angenot17.htm (accessed April 6, 2018).

"Awards." Blizzard Entertainment. Available online: http://us.blizzard.com/en-us/c ompany/about/awards.html#sc2 (accessed April 6, 2018).

Bagge, C. and A. Manning (2007), "Grammar and translation: The noun + noun conundrum," *Meta: Journal Des Traducteurs*, 52 (3): 556–67.

Balyan, R. and N. Chatterjee (2015), "Translating noun compounds using semantic relations," *Computer Speech & Language*, 32 (1): 91–108.

Béciri, H. (2003), "Néologie et transmission des connaissances," in J.F. Sablayrolles (ed.), *L'innovation lexicale*, 59–78, Paris: Honoré Champion Éditeur.

Berman, A. (2000), "Translation and the trials of the foreign," in L. Venuti (ed.), *The Translation Studies Reader*, 284–97, London: Routledge.

Bernal-Merino, M.A. (2006), "On the translation of video games," *The Journal of Specialised Translation*, 6. Available online: http://www.jostrans.org/issue06/art _bernal.php (accessed April 6, 2018).

Bernal-Merino, M.A. (2007), "Challenges in the translation of video games," *Revista Tradumàtica*, 5. Available online: http://www.fti.uab.es/tradumatica/r evista/num5/articles/02/02.pdf (accessed April 6, 2018).

Bernal-Merino, M.A. (2011), "A brief history of game localisation," *Trans: Revista de Traductología*, 15: 11–17.

Bernal-Merino, M.A. (2015), *Translation and Localisation in Video Games*, New York: Routledge.

Berns, A., A. González-Pardo, and D. Camacho (2011), "Designing videogames for foreign language learning," in *International Conference ICT for Language Learning (Florence, Italy)*, October 20–21, 2011, Milan: Simonelli Editore University Press. Available online: https://www.researchgate.net/profile/Anke_ Berns/publication/249993419_Designing_videogames_for_Foreign_Language_ Learning/links/02e7e51e8255cebbc2000000/Designing-videogames-for-Foreign- Language-Learning.pdf (accessed April 6, 2018).

Blizzard Entertainment (2010), *Starcraft II: Wings of Liberty* [Video game], Irvine, California: Blizzard Entertainment.

Carlson, R. and J. Corliss (2010), "Imagined commodities: Video game localisation and mythologies of cultural difference," *Games and Culture*, 6 (61). Available online: http://gac.sagepub.com/content/6/1/61 (accessed April 6, 2018).

Centre National de Ressources Textuelles et Lexicales [CNRTL] (n.d.). *CNRTL.* Available online: http://www.cnrtl.fr/definition/ (accessed April 6, 2018).

Chandler, H.M. and S.O. Deming (2012), *The Game Localisation Handbook*, London: Jones & Barlett Learning.

Delisle, J., H. Lee-Jahnke, and M.C. Cormier (eds.) (1999), *Translation Terminology*, Philadelphia: John Benjamins Publishing.

Di Marco, F. (2007), "Cultural localisation: Orientation and disorientation in Japanese video games," *Revista Tradumàtica*, 5. Available online: http://www.fti. uab.es/tradumatica/revista/num5/articles/06/06.pdf (accessed April 6, 2018).

Dictionnaire Des Éléments de Formation [DEF] (n.d.). *Le Robert.* Available online: https://robert-illustre.lerobert.com/pdf/dictionnaire-des-elements-de-formation. pdf (accessed April 6, 2018).

Dictionnaire Des Suffixes Du Français [DES] (n.d.). *Le Robert.* Available online: https://robert-illustre.lerobert.com/pdf/dictionnaire-des-suffixes.pdf (accessed April 6, 2018).

Dietz, F. (2007), "'How difficult can that be?' The work of computer and video game localisation," *Revista Tradumàtica*, 5. Available online: http://www.fti.uab. es/tradumatica/revista/num5/articles/04/04.pdf (accessed April 6, 2018).

Dozois, G. (1976), "Living the future: You are what you eat," in Charles L. Grant (ed.), *Writing and Selling Science Fiction*, Cincinnati: F&W Publications.

Ensslin, A. (2012), *The Language of Gaming*, New York: Palgrave Macmillan.

Ermi, L. and F. Mäyrä (2007), "Fundamental components of the gameplay experience: Analyzing immersion," in S. de Castell and J. Jenson (eds.), *Worlds in Play: International Perspectives on Digital Games Research*, 37–54, New York: Peter Lang.

Forsans, E. (2013), "Le marché du jeu vidéo en 2012—Synthèse," *Agence Française pour le Jeu Vidéo*. Available online: https://www.afjv.com/news/2702_le-marche -du-jeu-video-en-2012-synthese.htm. (accessed April 6, 2018).

Guttfeld, D. (2008), *English-Polish Translations of Science Fiction and Fantasy: Preferences and Constraints in the Rendering of Cultural Items*, Toruń: Grado.

Heinlein, R.A. (1977), *Starship Troopers*, London: NEL.

"*Hellion*" (n.d.), Blizzard Entertainment. Available online: http://us.battle.net/sc2/ en/game/unit/hellion (accessed April 6, 2018).

Humbley, J. (2005), "La néologie: Interface entre ancien et nouveau," in R. Greenstein (ed.), *Langues et cultures: Une histoire d'interface*, 91–103, Paris: Publications de la Sorbonne.

Kelly, A. (2014), "Alien: Isolation review—Giger's creature gets the game it deserves," *The Guardian*, October 3. Available online: http://www.theguardi an.com/technology/2014/oct/03/alien-isolation-review-giger-game-ridley-scott (accessed April 6, 2018).

Laguiller, A. (1974), *Arlette Laguiller 'Travailleuses Travailleurs'—Archive Vidéo INA*, Ina Politique. Available online: https://www.youtube.com/watch?v=fVi Hry4Rilk (accessed April 6, 2018).

Lieber, R. and P. Stekauer (2014), *The Oxford Handbook of Derivational Morphology*, Oxford: Oxford University Press.

Makuch, E. (2015), "Alien: Isolation sells 2.1 million," *GameSpot*, May 11. Available online: https://www.gamespot.com/articles/alien-isolation-sells-2-1-million/1100-6427238/ (accessed April 6, 2018).

Mangiron, C. and M. O'Hagan (2006), "Game localisation: Unleashing imagination with 'restricted' translation," *The Journal of Specialised Translation*, 6: 10–21.

Moldovan, D., A. Badulescu, M. Tatu, D. Antohe, and R. Girju (2004), "Models for the semantic classification of noun phrases," in *CLS '04: Proceedings of the HLT-NAACL Workshop on Computational Lexical Semantics*, 60–67, Stroudsburg, PA, USA: Association for Computational Linguistics.

"Navigation (Deck) Officer" (n.d.), *Careers at Sea*. Available online: http://www.careersatsea.org/career-development/navigation-officer/ (accessed April 6, 2018).

"Navigation Officer" (n.d.), *BTB Termium*. Available online: http://www.btb.termiumplus.gc.ca/tpv2alpha/alpha-fra.html?lang=fra&i=1&srchtxt=NAVIGATION+OFFICER&index=alt&codom2nd_wet=1#resultrecs (accessed April 6, 2018).

Noailly, M. (1999), *L'adjectif en français*, Paris: Ophrys.

O'Brien, L. (2014), "Alien: Isolation is the bravest game of 2014," *IGN*, November 2014. Available online: http://www.ign.com/articles/2014/11/23/alien-isolation-is-the-bravest-game-of-2014 (accessed April 6, 2018).

O'Hagan, M. (2007), "Video games as new domain for translation research: From translating text to translating experience," *Revista Tradumàtica*, 5. Available online: http://www.fti.uab.cat/tradumatica/revista/num5/articles/09/09art.htm (accessed April 6, 2018).

O'Hagan, M. and C. Mangiron (2004), "Games localisation: When 'arigato' gets lost in translation," in *Proceedings of New Zealand Game Developers Conference 2004*, 57–61, Dunedin: New Zealand Game Developers Association.

Onyett, C. (2010), "StarCraft II Wings of Liberty review," *IGN*, August 3, 2010. Available online: http://www.ign.com/articles/2010/08/03/starcraft-ii-wings-of-liberty-review (accessed April 6, 2018).

"*Overlord*" (n.d.), Blizzard Entertainment. Available online: http://us.battle.net/sc2/en/game/unit/overlord (accessed April 6, 2018).

Prucher, J., ed. (2009), *Brave New Words: The Oxford Dictionary of Science Fiction*, Oxford: Oxford University Press.

Przybylski, A.K., C.S. Rigby, and R.M. Ryan (2010), "A motivational model of video game engagement," *Review of General Psychology*, 14 (2): 154–66.

RPS (2014), "The bestest best horror of 2014—Alien: Isolation," *Rock, Paper, Shotgun*, December 2014. Available online: https://www.rockpapershotgun.com/2014/12/17/best-horror-games-2014-alien-isolation/ (accessed April 6, 2018).

Saint-Gelais, R. (1999), *L'empire du pseudo: Modernités de la science-fiction*, Québec, Canada: Editions Nota bene.

Sheehan, M.J. (2008), *Word Parts Dictionary: Standard and Reverse Listings of Prefixes, Suffixes, Roots and Combining Forms*, 2nd edn., Jefferson, USA: McFarland & Company.

Shuttleworth, M. and M. Cowie (2014), *Dictionary of Translation Studies*, New York: Routledge.

A. and A.M. Zwicky, eds. (1998), *The Handbook of Morphology*, Oxford: kwell Publishers.

;on, A., ed. [ODE] (2010), *Oxford Dictionary of English*, 3rd edn., Oxford: ford University Press.

ovelgy. *Glossary of Science Fiction Ideas and Inventions*. Available online: tp://www.technovelgy.com/ct/ctnlistalpha.asp (accessed June 9, 2018).

Creative Assembly (2014), *Alien: Isolation* [Video game], Horsham, England: Sega.

"Top 100 Blockbusters" (2011), *AMC*. Available online: http://www.amc.com/m ovie-guide/top-100-blockbusters (accessed April 6, 2018).

"Top 100 Des Meilleurs Blockbusters" (n.d.), *Sens Critique*. Available online: https://www.senscritique.com/top/resultats/Les_meilleurs_blockbusters/272024 (accessed April 6, 2018).

Tournier, J. (1985), *Introduction descriptive à la lexicogénétique de l'anglais contemporain*, Genève: Champion-Slatkine.

Triclot, M. (2011), *Philosophie des jeux vidéo*, Paris: Zones.

Tringham, N. (2015), *Science Fiction Video Games*, Boca Raton: CRC Press.

Valeontis, K. and E. Mantzari (2006), "The linguistic dimension of terminology: Principles and methods of term formation." Available online: https://www.res earchgate.net/profile/Elena_Mantzari/publication/253024944_THE_LINGUIS TIC_DIMENSION_OF_TERMINOLOGY_PRINCIPLES_AND_METHODS_ OF_TERM_FORMATION/links/54536ee00cf2bccc490a4364/THE-LINGU ISTIC-DIMENSION-OF-TERMINOLOGY-PRINCIPLES-AND-METHODS-OF- TERM-FORMATION.pdf?origin=publication_detail (accessed April 6, 2018).

Vermes, A.P. (2003), "Proper names in translation: An explanatory attempt," *Across Languages and Cultures*, 4 (1): 89–108.

Vinay, J.P. and J. Darbelnet (1958), *Stylistique comparée du français et de l'anglais: Méthode de traduction*, Paris, France: Didier.

Welsh, O. (2010), "StarCraft II: Wings of Liberty," *Eurogamer*, August 2010. Available online: http://www.eurogamer.net/articles/starcraft-ii-wings-of-libe rty-review (accessed April 6, 2018).

Wolf, Mark J. P. (2003), *The Video Game Theory Reader*, New York/London: Routledge.

Wozniak, M. (2014), "Future imperfect: Translation and translators in science- fiction novels and films," in K. Kaindl and K. Spitzl (eds.), *Transfiction: Research into the Realities of Translation Fiction*, 345–62, Amsterdam: John Benjamins Publishing Company.

5

End-user Agreements in Videogames

Plain English at Work in an Ideal Setting

Miguel Ángel Campos-Pardillos

1 Introduction: Online end-user license agreements

Before a person installs and uses software in a computer, he or she is asked to accept a legally binding agreement called an "End User License Agreement" (EULA). Such EULAs, as a legal instrument and as a form of communication, have attracted the attention of many scholars, not only those from the legal field but also sociologists and linguists, because of their legal characteristics and the special communicative conditions in which they are executed.

From the legal point of view, EULAs are a form of what is known as the *standard contract*. A logical product of the capitalist system and of twentieth-century mass distribution, the standard contract is a preformulated agreement developed by companies and used in every transaction for the same product or service. Historically, this has been grounds for criticism; as pointed out as early as 1943, it eliminates the "individuality" of the parties, especially of the consumer, since it leaves the consumer with no bargaining power, either because the author has a monopoly, or, as Kessler aptly describes, "because all competitors use the same clauses. His [the weaker party's] contractual intention is but a subjection more or less voluntary to terms dictated by the stronger party, terms whose consequences are often understood in a vague way, if at all" (Kessler 1943: 632).

These contracts are called *contracts of adhesion* or *standard contracts*; the first label describes the way they are entered (i.e., the adherent merely accepts without any negotiation), whereas the second one refers to the fact that they follow a pre-existing structure which does not vary as long as the offeror remains the same. In the academic sphere, legal scholars have paid attention to the enforceability of these agreements (see, for instance, Terasaki 2014), while others have mentioned the so-called "no-read phenomenon" (Ayres and Schwartz 2014), that is, that people do not usually read before accepting the conditions. As a rule, it is commonly accepted that they tend to favor the company, not only because of their nonnegotiable nature, but also because of the way the terms and conditions are drafted. In fact, they have been subject to frequent criticism, because of their "take it or leave it" assumption (i.e., one may not negotiate the terms, and if one desires to use the software, one must accept the agreement as it is). In the case of games, as Kunze (2008: 104) has pointed out, "[t]he user ordinarily has no real choice: either accept the license agreement as is, or forfeit the ability to play."

Indeed, it has been objectively proved that the conditions are tilted toward the seller, as is always the case in standard agreements. Marotta-Wurgler (2007, 2008) analyzed a sample of 647 software licenses, considering the bias toward one of the parties, and was able to prove that the EULAs were invariably more favorable to the seller, although, quite interestingly, no evidence was found that such bias was more serious in the case of products oriented toward the general public than those intended for businesses, or that it had any correlation with the degree of competition between companies.

In the world of software in general and videogames in particular, given the practical impossibility of any other type of relationship occurring between the parties, these contracts have been given great importance. This can be easily understood, on the one hand, considering the financial weight of videogames and the number of users (by June 2016, according to Microsoft, *Minecraft* (Mojang 2011) alone had sold over 100 million copies); not surprisingly, disputes over copyright and the interpretation of these contracts have become landmark cases (see, for instance, Shikowitz 2009 regarding *World of Warcraft* [Blizzard Entertainment 2004]). Also, on a somewhat darker note, these agreements must also consider privacy issues, since the platforms allow interaction between users which sometimes entail danger to individuals, especially children, who are exposed to pedophiles.[1]

All these reasons explain why the language used in contracts merits careful attention, as we shall see below. More specifically, this essay will explore the role of plain language in the EULA, not only from the point of

view of comprehensibility, but also as a strategic tool in order to create a
"community of users" as a promotion strategy.

2 Plain language in contract law

The fight for clarity and against obscurity in legal language is by no
means a recent one. In the second half of the twentieth century, there were
movements on both sides of the Atlantic trying to empower citizens so that
they could be aware of their rights and obligations (Adler 2012). Decades
after the famous Citibank's promissory note in plain language in the United
States, or the fight against gobbledygook and jargon launched by the Plain
English Campaign, legislators have now incorporated clarity as a statutory
requirement in texts addressed at the general public, which has led to some
success stories, such as the 2010 US Plain Writing Act (Williams 2015). The
regulations become even more demanding when the "public" may include
children. For instance, the General Data Protection Regulation (GDPR) (EU)
2016/679, in force since May 25, 2018, explicitly refers to understandability
and children's protection ("any information and communication, where
processing is addressed to a child, should be in such a clear and plain
language that the child can easily understand").

However, plain language is not only a legal requirement, but also a
social one. The corporate world has reacted to this need, sometimes due
to such legal obligation, but also for practical reasons. On the one hand,
clarity is also beneficial for the firms themselves, as it may help to warn
against infringements and thus avoid costly litigation (Gomulkiewicz
and Williamson 1996: 366). On the other, there are additional reasons
to use plain and clear language in contracts, which have more to do with
psychological factors. First and foremost, using obscure language is felt to
be not only illegal in some cases, but also unethical, even if it may have
been commonplace in the past (Felch 1985: 17) listed it as what is known
as "sharp" practices, "short of fraud, but unscrupulous"). In recent times,
clarity in the relationship between firms and stakeholders is seen as part of
"corporate social responsibility," which implies something else than "mere
compliance with laws and regulations" (Carroll 1991: 41), and the use of
clear contract terms is frequently included as one of the parameters of such
responsibility (Carter 2000), although, as discussed below, the concepts
of "clear" or "simple" may be, at least, subject to debate. On the other,
"consumer friendliness" and "ethics" are a source of competitive advantage,
a selling point: clear communication with consumers will not be the sole
factor in selling a product, but it may clearly add to the marketing mix.

In the software field, there is also a strong ideological component, especially
regarding the difference between "major players" and "independent
developers," the latter being associated with authenticity and with the

spirit of gamers themselves (see, for example, Lipkin 2013). In this respect, colloquial language may be seen as a way to depart from mainstream games, as we shall see below for the case of *Minecraft*.

Once the need for clarity has been discovered and taken on board by the corporate world, scholars have had the occasion to study the specific changes made to legal language toward simplification, concerning not only terminology and syntax, but also document planning, organization, and even testing on typical readers (Kimble 1996–97). Empirical studies have proved that the simplification of legal language enhances understanding (Masson and Waldron 1994), and, which may be more interesting considering the marketing factors mentioned earlier, that there is increased trust when clearer language is used (as proved by Van Boom et al. 2016, who compared two versions of an insurance contract with a C1 and a B1 CEFR language level). It is true, however, that the area of contract law seems to be resistant to change; this could be due to a number of reasons, among which Adams (2013: 60ff) lists general inertia (contracts are usually constant re-elaborations of previous ones), or fear of losing accuracy. An additional factor, as pointed out by Lemens and Adams (2015), is the importance of precedent, since established case law provides certainty regarding the interpretation of traditional terms and phrases, whereas simpler words or expressions are "untrodden ground" where corporate lawyers fear to tread.

As a rule, it may be acknowledged that the virtual economy (involving not only EULAs, but also online contracts in general) is causing contract language to evolve toward greater simplicity. One visible difference concerns the way the parties are mentioned; as Brunon-Ernst (2016: 45) has pointed out, in e-agreements, users are addressed as "you," which implies less distance and ambiguity than the third person ("the Buyer," "the Employee"). However, this apparent proximity also has its critics; for instance, given that "you" is an individual and "we" is a collective, there is still an imbalance between the parties (Bélanger and van Drom 2012).

Nevertheless, once the need for clear language has been established, there remains the problem of what constitutes "plain language" beyond traditional lists of words which may be advisable to avoid. One specific, objective way of measuring readability is the use of readability tests, a number of which have been proposed over the past decades, such as the Flesch index (Flesch 1948), the Gunning fog test (Gunning 1952), and many others (for a review of these traditional tests, see Tekfi 1987; for discussion and critique, see Bailin and Grafstein 2001). The criteria underlying these indexes are based on word frequency (as measured by word lists), word length, number of syllables, and sentence length. However, readability formulae are by no means an automatic instrument, and they are not without their defects. For instance, those based on syllable or letter count ignore that the principle "word length equals word difficulty" does not necessarily apply; for instance, the word *escrow* is shorter than *account* (and yet less frequent in English), and *affidavit* is shorter than *statement of truth* (although the English Civil Procedure rules have eliminated *affidavit* because of its obscurity). Another quantitative parameter, sentence length, does not necessarily entail difficult processing, as these formulae tend to—wrongly—consider it synonymous with syntactic complexity (see, for example, Bailin and Grafstein 2001: 290).

3 *Minecraft* in the world of videogames

One of the types of videogame which has become most popular over the past years, thanks to easy, affordable access to the internet, is what is known as a *massive multiplayer online game* (MMOG). The topics for these games are greatly varied, mainly depending on the type of activity and the scenario (e.g., building, flight simulation, first-person shooting, role-playing, etc.), and appeal to persons of all ages, building communities of sizes equivalent to small nations (Webber 2014). In fact, the term *game*, though certainly fit for the original motivation itself, is giving way to other more descriptive terms such as *worlds*, *communities*, or *lifeworlds* (see, for instance, Taylor 2006).

Among these, *Minecraft* (Mojang 2011) is probably one of the most successful games ever. Before its official release in November 2011, it had already sold over 3 million copies and had over 10 million players, which had increased to 40 million by June 2016 and 55 million monthly users by February 2017 (www.statista.com). The success was not affected by some apparent comparative disadvantages, such as a very simple, cubic design or the lack of an instruction manual, or the fact that this was an "indie" game, created by Markus ("Notch") Persson, and not supported by a large games company (or at least, until Microsoft bought Mojang in 2014). It may be argued, with Lastowka (2012), that some of these features have reinforced its "community" nature (for instance, players must look for instructions in Wikis, videos, and fora, which creates an online community).

Regarding the type of game, it may be defined as a "sandbox" game, one in which the player creates everything needed to progress and create his or her own virtual world; this category would include other games like *7 Days To Die* (The Fun Pimps 2013) or *Starbound* (Chucklefish 2016), among others. The game was first released in 2011 by Mojang AB, and at present is in Version 1.12.2 (according to the *Minecraft* Wiki). Regarding user profile, The Entertainment Software Rating Board (ESRB), a nonprofit body assigning ratings to videogames, has rated *Minecraft* as Everyone 10+, which according to the ESRB page, means "[c]ontent is generally suitable for ages 10 and up. May contain more cartoon, fantasy or mild violence, mild language and/or minimal suggestive themes."[2]

One of the distinguishing features of *Minecraft* is that it is probably one of the best accepted games by educators in general. The game has not only managed to avoid the type of criticism aimed at other games described as "risk-" or "violence-glorifying" (see, for instance, Anderson and Warburton 2011; or Hull et al. 2012), but it has even been incorporated into educational settings. For instance, it has been applied to the teaching of mathematics (Bos et al. 2014), a variety of science subjects (Short 2012), or even art (Overby and Jones 2015). *Minecraft* has capitalized on this acceptance and currently offers, through its https://education.minecraft.net website (formerly Minecraft.edu), a special version of the game called "Minecraft: Education Edition" intended for schools, which, according to the website, may be used with all educational levels, even from age 3, which makes it unique in the variety of ages it may appeal to.

Coming to our specific subject matter, the end-user license agreement, the *Minecraft* EULA certainly stands on its own as compared to other videogames. This has not escaped the attention of Lastowka (2012), who quotes fragments from an early 500-word version of the "Terms of Use," and remarks that it differs greatly from other EULAs in terms of register. According to this author, the approach by the authors of the document (signed by "Marcus Persson and friends") is greatly influenced by their ideological stance, which, at least at first sight, is less in favor of strict copyright protection. One cannot forget that, in 2011, Persson did say, in front of a large audience (Tassi 2011), that "Piracy is not theft" and, for good measure, added "If you steal a car, the original is lost. If you copy a game, there are simply more of them in the world."

Such remarks, and the way they are reflected in the terms of use, reflect, according to Lastowka, an interesting struggle between supporting user creativity and freedom, on the one hand, and the fact that, after all, developers create products from which they—quite legitimately—attempt to make a profit. As we shall see in the following section, these two conflicting principles are present in the *Minecraft* EULA.

4 The study

My working hypotheses will be as follows:

H1: The *Minecraft* EULA (hereinafter, E-*Minecraft*) differs from other EULAs in that an attempt is made by the drafters to make the language more comprehensible than and less similar to what may be found in traditional EULAs.

H2: In order to make the language more comprehensible, some linguistic strategies are used which visibly depart from what would be expected in legal language, even in the case of EULAs and videogames.

H3: All the linguistic strategies deployed are also aimed at creating proximity between the company and the user.

In order to test my hypotheses, I analyze the E-*Minecraft* in detail, comparing it to examples from other online MMOGs, two of a relatively simple nature, such as *7 Days To Die* and *Starbound*; and one of a slightly more sophisticated nature, *World of Warcraft* (a massive multiplayer online role-playing game [MMORPG]). I shall be using not only objective tools (readability indexes) but also a qualitative analysis, since it is our opinion that there are some features that may be better compared manually, or, in some cases, by allowing the text to speak for itself in order to comment on its features.

5 Analysis

5.1 Traditional readability measurements

In order to measure readability according to the established parameters, an online readability measurement tool, "Readability Calculator" (https://www.online-utility.org/english/readability_test_and_improve.jsp), was used; this tool was selected not only for its simple, straightforward interface, but also because it combined the most frequent readability indexes. The E-*Minecraft* was compared to other EULAs for videogames, such as *7 Days To Die*, *Starbound*, and *World of Warcraft* (hereinafter, "E-7Days," "E-*Starbound*" and "E-*WoW*,") which offered the following rough figures (Table 5.1).

In terms of pure brevity, as may be observed, the *Minecraft* EULA is shorter overall than the other three, and contains fewer sentences and shorter words in terms of both characters and syllables; the only potentially

TABLE 5.1 *Word, syllable, and sentence count for EULAs*

	No. of words	No. of sentences	Avg. characters per word	Avg. syllables per word	Avg. number of words per sentence
E-*Minecraft*	2,554	115.00	4.35	1.52	22.21
E-7Days	3,284	134	5.20	1.81	24.51
E-*Starbound*	4,292	252	4.75	1.70	17.03
E-*WoW*	7,835	478	4.94	1.74	16.39

TABLE 5.2 *Readability indexes for EULAs*

	Readability indexes					
	Gunning-fog	Coleman-Liau	Flesch-Kincaid	ARI	SMOG	Flesch reading ease
E-*Minecraft*	12.45	8.48	10.96	10.17	11.95	56.04
E-*7Days*	16.64	13.58	15.34	15.30	15.98	28.76
E-*Starbound*	12.81	10.44	11.15	9.47	23.02	45.44
E-*WoW*	12.05	11.47	11.36	10.03	12.47	42.83

"negative" figure is the number of words per sentence, which, as I mentioned earlier, is not necessarily indicative of difficulty. For instance, compare these two sentences, dealing with the same content:

- This license is a legal agreement between you and us (Mojang AB) and describes the terms and conditions for using the Game. (E-*Minecraft*)
- This Agreement sets forth the terms and conditions under which you are licensed to install and use the Platform. (E-*WoW*)

Apparently, the E-*Minecraft* sentence contains 22 words, whereas the E-*WoW* one contains 19. However, the comparison is clear; for instance, E-*WoW* contains a relative clause ("under which"), a passive sentence, and the archaic "set forth" (as compared to "describe"). In the light of examples like this, it appears that the readability indexes must be interpreted considering the weight attached to sentence length (Table 5.2).

As predicted, however, the length of the sentences, or rather, the punctuation, penalizes the readability index, since the average number of words per sentence is 22.21, which leads to a Gunning fog index of 12.45, that is, the reader requires 12–13 years of formal education to read the text. Other indexes which give slightly lower results are Flesch-Kincaid (11th grade, index = 10,96), SMOG (12th grade, index = 11.95), and Automatic Readability Index (10th grade, index 10.17), whereas, curiously enough, Coleman-Liau attributes a level of 8.48.

5.2 Beyond readability: Register and colloquialism

In this second part of our analysis, I shall perform a qualitative study of specific elements which not only intentionally move away from what is expected in EULAs and improve readability but may also have other

additional positive effects. In order to do so, it may suffice to compare the
initial paragraph of the *Minecraft* EULA with that of *World of Warcraft* to
see that the E-*Minecraft* belongs to a class of its own:

Minecraft	*World of Warcraft*
In order to protect *Minecraft* (our "Game") and the members of our community, we need these end-user license terms to set out some rules for downloading and using our Game. This license is a legal agreement between you and us (Mojang AB) and describes the terms and conditions for using the Game. We don't like reading license documents any more than you do, so we have tried to keep this as short as possible. If you break these rules we may stop you from using our Game. If we think it is necessary, we might even have to ask our lawyers to help out.	Thank you for your interest in Blizzard Entertainment, Inc.'s online gaming platform (formerly known as "Battle.net") and interactive games, and the interactive games from other developers ("Licensors") who make their games available for purchase and use on and through the Platform (collectively, the "Games"). Except as otherwise provided below, if you reside in the United States, Canada, or Mexico, use of the Platform is licensed to you by Blizzard Entertainment, Inc., a Delaware corporation, and if you are not a resident of the United States, Canada, or Mexico, use of the Platform is licensed to you by Activision Blizzard International B.V., Beechavenue 131 D, 1119 RB Schiphol-Rijk, the Netherlands (Blizzard Entertainment, Inc., and Activision Blizzard International B.V. are referred to herein as "Blizzard," "we," or "us"). This Agreement sets forth the terms and conditions under which you are licensed to install and use the Platform. The term "Platform," as used in this Agreement, means and refers collectively, and at times individually, to (1) the Blizzard App Client software (formerly known as the "Battle.Net" Client), (2) the gaming services offered and administered by Blizzard in connection with the Blizzard App Client and the Games, (3) each of the Games (including any authorized mobile apps relating to the Games), (4) Blizzard's Game-related websites and their associated forums, and (5) all features and components of each of them, whether installed or used on a computer or mobile device. IF YOU DO NOT AGREE TO THE TERMS OF THIS AGREEMENT, YOU ARE NOT PERMITTED TO INSTALL, COPY, OR USE THE BLIZZARD PLATFORM. IF YOU REJECT THE TERMS OF THIS AGREEMENT WITHIN FOURTEEN (14) DAYS AFTER YOUR PURCHASE OF A GAME FROM BLIZZARD, YOU MAY CONTACT BLIZZARD AT 1-800-592-5499 TO INQUIRE ABOUT A FULL REFUND OF THE PURCHASE PRICE OF THAT GAME. IF YOU PURCHASED A GAME AT RETAIL, YOUR RIGHT TO RETURN THE GAME IS SUBJECT TO THE RETAILER'S RETURN POLICY.

The initial paragraph for *Minecraft* sets the tone from the very beginning. The differences are quantitative (105 words vs. 317 for *WoW*) and objective (e.g., no numerals, addresses, only one defined term), but, most importantly, there is an attempt at complicity, an *excusatio* ("We don't like reading license documents any more than you do..."), which is a blatant infringement of the unwritten rules of contract writing: a contract does not describe itself as an unpleasant experience. E-*Minecraft* is presented as something that the developers would have preferred not to do ("we **need**[3] these end-user license terms to set out some rules for downloading and using our Game"). This, of course, does not mean that *Minecraft* is not serious about legal action, and such is said throughout, but the threat is toned down by "might even," and by the intimation that this could be an undesired measure by Mojang ("even *have* to ask our lawyers"). After all, Mojang does not want legal actions ("If you and we ever have a dispute in court (and **we hope that won't happen** just as much as you do").

As was noted earlier, *Minecraft* was developed from an "indie" point of view by a developer who favors creativity and is not inherently opposed to "piracy," or at least claims not to share the approach to piracy common in the industry. This has a direct impact on both the content and the register used in the E-*Minecraft*, which presents a number of colloquial and conversational features not commonly found in these contracts. All of these are perfectly aligned with the intention to create a "collaborative" atmosphere and the sense of an "online community," in order to overcome a feeling of mistrust, or even plain hostility, between licensors and software users (Stern 1985). Indeed, the agreement, as we shall see below, is not only a two-party relationship between a company (Mojang AB) and players, but also reproduces a friendly "dialogue."

In order to analyze how this "dialogue" is created, I shall comment on some stylistic features that may be observed in the *Minecraft* EULA, which are typical of colloquial and/or conversational discourse and clearly avoid what would be expected in "EULA legalese": colloquialisms and avoidance of legalese, contractions, personal pronouns, and other pragmatic and "friendly" markers.

5.2.1 Colloquialisms and avoidance of legalese

In terms of register, for instance, one may compare these clauses related to user-created content:

- If you are going to make something available on or through our Game, it must not be offensive to people or illegal, it must be honest, and it must be your own creation. Some examples of the types of things you must not make available using our Game include: posts that include racist or homophobic language; posts that are bullying or trolling; posts that are offensive or that damage our or another person's reputation; posts that include porn or someone else's creation or image; or posts that impersonate a moderator or try to trick or exploit people. (E-*Minecraft*)

- You further represent and warrant that you will not use or contribute User Content that is unlawful, tortious, defamatory, obscene, invasive of the privacy of another person, threatening, harassing, abusive, hateful, racist or otherwise objectionable or inappropriate. (E-*WoW*)

Although the content and the purpose is the same, the language is not, beyond the mere word count; in fact, this fragment is good evidence that "shorter" does not necessarily mean "clearer." The *WoW* agreement includes some of the classic features of legal language, for example, the use of "otherwise" to cover all possible cases, or binomials such as "represent and warrant." For its part, the E-*Minecraft* uses colloquial clippings ("porn" vs. the more legal "obscene") and periphrases such as "that damage our or another person's reputation," compared to the legalese "defamatory"; also, *Minecraft* prefers "illegal" to "unlawful," and instead of adjectives such as "abusive" or "hateful," *Minecraft* prefers to even resort to in-group language, including "trolling," an activity which may not be strictly against the law but represents probably one of the most serious threats to bona fide online activity.

As one would expect, E-*Minecraft* avoids the "usual suspects" of legalese. Concerning "here-" or "there-" compound adverbs, there are no single instances of them, unlike the other agreements (Table 5.3):

TABLE 5.3 *Compound adverbs in EULAs analyzed*

	E-*Minecraft* (2,554 words)	E-*7Days* (3,284 words)	E-*Starbound* (4,292 words)	E-*WoW* (7,835 words)
hereby	-	6	6	5
thereby	-	4	1	1
hereof	-	-	-	1
thereof	-	1	2	9
herein	-	-	1	4
therein	-	-	1	2
hereunder	-	-	1	-
therefore	-	-	1	-
Total	-	11	13	22

For its part, and although most guidelines remind that "shall" is ambiguous and should be replaced by "must" if it means "obligation," it is still used in the other three EULAs (although it is true that it alternates with "will" in some of them), whereas there is no single instance in E-*Minecraft*, which prefers "must" in case of obligation and "will" for plain reference to the future. For instance, let us compare what would happen in case of disputes:

- If you and we ever have a dispute in court..., the exclusive forum (that is, the place it will be handled) **will** be a state or federal court in King County.... (E-*Minecraft*)
- This Agreement **shall** be construed (without regard to conflicts or choice of law principles) under the laws of the State of Texas. (E-*7Days*)

There are many other cases where legalese is omitted. In order to illustrate this, I have selected some of the "words/structures to be avoided" from the plainlanguage.gov website, showing how E-*Minecraft* compares favorably (Table 5.4).

Another thing that E-*Minecraft* does is, when the legal term is absolutely unavoidable, is to explain its meaning. In fact, the agreement contains eight instances of rephrasing (1 "that is" and 7 "mean," most related to words or phrases which are seldom, if ever, explained in EULAs):

If you and we ever have a dispute in court (and we hope that won't happen just as much as you do), the exclusive forum (**that is**, the place it will be handled) will be a state or federal court in King County, WA [...]

Within reason you're free to do whatever you want with screenshots and videos of the Game. **By "within reason" we mean that** you can't make any commercial use of them or do things that are unfair [...]

Nothing we say in these terms will affect those legal rights, even if we say something which sounds like it contradicts your legal rights. **That's what we mean when we say** "subject to applicable law."

It is, arguably, unavoidable to include some traditional contract terms in these agreements, especially since they have a well-established meaning in contract law. One of them is "as is," which limits liability on the part of the supplier (for a recent review, see Marks 2017). E-*Minecraft* does include this phrase but escapes boilerplate language and attempts to give an explanation, while others stick to the traditional formulae, in spite of their verbosity and the "unfriendly" capitalization:

TABLE 5.4 *Use of archaisms in EULAs analyzed*

	E-Minecraft (2,554 words)	E-7Days (3,284 words)	E-Starbound (4,292 words)	E-WoW (7,835 words)
and/or (*a or b or both*)	0	7	0	22
accompany (*go with*)	0	6	0	0
additional (*added, more, other*)	0	1	3	14
appropriate (*proper, right*)	0	1	1	5
constitute (*is/are, forms*)	1	2	1	3
fail to (*do not, did not*)	0	0	1	1
notwithstanding (*in spite of*)	0	1	0	1
permit (*let*)	1	4	7	7
represent (*is*)	0	1	0	3
such (*this, that*)	0	12	11	21
set forth (*list, describe*)	0	7	0	9

- When you get a copy of our Game, we provide it "**as is**". Updates are also provided "**as is**". **This means that** we are not making any promises to you about the standard or quality of our Game, or that our Game will be uninterrupted or error free. (E-*Minecraft*)
- THE GAME IS PROVIDED "**AS IS**" WITHOUT WARRANTY OR GUARANTEE OF ANY KIND, EITHER EXPRESS OR IMPLIED, INCLUDING, WITHOUT LIMITATION, THE IMPLIED WARRANTIES OF SATISFACTORY QUALITY, MERCHANTABILITY, FITNESS FOR A PARTICULAR PURPOSE OR NON-INFRINGEMENT. (E-*Starbound*)

- THE PLATFORM, ACCOUNTS, AND THE GAME(S) ARE
  PROVIDED ON AN **"AS IS"** AND "AS AVAILABLE," BASIS FOR
  USE, WITHOUT WARRANTY OF ANY KIND, EITHER EXPRESS
  OR IMPLIED, INCLUDING WITHOUT LIMITATION ANY
  IMPLIED WARRANTIES OF CONDITION, UNINTERRUPTED
  OR ERROR-FREE USE, MERCHANTABILITY, FITNESS FOR A
  PARTICULAR PURPOSE, NONINFRINGEMENT, TITLE, AND
  THOSE ARISING FROM COURSE OF DEALING OR USAGE OF
  TRADE. (E-*WoW*)

There are other examples of colloquialism; for instance, a section in
E-*Minecraft* is literally titled "GENERAL STUFF," and the colloquial,
but probably ambiguous, term "things" occurs 6 times (users do/make/
create "things"). The EULA also contains phrasal verbs ("help out," "play
around"), adverbs typical of oral discourse ("too") and even idiomatic and
metaphorical expressions:

> [...] sometimes the law changes or someone does something that affects
> other users of the Game and we therefore need to **put a lid** on it.

> Any Mods you create for the Game **from scratch** belong to you (including
> pre-run Mods and in-memory Mods).

> Below we also give you limited rights to do other things but we have to
> **draw a line** somewhere or else people will go too far.

> And so that we are **crystal clear**, "the Game" or "what we have made"
> includes, but is not limited to, the client or the server software ...

Probably the most extreme attempt at oral-conversational discourse is the
use of "oh," which implies that one has forgotten to say something and
then remembers (completely incompatible with written language, let alone
a contract):

> **Oh** and if the law expressly allows it, such as under a "fair use" or "fair
> dealing" doctrine then that's ok too—but only to the extent that the law
> applicable to you says so.

In order to finish this section, mention may be made of the register used in order
to warn users about what they are or are not allowed to do. Infringements
are described in terms of "prohibited" or "acceptable" in all other EULAs,
but E-*Minecraft* prefers to talk about things which are "okay" or "(no) fun":

> Basically, Mods are **okay** to distribute; hacked versions or Modded
> Versions of the Game client or server software are not **okay** to distribute.

> Also, don't just rip art resources and pass them around, that's **no fun**.

5.2.2 *Contractions*

Minecraft contains contractions (a total of 54 in 2,526 words), which are nowhere to be found in any of the other EULAs I have reviewed:

> We **don't** like reading license documents any more than you do, so we have tried to keep this as short as possible.

> If you **don't** want to or **can't** agree to these rules, then you must not buy, download, use or play our Game.

Although it is felt that contractions are out of place in legal writing, advocates of plain English argue that they make documents more readable. Back in the 1940s, one of the developers of the readability formulae remarked that using contractions was "the most conspicuous and handiest device" for doing so (Flesch 1949: 82), an impressionistic comment which has been later supported by empirical evidence. Also, the use of contractions helps to achieve a relaxed tone showing confidence (Garner 2001: 49). Nevertheless, it should not be felt that this is an automatic, search-and-replace attitude, but rather a choice of natural contexts. As Flesch noted, "[s]ometimes, they [contractions] fit, sometimes they don't. It depends on whether you would use the contraction in speaking *that particular sentence*" (Flesch 1949: 83, original emphasis). E-*Minecraft* uses contractions in conjunction with personal pronouns (see above), whereas with impersonal subjects, it tends to use the full forms; compare the following sentences, with nonanimated subjects:

> And so that we are crystal clear, "the Game" or "what we have made" includes, but **is not** limited to, the client or the server software for our Game and includes *Minecraft* Pocket Edition on all platforms.

> We also agree that class action lawsuits and class-wide arbitrations **are not** allowed under the terms of the agreement.

As said earlier, this is a tendency; there are cases with a "you" subject with no contractions (e.g., "you are not buying the Game itself"); however, the opposite is true, that is, with only one exception (third-party tools), all the uses of contractions have a pronoun as a subject. Therefore, although contractions are a visible difference between the E-*Minecraft* and other agreements, they must be considered in conjunction with other colloquial features, that is, the use of a higher number of pronouns, as we shall see in the following subsection.

5.2.3 *Personal pronouns*

As we saw in the literature review, it is a fact that in the virtual economy users are more commonly referred to as "you" instead of "the user," "the purchaser," etc. (Brunon-Ernst 2016: 45). Indeed, in the cases studied, "you" is consistently used by all the EULAs, and "user" is mostly found in defined terms such as "user content" (E-*Starbound*), "end user" (E-*Minecraft*, *Starbound*), although it is quite remarkable that E-*WoW* still resorts to "user" in cases which clearly include the addressee ("The user is responsible for the costs of returning media to Blizzard"). However, in spite of the "you" mode of address, there are many cases in which the EULAs resort to impersonal constructions (especially nominalizations), whereas *Minecraft* maintains the "you" approach:

- Any **use, reproduction or redistribution** of the Game not in accordance with the terms of this EULA is expressly prohibited. (E-*Starbound*)
- **you** can't make any commercial use of them or do things that are unfair or adversely affect our rights […]. (E-*Minecraft*)

Unlike "you," "we" is not so frequent in EULAs. In the other agreements studied, the developers do not use "we" at all; instead, they use the passive or refer to themselves in the third person:

- **Chucklefish** hereby grants, and by installing the Game you thereby accept, a limited, non-exclusive right and license … (E-*Starbound*)
- To play Games on the Platform, you will need to add a Game license to an Account, which requires an authentication code generated by **Blizzard**. (*WoW*)
- **Licensor** hereby grants you the nonexclusive, non-transferable, limited right and license to use one copy of the Software […]. (E-*7Days*).

This is not the case with E-*Minecraft*, which starts very early (in the first paragraph) to use "we" and "you" for the contracting parties, to such an extent that, in a comparatively short EULA, there are as many as 128 occurrences of first-person plural indexicals (64 "we," 53 "our" and 11 "us"), that is, 5 % of the total. E-*Minecraft* even uses "we and you" where it feels necessary (whereas E-*Starbound*, for instance, prefers "Chucklefish and you"):

- This license is a legal agreement between **you and us** (Mojang AB) and describes the terms and conditions for using the Game. (E-*Minecraft*)
- You also acknowledge and agree that this EULA is the complete and exclusive statement of the agreement between **Chucklefish and you** and that this EULA supersedes any prior or contemporaneous agreement, either oral or written, and any other User Content between **Chucklefish and you**. (E-*Starbound*)

5.2.4 Other pragmatic markers and "friendly" insertions

In addition to the linguistic devices and strategies we have seen in previous sections, the creation of a friendly atmosphere is also reflected in the content of E-*Minecraft*. In previous sections, we saw an outstanding self-deprecation of the genre itself ("We don't like reading license documents any more than you do…"), which aims to create a "shared ideological space" between the developer and the user. In addition to these, there are quite a few comments which define a friendly space. For instance, the crucial decision to install the software or not, which implies acceptance of the term, may lead a potential user not to install the software, but the reasons vary from one game to another:

- IF YOU DO NOT AGREE WITH THE TERMS OF THIS EULA, YOU MAY NOT USE, DOWNLOAD OR INSTALL STARBOUND. (E-*Starbound*)
- If you do not want your information shared in this manner, then you should not use the Software. (E-*7Days*)
- If you don't want to or **can't** agree to these rules, then you must not buy, download, use or play our Game. (E-*Minecraft*)

E-*Minecraft* accepts that lack of consent may stem from free will ("don't want"), but also from principled dissent ("can't agree"). The underlying principle is that one may disagree with copyright, and that the user is to be protected.

At times, E-*Minecraft* "puts itself in the gamers' shoes," recognizes errors (rather than simply avoiding responsibility for them), and even justifies such errors, appealing to the gamers' desire for novelty:

You have to accept that we may release games well before they are complete and so they may (**and often will**) have bugs—**but we prefer to release these features early than make you wait for perfection.**

In other cases, there are appeals at solidarity between the developer and the user (*dura lex, sed lex*):

We're not going to be unfair about this though—but sometimes the law changes or someone does something that affects other users of the Game.

6 Conclusions

EULAs are a major genre of videogame metadiscourses which define and govern the relationships between major firms and millions of users, and they have relevant repercussions not only in financial terms, but also for the privacy of individuals. Therefore, the way they are drafted is capital, because, even if they may be ignored or read very quickly, they do set the rules for events that may arise from noncompliance or disagreement.

As this analysis has shown, the *Minecraft* EULA differs from other end-user license agreements, even within the realm of online games. Traditional readability measurements, in spite of their faults, do show that there is an attempt at enhanced readability, whereas an in-depth analysis of the language in the clauses shows that the drafters have gone to great lengths to avoid the defects of legalese and produce a "user-friendly" agreement. However, the text, as we have seen, is not only "friendly" in terms of readability, but also literally speaking, that is, there is a constant strategy to create solidarity between the developers and the users, which is aimed at increasing trust, but also, more likely than not, part of a marketing strategy.

The "you and us" approach taken by the E-*Minecraft* is aligned with its history and its spirit. *Minecraft* was born and purports to retain the spirit of an "indie" game, and its EULA represents a way to resist mainstream co-optation. In spite of the fact that Mojang AB has been bought by Microsoft, or "taken over by mainstream," its EULA still reflects its "indie" soul, as a way of what Lipkin calls incorporating the spirit "directly into content" (2013: 21). In this case, we have found many strategies pointing at a shared ideological space, which range from purely linguistic ("you and we") to ideological ones, which attempt to create the impression that the developer and the user share the same views regarding gaming and intellectual property. This approach may be greatly beneficial in many respects; for a start, users are empowered and allowed to understand exactly which terms they are accepting, especially in cases where intellectual property and privacy issues are involved. In addition to this, there is a further intangible benefit for game developers (and probably to any other firms following this example, beyond the world of computing software), as plain-language agreements help to create an image of a socially responsible firm, which may provide a competitive edge in an environment where corporate reputation is a major ideological component.

In the same way that I believe that this type of EULA points the way ahead and may set an example for others to follow, I am aware that this study has not only shown interesting facts but has also left questions unanswered. This analysis focuses on a very specific case, the *Minecraft* EULA, which has been shown not only to have very attractive characteristics but also to be a relatively isolated case within the environment of gaming software. Future studies might explore issues like (a) the potential impact this type of EULA may have on "informed" gamers (e.g., due to its massive success among young people from very early ages, there may well come a time when most gamers who download and install games have already been exposed to the *Minecraft* EULA, which may, or may not, set a comparative standard making them more demanding as "educated gamers"); (b) intermediate possibilities in user readability (as we have seen in this chapter, there are certainly some EULAs which use extreme legalese, where others, without going as far as *Minecraft*, do try to simplify their language); or even (c) whether these simplified EULAs lead to more or less litigation, which might be crucial in encouraging other software developers to join the trend and simplify their agreements or, conversely, retreat to the safer ground provided by established case law. I am sure that both these studies and the one we have conducted here may be of great interest not only to academics, but also to developers and gamers alike, as they deal with a legal relationship that practically every individual on this planet is likely to enter.

Notes

1   http://www.itechpost.com/articles/96096/20170417/pedophiles-now-targeting-young-players-from-world-of-warcraft-minecraft-and-avakin-life.htm.

2   https://www.esrb.org/ratings/ratings_guide.aspx (accessed December 30, 2017).

3   Unless otherwise stated, emphases are mine.

References

Primary sources

7 Days to Die. Available online: www.7daystodie.com/eula (accessed December 30, 2017).

Minecraft. Available online: https://account.mojang.com/documents/*Minecraft*_eula (accessed December 30, 2017).

Starbound Terms and Conditions. Available online: https://playstarbound.com/eula/.

World of Warcraft Terms of Use. Available online: http://us.blizzard.com/en-us/company/legal/eula (accessed December 30, 2017).

Secondary sources

Adams, K.A. (2013), *A Manual of Style for Contract Drafting*, 3rd edn., Chicago, IL: American Bar Association.

Adler, M. (2012), "The Plain English movement," in P.M. Tiersma and L.M. Solan (eds.), *The Oxford Handbook of Language and Law*, 67–83, New York: Oxford University Press.

Anderson, C.A. and W.A. Warburton (2011), "The impact of violent video games: An overview," in W. Warburton and D. Braunstein (eds.), *Growing Up Fast and Furious: Reviewing the Impacts of Violent and Sexualised Media on Children*, 56–84, Annandale, NSW, Australia: The Federation Press.

Ayres, I. and A. Schwartz (2014), "The no-reading problem in consumer contract law," *Stanford Law Review*, 66: 545–610.

Bailin, A. and A. Graftstein (2001), "The linguistic assumptions underlying readability formulae: A critique," *Language and Communication*, 21: 285–301.

Bélanger, A. and A. van Drom (2012), "A dialogical and polyphonic approach to contract theory," in V.K. Bhatia, C.A. Haftner, L. Miller, and A. Wagner (eds.), *Transparency, Power and Control, Perspectives on Legal Communication*, 85–107, Farnham: Ashgate.

Blizzard Entertainment (2004), *World of Warcraft*, Irvine, CA: Blizzard Entertainment Inc.

Bos, B., L. Wilder, M. Cook, and R. O'Donnell (2014), "Learning mathematics through *Minecraft*," *Teaching Children Mathematics*, 21 (1): 56–59.

Brunon-Ernst, A. (2016), "The fallacy of informed consent: Linguistic markers of assent and contractual design in some e-user agreements," *Alicante Journal of English Studies*, 28: 37–58.

Carroll, A.B., (1991), "The pyramid of corporate social responsibility: Toward the moral management of organizational stakeholders," *Business Horizons*, 34 (4): 39–48.

Carter, C.R. (2000), "Ethical issues in international buyer–supplier relationships: A dyadic examination," *Journal of Operations Management*, 18: 191–208.

Chucklefish (2016), *Starbound*, London: Chucklefish Limited.

Felch, R.I. (1985), "Standards of conduct: The key to supplier relations," *Journal of Purchasing and Materials Management*, 21: 16–18.

Flesch, R. (1948), "New readability yardstick," *Journal of Applied Psychology*, 32 (3): 221–33.

Flesch, R. (1949), *The Art of Readable Writing*, New York and Evanston: Harper & Row Publishers.

Garner, B.A. (2001), *Legal Writing in Plain English: A Text with Exercises*, Chicago and London: University of Chicago Press.

Gomulkiewicz, R.W. and M.L. Williamson (1996), "A brief defense of mass market software license agreements," *Rutgers Computer and Technology Law Journal*, 22: 336–67.

Gunning, R. (1952), *The Technique of Clear Writing*, New York: McGraw-Hill.

Hull, J.G., A.M. Draghici, and J.D. Sargen (2012), "A longitudinal study of risk-glorifying video games and reckless driving," *Psychology of Popular Media Culture*, 1 (4): 244–53.

Kessler, F. (1943), "The contracts of adhesion—Some thoughts about freedom of contract role of compulsion in economic transactions," *Columbia Law Review*, 43 (5): 629–42.

Kimble, J. (1996–97), "Writing for dollars, writing to please," *The Scribes Journal of Legal Writing*, 6: 1–38.

Kunze, J.T. (2008), "Regulating virtual realms optimally: The model end user license agreement," *Northwestern Journal of Technology and Intellectual Property*, 7 (1): 102–18.

Lemens, C. and K.A. Adams (2015), "Fixing your contracts: What training in contract drafting can and can't do," *Docket*, September 22. Available online: http://www.accdocket.com/articles/fixing-your-contracts.cfm (accessed April 7, 2018).

Lipkin, N. (2013), "Examining indie's independence: The meaning of 'indie' games, the politics of production, and mainstream co-optation," *Loading… The Journal of the Canadian Game Studies Association*, 7 (11): 8–24.

Marks, C.P. (2017), "Online and 'as is'," *Pepperdine Law Review*, 45 (1): 1–53.

Marotta-Wurgler, F. (2007), "What's in a Standard Form Contract? An empirical analysis of software license agreements," *Journal of Empirical Legal Studies*, 4 (4): 677–713.

Marotta-Wurgler, F. (2008), "Competition and the quality of Standard Form Contracts: The case of software license agreements," *Journal of Empirical Legal Studies*, 5 (3): 447–75.

Masson, M.E.J. and M.A. Waldron (1994), "Comprehension of legal contracts by non-experts: Effectiveness of plain language redrafting," *Applied Cognitive Psychology*, 8 (1): 67–85.

Mojang (2011), *Minecraft*, Sweden: Mojang AB.

Overby, A. and B.L. Jones (2015), "Virtual LEGOs—Incorporating *Minecraft* into the art education curriculum," *Art Education*, 68 (1): 21–27.

Shikowitz, R. (2009), "License to kill: MDY v Blizzard and the battle over copyright in *World of Warcraft*," *Brook L. Rev.*, 75 (3): 1015–53.

Short, D. (2012), "Teaching scientific concepts using a virtual world—*Minecraft*," *Teaching Science*, 58 (3): 55–58.

Stern, R.H. (1985), "Shrink-wrap licenses of mass marketed software: Enforceable contracts or whistling in the dark," *Rutgers Computer and Technology Law Journal*, 11: 51–92.

Tassi, P. (2011), "*Minecraft*'s Notch: 'Piracy is not theft'," *Forbes*, March 4. Availiable online: https://www.forbes.com/sites/insertcoin/2011/03/04/minecrafts-notch-piracy-is-not-theft/

Taylor, T.L. (2006), *Play between Worlds: Exploring Online Game Culture*, Cambridge, MA/London: The MIT Press.

Tekfi, C. (1987), "Readability formulas: An overview," *Journal of Documentation*, 3: 261–73.

Terasaki, M. (2014), "Do end user license agreements bind normal people?" *Western State University Law Review*, 41 (2): 467–89.

The Fun Pimps (2013), *7 Days to Die*, Allen, TX: The Fun Pimps Entertainment LLC.

Van Boom, W.H., P. Desmet, and M. Van Dam (2016), "'If it's easy to read, it's easy to claim'—The effect of the readability of insurance contracts on consumer expectations and conflict behaviour," *Journal of Consumer Policy*, 39: 187–97.

Webber, N. (2014), "Law, culture and massively multiplayer online games," *International Review of Law, Technology*, 28 (1): 45–59.

Williams, C. (2015), "Changing with the times: The evolution of plain language in the legal sphere," *Alicante Journal of English Studies*, 28: 183–203.

Minecraft Wiki, "Version History." Available online: https://*Minecraft*.gamepedia.com/Version_history (accessed December 30, 2017).

PART TWO

Player Interactions: (Un)Collaboration, (Im)Politeness, Power

6

Bad Language and Bro-up Cooperation in Co-sit Gaming

Astrid Ensslin and John Finnegan

1 Introduction

This chapter examines the quasi-paradoxical relationship between cooperative and impolite verbal behavior in co-situated ("co-sit") player communication. In doing so, it takes a broad paratextual, interactional approach, locating itself in the overlapping territory between videogame textuality research (see Part Three of this volume) and verbal and multimodal player interaction. More specifically, we seek to investigate the role of BLE ("Bad Language Expressions," see McEnery 2006) as part of the discourses of cool and fun (Ensslin 2012) based on participants' use of swear words and other types of BLE, as well as their own assessment of the pragmatic functions and psychological and social motivations of these overused structures. This will be juxtaposed with an examination of the distinctly cooperative, mutually supportive and ultimately polite "bro-up" behavior displayed by co-sit players to secure and maintain bonding and cooperation. The data is methodologically triangulated from transcribed and BLE-annotated video/audio recordings and semistructured paired interviews. We will show that the metaludic and metacommunicative information garnered from the interviews lifts the appearance of paradoxical behavior and instead refigures bad language as face-saving and ultimately prosocial tool in player communication.

2 Co-sit interaction as paratext

Research into videogame paratexts has proliferated in the past five years or so, triggered partly by Mia Consalvo's seminal book, *Cheating: Gaining Advantage in Videogames* (2009), which first considered paratextual media (industries) such as game magazines, guides, and walkthroughs and how they blur the boundaries between game world and player world. Paratexts, according to Genette (1991), are "verbal and other productions" "beyond the naked state" of the text (Genette 1991: 261). Elements like the author's name, preface, illustrations, blurbs, interviews, and reviews "surround or extend" (Genette 1991: 261) the core text of a book, which was Genette's medial focus. Clearly, however, no matter the medium, in contemporary participatory and fan culture, the text itself becomes the material starting point for multilayered user interactions and communications, from debates in online discussion fora to cosplay (costume play) conventions and oral after-school chit-chat. It may thus be argued that, without the paratext, the text itself cannot come into being and that, in modern transmedia, "we can only approach texts *through* paratexts" (Gray 2010: 25). Games are an extreme case in point as their narratives are emergent and experiential (Aylett 1999; Jenkins 2004) and do not come into being without the player's "autotelic enactment" (Ryan 2004: 348), that is, their self-motivated decision-making and interactions in the game world. Thus, paratexts like walkthroughs, after-action reports, and "let's plays" (online videos that "feature gameplay footage accompanied by simultaneous commentary recorded by the player," Burwell and Miller 2016: 109) are discursive manifestations of individualized player experiences that generate and simultaneously individualize players' experiences of the "orthogame" (Carter et al. 2012) itself. Souvik Mukherjee (2015: 106) goes as far as to argue that "to construct any textuality for the videogame, the paratextual needs to be considered first."

In this chapter, we propose to take videogame paratextuality one decisive step further and argue that unplanned, spontaneous player communication during physically co-located gameplay, which we call co-situated (or "co-sit") player interaction, is a form of paratext as well, comparable in its immediacy and meaning-making negotiations to, for instance, the conversations between a parent and a child during a joint reading session. Indeed, co-sit gaming might be seen as one of the most intrinsic forms of metaludic communication as it happens in direct response to on-screen action and reflects the experiential aspects of videogames' mechanics and narrativity with a maximum degree of immediacy and emotional rigor.

Co-sit interaction can be differentiated from other, seemingly spontaneously occurring oral videogame paratexts such as let's plays. Both genres emerge synchronously during gameplay, and although let's plays do occasionally feature more than one player, they are essentially videos posted on channels like YouTube, while co-sits are essentially private gaming sessions and not primarily intended for publication. While some players do record their co-sit play and put it on the PlayStation network, for example, compared to performance-oriented and often (partly) scripted and rehearsed let's plays, co-sits tend to be unplanned and primarily socially rather than public-performance or even—in the case of professional YouTube channels—commercially oriented let's plays. Let's plays are also conceptually monologic, with (typically) one player commenting on their gameplay, whereas co-sits are conceptually dialogic in response to screen action, bringing relationships between players to the fore. In other words, let's plays and other forms of public metagame paratexts are essentially performative in function, addressed to an audience of fans, followers, or followers-to-be, and the conative nature of let's plays is a lot more pronounced than the more competitive, inward-looking, interactional intent behind co-sits.

3 Subverting the discourse of power

Johan Huizinga's (1950) concept of the *magic circle* has been widely debated and partly subverted by contemporary ludologists (e.g., Consalvo 2009). It describes play in terms of being in "a temporary world within the ordinary one" (Dippel and Fizek 2016: 1), a world that we enter when we embark on gameplay and that makes us feel immersed and separated from the actual world. The magic circle allows or even forces players to adopt what Bernard Suits (2005) has called a *lusory* attitude, a psychological state that makes players accept and embrace the rules of the game they are engaged in, including its embedded values and ethical implications, no matter how unethical those rules and conditions may be outside the magic circle. What this implies is that, while in the magic circle, players dispense with many of their actual-world rules and values and replace them with alternative value and rule sets.

Player groups conjoin the idea of lusory attitude with that of *community of practice*, which, according to cognitive anthropologists Jean Lave and Etienne Wenger (1991), refers to a group of people who share an interest, a craft, and/or a profession. The group can evolve naturally because of its members' common interest in a particular domain or area, or it can be created specifically with the goal of gaining knowledge related to their field. Importantly, these shared practices involve various social actions, including specific linguistic and communicative practices. Furthermore, the tacit knowledge shared by members of a specific community of practice is akin to, or contains within itself, the linguistic concept of pragmatic presupposition (Stalnaker 1974), which refers to the shared assumptions about the context in which communication takes place and about the interlocutor and their knowledge and world view. Habitual game players are fully aware of the shared world views and social practices afforded or even demanded by specific gaming situations, and while gameplay itself is understood in terms of a general fun- and entertainment-oriented activity, player expectations and social practices can vary widely depending on the game genre or even the game itself.

Previous work by Ensslin (2012) on player discourses has documented the frequent use of swearing and other types of strong language, which reflects highly specific social contexts and communicative embeddings. In *Swearing in English*, Tony McEnery (2006) argues that swearing subverts the discourse of power, a purist language ideology pursued by powerful social groups who have the cultural capital (Bourdieu 1986) to determine what is appropriate language use for any given situational context. The discourse of power works hegemonically, that is, it is reconfirmed and perpetuated by anybody following and adapting to it. We would argue that this is exactly what most gamers refuse to do: they subvert the discourse of power (in the sense of privileged, elitist social codes) by communicating in ways that are not considered "refined," that are regularly censored in prewatershed public broadcasting, and that are typically scolded by powerful stakeholders such as parents, teachers, and politicians in the media.

Subverting the discourse of power happens at many levels of metaludic communication and player culture more generally:

> Player behaviors can often resist or even challenge the underlying social order [imposed by rules and subscribed by players to be able to play the game in the first place]. This includes technically specialized interventions, such as hacking and modding, as well as widespread player practices such as cheating, technological appropriations, subversive readings, interpersonal relationships, and the production of unofficial game "paratexts" (such as fan fiction, walkthroughs, etc.). (Grimes and Feenberg 2009: 108)

Interestingly, however, as our ensuing data analysis will demonstrate, what may prima facie and plausibly be seen as subversive behavior is not necessarily understood as such by players themselves—at least not in their conscious, explicitly rendered perceptions.

4 Player discourses

In *The Language of Gaming*, Ensslin (2012) outlines three dominant player discourses that can be found across videogame paratexts: the *Discourse of Fun*, the *Discourse of Cool*, and the *Discourse of Appreciation*. All three can be viewed under the broad umbrella of affective discourse because they all express emotions in more or less explicit ways: fun is inherent in the experience of play, although, of course, the game itself does not necessarily have to be funny or even evoke positive feelings (think of horror games); cool is a concept that paradoxically and effectively blends detachment and engagement; and appreciation involves evaluative discourses relating to game ratings and other forms of quality assessment.

In this chapter, we align players' use of so-called Bad Language Expressions (BLE, adapted from McEnery's more restrictive term, "Bad Language Words" [BLW], 2006) with the Discourses of Fun (DoF) and Cool (DoC) in particular. The DoF reflects players' extreme emotional investment in the games they play and the activities they perform in and around gameplay. This emotional investment, combined with a strong dedication to ritualized performance, typically manifests in the form of a vast diversity of expressive interjections, including response cries (Conway 2013; Goffman 1981) and laughter. Interjections signaling enjoyment can adopt different emotional connotations, such as malicious and/or supportive amusement, awe, shock, or bawdiness.

The DoC is slightly more elusive and challenging to capture, but it generally revolves around the paradox of detached engagement; on the one hand, cool is about projected emotional detachment, while on the other, it is about the performance of playing along as a mode of social engagement and emotional bonding. "Cool both enables and prevents relationships developing" (Nicholson and Hoye 2008: 28), and it is exactly this paradoxical tension that makes it so socially attractive. Put differently, cool detachment serves the purpose of trying to fit in and being accepted by others through respect rather than overt expressions of personal interest or emotional engagement. Cool assumes a sense of propriety, which involves knowing just how close one needs to get in order to both respect and gain respect from the other. The ideal cool distance cannot be guaranteed in advance, but it needs to be neither too far nor too close, ensuring the other's retention of effort in the relationship as well as the potential benefits of what a good relationship with the other can offer (Blackshaw and Long 2005: 253). In player discourse, then, the DoC marks the playful interception between social and antisocial communication, for example through mock bravado and instances of clipped lexis and very short, staccato syntax in players' turn-taking, as well as an overuse of expletives and other forms of BLE.

5 Methodology

This particular study was motivated by the anecdotal observation that swearing in videogame paratexts, and especially spoken paratexts, is abundant. Our aim was therefore to find out more about (1) the actual patterns of bad language (BL) use in player communication and, more specifically, spoken, synchronous turn-taking during co-sit play; and (2) how players themselves explain their linguistic behavior and, specifically, their motivations for using BL, following an actual co-sit session. For the first goal, we first defined a BLE tagset for corpus annotation, based on McEnery's (2006: 25; see also McEnery et al. 2006: 264) categories of BLW. McEnery distinguishes between actual expletives, or swear words related to sex, defecation, and other body parts and fluids generally known to trigger revulsion (FUCK, PISS, SHIT, CUNT, BLOODY); animal terms of abuse (PIG, COW, BITCH, ASS); sexist terms of abuse (BITCH, WHORE, SLUT); intellect-based terms of abuse (IDIOT, PRAT, IMBECILE, DORK); racist terms of abuse (PAKI, NIGGER, CHINK); religious swear words (JESUS, GOD, HEAVEN, HELL, DAMN, HOLY); and homophobic terms of abuse (QUEER, FAG, POOF, DYKE). There is a certain degree of overlap in some terms of abuse (e.g., BITCH and COW are both sexist and animal-related), and they often combine into expressions like HOLY SHIT.

Data collection took place in summer 2015 and resulted in a corpus of four 30-minute video- and sound-recorded co-sit gaming sessions featuring undergraduate students (two per co-sit session) at Bangor University, where both authors were based at the time. Six out of our eight participants were male (we did not follow a gender critical approach in this study and therefore followed a logistically feasible rather than gender-balanced recruitment strategy), and all participants were native (n = 7) or near-native (n = 1) speakers of British English. All participant pairs were gaming "buddies," that is, they were habitual co-sit players and were used to playing with each other. To further support their familiarity levels, they were allowed to play a game of their choice. The games they chose were *Helldivers* (Arrowhead Game Studio 2015; PS4 version; P1/P2), *Streetfighter X Tekken* (Capcom 2012; Windows with Fightstick; P3/P4), *Halo 4* (343 Industries; Xbox 360; P5/P6), and *Yoshi's Woolly World* (Good-Feel 2015; Wii U; P7/P8). Each session was immediately followed by semistructured interviews with each co-sit pair. The corpus contains 10,287 tokens. The transcripts were analyzed as raw data and subsequently tagged in terms of BLE types as outlined above. We then analyzed the corpus data in terms of frequency lists, keywords (using the BNC Spoken as reference corpus), concordances and keywords in context, collocations and clusters, using WordSmith 6.0. The interviews were coded for thematic nodes in NVivo 11 with a view to distilling some grounded theory relating to habits and motivations for bad language.

6 Data and findings

6.1 Corpus analysis

A look at the keyword lists of the co-sit transcripts strongly suggests overuse of the expletives SHIT (0.21%, keyness = 101.06; used particularly often in the phrases OH SHIT and ARR SHIT, and directly collocated by religious

BLEs such as OH GOD—see below), CRAP (0.12%, keyness = 77.01), and FUCK (0.07%, keyness = 24.37, also used as prepositional verb, FUCK OFF, and in the formulaic verb-complement phrase, FUCK ME)—and these values only reflect the represented core forms themselves, excluding derivatives such as CRAPPY (used for example in CRAPPY LITTLE ELF). Also frequent were religious BLEs (GOD, 0.45%; keyness = 557.12; used specifically in OH MY GOD and OH GOD and in combination with expletive BLEs—see above).

Overall, the distribution of BLEs across the thus annotated corpus is shown in Figure 6.1 and their normalized frequencies in Table 6.1. The vast majority of BLEs in the corpus are religious terms of abuse and actual expletives, and there is a small percentage of animal, intellect-based, and sexist terms of abuse. Racist and homophobic BLEs are not represented in the current dataset.

In addition to overusing[1] certain BLE categories, however, players also overused certain terms of endearment and politeness, such as THANKS (0.15%, keyness = 58.52) and SORRY (0.35%, keyness = 70.06). A closer look at the concordance of THANKS reveals that it is often used cynically, as in THANKS, MATE or THANKS FOR THE REASSURANCE. The cynical reading of these phrases is reconfirmed by a multimodal analysis of the audiovisual data recordings, which took into account intonation and body language. Terms of gratitude were often used ironically with the intention to tease the other player about their antisocial behavior in the game, for example, or about an error they had just made via their player

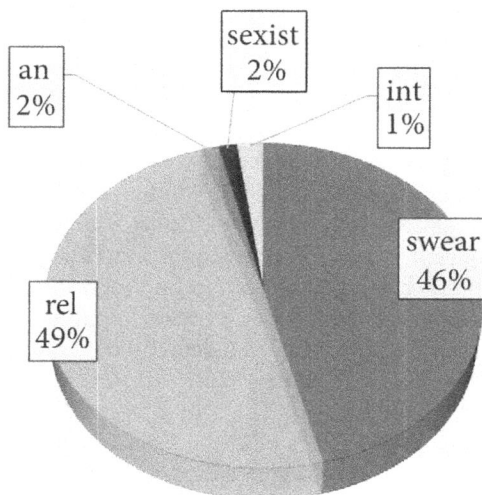

FIGURE 6.1 *Relative distribution of BLEs across corpus.*

TABLE 6.1 *Normalized frequencies of BLE categories with types occurring in corpus*

Category of BLE	Types	Frequency (normalized) relative to entire corpus	Percentage of all BLEs
rel (religious terms of abuse)	GOD, JESUS (CHRIST), JEEZ, DARN, (BLOODY) HELL, GODDAMMIT, HOLY LORD	0.00651	49%
swear (expletive), including euphemisms	CRAP, CRAPPY, FUCK (OFF), FUCKIN(G), FECKING, ASS, SUGAR, BLOODY (HELL), SHIT, SHITE, BOLLOCKS, BUM	0.00612	46%
an (animal terms of abuse)	(SON OF A) BITCH	0.00029	2%
sexist (terms of abuse)	(SON OF A) BITCH	0.00029	2%
int (intellect-based terms of abuse)	TWIT	0.00019	1%
ALL		1.33%	100%

character that had harmed either the speaker's player character or both player characters. Terms of apology, on the other hand, were commonly used sincerely, typically to exculpate the speaker from making a perceptually grave blunder (e.g., "Aww gosh, I'm really sorry. That was my fault there."). In turn, the player at whom the apology was directed would often respond with phrases containing another overused lemma, FINE, as in "No, hum, it's fine. You carry on." Further reconfirming polite interaction is the fact that FINE was used more generally in utterances signaling mutual reassurance ("It's fine. You can do this" or "Are you all fine there?").

The above observation underlines previous findings made by Ensslin (2012) suggesting a mixture of emotional detachment and engagement, which is characteristic of the DoC. Unsurprisingly, COOL itself was overused by participants (0.08%, keyness = 44.78), which is partly due to it frequently occurring in reduplicative clusters (COOL, COOL!). Examining the actual uses of COOL allows a more complex picture. In the following passage, the speaker's emotions in response to on-screen action switch rapidly between controlled "cool" to a panicky expletive, reflecting the realization that death was imminent for a split second. The end of the passage ("Right, okay") echoes the restoration of the speaker's self-control, or at least a level of acceptance that marks willingness and ability to move ahead in the game. This restoration of self-control can be understood in the context of aiming to keep one's "cool" in the game and its metaludic discourse, and thus to maintain positive rapport between the players.

> P2: Cool. Oh shit, I almost got lazered by you then. Right, okay.

The following exchange between Player 1 (P1) and Player 2 (P2) sees P2 announcing a rather inglorious yet somewhat gentlemanly choice of weapon, compared to the high-tech armor typically used in the *Helldivers* universe. The detached way in which P2 talks about how he will kill the final enemies without seemingly having to fear counterstrike or self-destruction is reflected in P1's question referring to P2's detached engagement in the termination process, and P2's ensuing unambiguously affirmative answer, "Yeah."

> P2: I'm gonna end this mission with a pistol.
> P1: Wow. Don't you feel cool? <giggles>
> P2: Yeah.

6.2 Interviews

To gain a more profound understanding of participants' motivation for overusing BLEs, we conducted semistructured interviews following their co-sit sessions. They were asked more generally about their linguistic habits during gameplay and particularly about how they accounted for the strong presence of bad language and expletives in particular in player discourse. One question related to McEnery's (2006) argument linking swearing to subversive behavior specifically asked them actually about subversive behavior, which none of them seemed to agree with, at least not in the sense of being the main motivation behind swearing in player discourse.[2] Instead, they explained their BLE motivations and behavior in terms of the following hierarchical list:

(1) *Immersion, spontaneity, control,* with strong levels of immersion in a game taking away any linguistic inhibitions by which participants would be impacted in other situations in everyday life;

(2) *Strong feelings,* including frustration, anger, fear, freedom, emotional investedness in serious play, self-expectations, and ambitions;

(3) *Communities of practice* and their imposition of copycat behavior and social norms, which cause players to apply certain medium- and context-specific social and linguistic behaviors as part of individual personality management;

(4) *Metacommunication* (Bateson 1955), commanding players to exhibit certain degrees of fake emotions, irony, and sarcasm;

(5) *Taboo removal*, specifically in relation and opposition to norms followed and imposed by older generations and culture-specific norms;

(6) *Player, platform and game specificity*, with specific linguistic choices such as the amount and selection of BLE being contingent on co-player, platform, or even the game itself;

(7) *Behavior linked to online culture*, which tends to give players a sense of safety and privacy regarding their social and linguistic behavior but also generates the need to express feelings more strongly through verbal means than in face-to-face communication.

Blending motivations (1) and (2), one participant pair described the immersion-related removal of inhibition in very material terms, as a profanity filter that evaporates in an emotionally heightened, immersive state (Pn refers to participant number 1–8; R refers to the interviewing researcher; // marks partial overlap with next speaker's turn; <.> marks one-second pause; <..> marks pause of two to three seconds[3]):

P1: *...you get so immersed*
P2: *Yeah you get kind of immersed*
P1: *Because it's like ah, shit, if we die we've failed//*
P2: *You get excited and yeah you feel like it's gonna be like a failure if you er if you mess up so <.> I don't know, you kind of just lose your filter a little bit I guess. <..> The profanity filter just sort of//*
P1: *It evaporates <giggles>*

When referring to community of practice-related behavior, another participant pair expressed themselves thus:

P3: *But yeah, I'd say that, it's a more relaxed community.*
P4: *Yeah.*
P3: *Gamers in general. Er, I don't think there's so much hang-up on politeness, erm, <..> yeah like <..>*
P4: *There's almost like a desensitization to swearing in general, like it doesn't hold the same weight, like the words don't hold the weight that they usually mean or or do with other people, I think, within the circle of people playing games.*

Thus, players' community of practice is described as less restrictive and socially rule governed than other communities surrounding P3 and P4. Furthermore, expletives seem to be so common in players' community of practice that members have become "desensitized" to the social taboos associated with them in other communities and situations of everyday life.

Participant utterances represented in the following passage relate to metacommunication in play, which refers to Bateson's (1955) observation that play behavior, and particularly play-fighting, communicates its simulative nature, involving "deception, threat, and imitation" (Mitchell 1991: 73). This is reflected in P1's description of trying to come to terms with frustration that, if seen rationally, may not be legitimized by the "joke" nature of the game because players are bound to "deluded" into being overly affected emotionally. In other words, rationality should tell them not to get emotionally involved in the fictional interactions represented in and by gameplay. At the same time, however, play is perceived as extremely serious and real ("the stakes are so high"), which gives rise to exclamatory expressions of frustration or anger in moments of failure, such as "arr, fuck" and other expletives:

> P1: *I think it's because in the moment, so when you mess something up, like, you can vent your frustration although you're like you're not really frustrated but you kinda like oh well first it's only a joke, isn't it, and then "arr, fuck." <last two words whispered>*
>
> P2: *Yeah it's sort of like obviously, ha, we're not deluded, we know it's a a game but like in your head, when you're playing, it kind of like the stakes are so high.*

Incidentally, the fact that the quoted expletives at the end of P1's turn are whispered rather than spoken at the regular volume of interaction signals that the social context of the interview is perceived as too formal to utter the expletives in the same way he would in a gaming context. This underlines the discrepancy in perceived appropriateness between both communicative situations.

The topic of taboo removal linked to generational etiquette changes and culture specificity was mentioned in particular by P3 and P4 (who had been playing *Streetfighter IV*):

> P4: *One of [the reasons] is the generation that like sort of is the main players of videogames like it's normally the younger generation, if you don't have that sort of built-in taboo reservation for swearing*
>
> P3: *Yeah, I think it's generational because I erm I did spend a year in France, for a year, and I found myself like learning the swearwords in that language and using them fairly frequently and then people would come up to me and say "Why do you swear so much?", and erm*
>
> R: *What kind of people would ask you that?*

P3:   *I was at university over there.*
R: *Okay.*
P3:   *So students about the same age as me. They were just picking me up for swearing so much because I assumed they'd do it at the same frequency as people in English.*

The fact that the above passages refer to swearing in situations other than videogames provides further context for the dominance of BLE in player communication. However, the use of BLE in terms of intensity, frequency, and/or linguistic choices can also be linked to the specific game or game genre being played, the platform being used, or the types of games produced by specific companies and the designs they stand for:

P8:   *Yeah it's sort of like a control thing.*
R: *Would you say that is true, does that make sense?*
P8:   *For the most part yeah, especially when it comes to like it depends like console to console as well. ... So it's stuff like Sony and Microsoft, you will get a lot of people like they will play the games and they will swear a lot but if you're like Nintendo, you won't get it as much.*
R: *Uhum.*
P8:   *Like it depends on like, the different companies like how their games are and like if you've got a bright, colourful game you don't really wanna be like really like rude, do you?*
R: *Uhum.*
P8:   *But if you've got something like a war game or something, you're gonna be like, you're gonna get the emotions getting angry and stuff like because it just reflects how the games are.*

Interestingly, here P8 (who played Nintendo's *Yoshi's Woolly World* in the preceding co-sit session) links the use of BLEs to a need to express control, or power, in relation to their abilities or experience with a specific technology—a console or type of game. Yet, predictably, BLE habits can also vary massively from player to player, as P5 (one of the *Halo 4* players) confirmed in his interview: "I don't usually swear at all," and in whose co-sit transcript there really was no instance of expletives to be found, and only a few instances of religious BLEs.

Finally, participants related the frequent use of BLEs to online identity management:

P4:   *I think a big part of [BLE use] comes as well from the online aspect of not being there in the same room so the best way to like show your expression is sometimes like to overreact and that means like to swear you head off all the time.*

he above-mentioned corpus data pointing at somewhat paradoxical
ehavior mixing with the use of BLEs were complemented by
wees explaining that apologetic or other types of polite behavior
deed quite common in their co-sit communication:

> And what struck me …, you apologized quite a lot as well.
> /:   Yeah, like "sorry, I just ate you." <all laugh>
> P7:   "Sorry to carry you around as an egg now."
> P8:   <laughs> Just not normal phrases but in the context of the
> videogames it is.
> R: Okay, and would this be quite typical, like is this usual with you
> more or less?
> P7:   Yeah//
> P8:   Yeah.

Importantly, polite expressions are here ridiculed or parodied as they are
combined with fictional lexis relating to the fantasy world of *Yoshi's Woolly
World*. An important motivation for polite behavior in heated and often
verbally abusive player communication is given by P1 and P2 (the *Helldivers*
players):

> P2:   I think we also tend sort of er <.> I can't think of a better phrase
> to describe it, so like "bro-up"
> P1:   Bro-up, yeah.
> R: Bro-up?
> P2:   Yeah//
> P1:   Yeah.
> P2:   Like stick together sort of thing.
> P1:   Yeah like
> P2:   We'll we'll scate, scathe each other but … [it] pretty much adds to
> the humour for us though.
> R: Yeah I thought that was really interesting because you were
> using a lot of polite phrases as well. You would say like
> "thank you" or "you're welcome," or you would apologize or
> something, and then all of a sudden there was a swear word like
> "shit" like.
> P1:   <laughs> That's pretty much what we do <laughs hysterically>
> <All laugh>
> P1:   It's like oh, we're professional, er. <laughs>

Thus, the need to "bro-up" and to create an atmosphere relaxed and amused
enough for both BLEs and polite language to contribute to fun and laughter
seems paramount to these and other co-sit players.

7 Discussion

Our corpus analysis showed that certain types of BLEs were overused in the co-sit player interactions we studied in this project. Particularly dominant were religious BLEs and actual expletives, with phrases containing GOD, SHIT, and FUCK featuring particularly high keyness values. This finding correlates with Ensslin's (2012: 94–95) finding that, generally speaking, expressive speech acts are dominant in player discourse, signaling the need to communicate a variety of feelings and emotional attitudes, such as joy, pain, desire, pleasure, and sorrow. When immersed in gameplay, players become emotionally involved as they need to re-prioritize their aims and objectives in order to succeed in the gameworld. Success and failure lie close together and tend to be represented in terms of extreme in-game metaphors such as, quite literally, life and death. Players' emotional responses to in-game events and circumstances therefore tend to be just as, if not more, heightened than actual life events, and these emotional outbursts manifest themselves in a variety of linguistic and paralinguistic expressions, such as *spill cries* (spontaneous, impulsive expressions of negative surprise or anger, typically after having been flawed in or by a game), *threat startles* (controlled expressions of negative surprise or fear), revulsion sounds, and laughter (Conway 2013). The boundary between spill cries and threat startles is of course a fluid one, as the extent to which an affective speech act is uttered in an impulsive or controlled fashion is not only subjective to speaker and listener but also a matter of degree rather than an absolute distinction. It can be argued, however, that fully verbalized expressive speech acts like OH GOD or FOR FUCK'S SAKE may indicate a more controlled (threat-startle) response than nonverbal interjections like WHOOA!

The predominance of religious BLEs in our corpus may be a reflection of the fact that "religious swearing [in English] is not as powerful as it once was—linguistic functions that once were performed by words and phrases that called on God or blasphemed him are now performed mostly by words for taboo human body parts and actions" (Mohr 2013: 10). Thus, religious BLEs have lost much of their original taboo stigma and are used commonly, not only in player discourse but also in everyday verbal interaction more generally. Conversely, the almost complete absence of other types of BLEs, namely animal, sexist, and intellect-based BLEs, may point to either personal preferences on the part of the admittedly small population under investigation here, or indeed the fact that they reflect a more general convention followed by (British) players' community of practice.

To further explain the observed patterns of BLE use in our corpus, our interview data suggested that the frequent use of expletives in player discourse is not only a reflection of the heightened and authentically felt emotions experienced by immersed players. A number of responses indicated that players' communities of practice operate according to different, more relaxed social norms than other communities surrounding our participants. Interestingly, participants did not link the community of practice motivation with a need to revolt against the discourse of power (McEnery 2006) but rather with the social safety offered by the magic circle (Huizinga 1950), which endorses the suspension of everyday social rules within its own psychological and ludonarrative confines. They indicated that the relaxed attitude toward profanity may further be related to generational shifts in assumptions about appropriate and inappropriate verbal behavior, which again reflects a general trend among young people to use BLEs, including expletives, frequently and habitually as part of their peer group's accepted, if not mandatory, code of communication, and specifically codes dictated and/or perpetuated by computer-mediated communication (Kiesler et al. 1984; Baron 2003).

Generally speaking, the range of motivations offered by our participants points at a complex and relativistic motivational framework for BLE use in gameplay. They mentioned the importance of metacommunicative play (Bateson 1955), which renders players' awareness of the fictional nature of their ludic interactions and indexes the importance of expressing the simulative nature of play through ludic behavior that marks itself as self-consciously ludic. The anonymity yielded by online culture was another contextualizing facture mentioned by participants, which adds an interesting virtual component to the physicality of co-situated gaming. Players seem to retain an awareness of the hypermediated landscape in which their specific co-sit experiences are embedded, thus leading to attitudes and assumptions that cross over from online to physically co-experienced communication— bearing in mind the increasingly fluid boundaries between the former and the latter. Finally, participants demonstrated a keen awareness of the platform-, genre-, game-, and brand-specificity that their communicative behavior is contingent upon, as well as the ways in which they adjust their verbal choices to individual co-players.

Our data further indicates that co-sit discourse is altrocentrically oriented and emerges in continuous mutual responses to what the other player has done or said vis-à-vis on-screen action, rather than being directed at the player's egocentric performance in let's play discourse, for example (see Ensslin 2016). We found that co-sit players in our study combined forms of verbal abuse and aggression with distinct and frequent forms of politeness (especially P1/2 and P7/8). There was a lot of polite discourse involving reciprocal thanking, apologizing, and reassurance-seeking between players, an mechanism described by one player pair as "bro-up" procedure. Importantly, this bro-up mechanism was perceived to increase the humorous side of co-sit gameplay, which in turn underscores the multilayered nature and social significance of the DoF in gameplay.

What is striking about co-sit interaction is the extreme affective polarity of players' expressed emotional states and the frequent and immediate shifts between extreme forms of affective discourse to rational composure. Thus, it can be argued that the discourse of co-sits follows important social norms related to detached engagement, or dictates of the DoC. It also became clear, however, that in each pairing, there was a hierarchy established by the cultural capital (Bourdieu 1986) of the stronger or more experienced gamer. This hierarchical relationship manifested itself in mechanisms of asking for and giving advice, but also in patterns of ridicule and directive speech acts.

8 Conclusions

This chapter offered an examination of co-sit player communication and focused in particular on patterns of BLE use and on what might be perceived as its paradoxical counterpart: politeness in otherwise heated and seemingly emotionally uncontrolled player communication. We found that this apparent paradox is subverted by the important context-specific ways in which the use of BLE can be seen as part of a community-of-practice-intrinsic code which contributes to perceived levels of fun and social bonding, as it is often used humorously and with the aim of increase psychological connectedness between players. In our dataset, there was a general pattern that religious and expletive terms by far outweighed other types of bad language, and that specific BLEs are key in the sense of being overused compared to common everyday oral interaction. Our corpus findings were relativized and contextualized through data garnered from participant interviews, which yielded a complex, multilayered, and relativist picture of the motivations behind BLE use and specifically swearing in (British English) co-sit discourse.

Clearly, the limited scope of this study cannot yield any statistically significant or generalizable results. Neither can the results gained from a spoken corpus that is just over 10,000 tokens in size tell a conclusive story beyond the data set investigated at the time and in its site-specific context. It may further be conceded that participant behavior and responses will likely have been affected by the Hawthorne effect, as the researchers were present during the video-recorded co-sit sessions in order to be able to refer to specific verbal behavior in the directly ensuing interviews. Nonetheless, the findings strongly corroborate anecdotal evidence of player behavior experienced frequently by both authors of this paper.

By ways of a final thought, the immediate communicative and spontaneously emergent nature of co-sits begs the question of whether co-sit discourse is indeed paratextual or whether it should rather be regarded as part of the actual text, or orthogame, itself (Carter et al. 2012). After all, game narrative does not emerge until players perform in-game interactions, and the conversations between co-sit players are at least partly a manifestation of the first-person enactment of their embodied roles as avatars. Players often speak in character, and their fictional dialogues alternate with metaludic dialogic commentary and response cries. This embodied enactment performs an important act of dual situatedness, thus corroborating social bonding and legitimizing verbal and other forms of abuse. Future research is clearly needed that will not only refine our understanding of male-identified "bro-up" sociality in co-sit gameplay, but that will also, importantly, cast a far more gender-inclusive picture of quasi-paradoxical cooperative ludic behavior.

Notes

1   We use the term *overuse* in relation to the reference corpus, the BNC Spoken.
2   Interestingly, some participants mentioned taboo removal as a motivation for swearing (see point 5 in the ensuing list). Hence, while they all seemed to agree that their swearing behavior was not subversive per se, they aligned it with a generational shift that they considered to be the sociolinguistic context within which a more general increase in swearing behavior has to be seen.
3   The oral discourse annotation conventions were adapted from Bousfield (2008: 8).

References

Arrowhead Game Studios (2015), *Helldivers*, San Mateo, CA: Sony Interactive Entertainment.

Aylett, R. (1999), "Narrative in virtual environments: Towards emergent narrative," in *Proceedings of AAAI Symposium on Narrative Intelligence*, 83–86.

Baron, N.S. (2003), "Language of the internet," in A. Farghali (ed.), *The Stanford Handbook for Language Engineers*, 1–63, Stanford, CA: CSLI Publications.

Bateson, G. (1955), "A theory of play and fantasy," *Psychiatric Research Reports*, II: 177–78.

Blackshaw, T. and J. Long (2005), "What's the big idea? A critical exploration of the concept of social capital and its incorporation into leisure policy discourse," *Leisure Studies*, 24: 239–58.

Bourdieu, P. (1986), "The forms of capital," in J. Richardson (ed.), *Handbook of Theory and Research for the Sociology of Education*, 241–258, New York: Greenwood.

Bousfield, D. (2008), *Impoliteness in Interaction*, Amsterdam: John Benjamins.

Burwell, C. and T. Miller (2016), "Let's Play: Exploring literacy practices in an emerging videogame paratext," *E-Learning and Digital Media*, 13 (3–4): 109–25.

Capcom (2012), *Streetfighter X Tekken*, Osaka: Capcom.

Carter, M., M. Gibbs, and M. Harrop (2012), "Metagames, paragames and orthogames: A new vocabulary," *Proceedings of the International Conference on the Foundations of Digital Games*, May 29–June 1, Raleigh, NC, USA, ACM.

Consalvo, M. (2009), *Cheating: Gaining Advantage in Videogames*, Cambridge, MA: MIT Press.

Conway, S. (2013), "Argh! An exploration of the response cries of digital game players," *Journal of Gaming and Virtual Worlds*, 5 (2): 131–46.

Dippel, A. and S. Fizek (2016), "Ludification of work or Labourisation of play? On work-play interferences," in *Proceedings of the 1st Joint Conference of DiGRA and FDG*. Availiable online: https://library.med.utah.edu/e-channel/wp-content/uploads/2017/02/paper_1681.pdf.

Ensslin, A. (2012), *The Language of Gaming*, Basingstoke: Palgrave Macmillan.

Ensslin, A. (2016), "The language of gaming: Affective discourse patterns in two videogame paratext genres," keynote, LEXESP 2016: Videogames and Language, May 5–6, 2016, University of Alicante.

Genette, G. (1991), "Introduction to the paratext," *New Literary History*, 22: 261–72.

Goffman, E. (1981), *Forms of Talk*, Oxford: Blackwell.

Good-Feel (2015), *Yoshi's Woolly World*, Kyoto: Nintendo.

Gray, J. (2010), *Show Sold Separately: Promos, Spoilers, and Other Media Paratexts*, New York: New York University Press.

Grimes, S.M. and A. Feenberg (2009), "Rationalizing play: A critical theory of digital gaming," *The Information Society*, 25: 105–18.

Huizinga, J. (1950), *Homo Ludens*, Boston, MA: Beacon Press.

Industries (2012), *Halo 4*, Redmond, WA: Microsoft Studios.

Jenkins, H. (2004), "Game design as narrative architecture," in N. Wardrip-Fruin and P. Harrigan (eds.), *First Person: New Media as Story, Performance, and Game*, 118–30, Cambridge, MA: MIT Press.

Kiesler, S., J. Siegel, and T.W. McGuire (1984), "Social psychological aspects of computer-mediated communication," *American Psychologist*, 39: 1123–34.

Lave, J. and E. Wenger (1991), *Situated Learning: Legitimate Peripheral Participation*, Cambridge: Cambridge University Press.

McEnery, T. (2006), *Swearing in English: Bad Language, Purity and Power from 1586 to the Present*, London: Routledge.

McEnery, T., R. Xiao, and Y. Tono (2006), *Corpus-Based Language Studies: An Advanced Resource Book*, London: Routledge.

Mitchell, R.W. (1991), "Bateson's concept of 'metacommunication' in play," *New Ideas in Psychology*, 9: 73–87.

Mohr, M. (2013), *Holy Sh*t: A Brief History of Swearing*, Oxford: Oxford University Press.

Mukherjee, S. (2015), *Video Games and Storytelling: Reading Games and Playing Books*, Basingstoke: Palgrave Macmillan.

Nicholson, M. and R. Hoye (2008), *Sport and Social Capital*, Burlington, MA: Butterworth-Heinemann.

Ryan, M.L. (2004), *Narrative across Media*, Lincoln, NE: University of Nebraska Press.

Stalnaker, R. (1974), "Pragmatic presuppositions," in M. Munitz and P. Unger (eds.), *Semantics and Philosophy*, 197–214, New York: New York University Press.

Suits, B. (2005), *The Grasshopper: Games, Life and Utopia*, Peterborough, ON: Broadview Press.

7

"Shut the Fuck up Re![1] Plant the Bomb Fast!"

Reconstructing Language and Identity in First-person Shooter Games

Elisavet Kiourti

1 Introduction

In contemporary capitalistic society, the proliferation of widely affordable and accessible internet connectivity has transformed, in many ways, how videogames are played. Millions of players globally connect to violent multiplayer first-person shooter games, for instance *Call of Duty: Black Ops III* (Activision 2018) and *Counter Strike: Global Offensive* (Valve 2018), and/or massively multiplayer online role-playing games such as *League of Legends* (Statista 2016) and *World of Warcraft* (Mahardy 2014) in their everyday lives. Yet, videogames are often associated with increased aggression and delinquent behavior in the players (Anderson et al. 2010). The American Psychological Association, for example, released a technical report of laboratory[2] experiments claiming a "consistent relation between violent video game use and increases in aggressive behavior, aggressive cognitions and aggressive affect, and decreases in prosocial behavior, empathy and sensitivity to aggression" (American Psychological Association 2015: 11).

This narrative, among other mass media releases (Christopher 2017) and statements from politicians linking videogames' content with mass school shootings in the United States (Salam and Stack 2018), tend to evoke moral panic[3] (Cohen 1980). Research across disciplines, though, indicates that gaming contexts are far more complex and demanding environments than publicly assumed: videogames are rich contexts for learning (Gee 2003, 2004, 2005a; Shaffer et al. 2005; Squire 2008). During gameplay, players are participating in social and literacy practices (Kiourti 2018), and they need to effectively interact with the game itself, their co-players, and their opponents linguistically and performatively. Talk-in-interaction in videogames is a highly performative event wherein those who take part are ascribed specific roles and participation membership; they follow certain rituals and rules, and they embed contextual knowledge and literacy practices.

Drawing on the framework of sociolinguistics (Bourdieu 1994; Fairclough 2001; Halliday 1985), theory of politeness (Allan and Burridge 2006), unified discourse analysis (Gee 2015), and frame analysis (Goffman 1974), this chapter follows a line of ethnographic research that focuses on a multimodal analysis of the linguistic strategies of swearing and bad language performed by four youth Cypriot gamers during gameplay of *Counter Strike: Global Offensive* (*CS:GO*). The data from the research reveal that videogames function as frame social spaces in which swearing and bad language practices have effective purposes. More specifically, players use swearing as a linguistic strategy with an aim of preventing individual or team-based performative face-loss when communicative violations occur during gameplay. Secondly, the use of swearing and expletives functions as linguistic strategy to cool stress and to ensure in-group bonding. Finally, players use swearing and expletives as fast language mechanisms to provide feedback to their co-players when they employ low performative actions during gameplay. Players, in other words, break the rules of politeness for the sake of effective gameplay and to protect and maintain their positive identity as gamers.

2 Linguistic and communicative patterns during gameplay

Socioeconomic contexts are constructed through social processes that identify the subjectivity of the members participating in them (Bourdieu 1994). Likewise, language as a semiotic system is shaped according to sociocultural environments, and it is used for representation (ideational metafunction), to enact roles, and to negotiate power (interpersonal metafunction) (Halliday 1985). Under this spectrum, our ideas, attitudes, beliefs, and practices and the linguistic choices we as human beings make are not neutral. They are embedded in what Fairclough (2001) refers to as *ideology*. Ideology in language interactions is formed based on an implicit knowledge and the interconnection of situative context, the interlocutors, and the hierarchies established by social context. Thus, the way we choose to use language reflects, constructs, permeates, and subverts sociocultural contexts, world views, identities, and human relationships in the sense of a "a way of signifying a particular domain of social practice from a particular perspective" (Fairclough 1995: 14). Moreover, the linguistic and communicative patterns we use incorporate fundamental requirements of theatricality (Goffman 1981). When using language and other semiotic practices (clothing, physiognomy, gestures, see Gee 2005b), we automatically perform "acts of identity" in the sense of conveying to other people who we are and how we want to be seen and understood.

Video gaming contexts as problem-solving digi-sociocultural spaces are similar to our societies in the sense of being designed with specific rules, constraints, and allowances in mind. For a gamer, it is crucial to know what to say and how to say it and to be aware of the social and cultural settings in which each communicative act is embedded. It is therefore not surprising that gameplay conversations, unlike other types of face-to-face conversations, exhibit a high frequency of short and long pauses, "for the sake of focused gameplay" (Ensslin 2012: 99).

In addition, players use language (words, grammar, and discourse) in certain ways to enact a socially significant identity. They use a distinctive style of language, a "social language" (Gee 2015: 38). Gee (Gee 2004) describes an example of *Yu-Gi-Oh*[4] players' social language. The *Yu-Gi-Oh* language, like all social languages and jargons (see Balteiro and Gledhill, this volume) "is a technical of specialist social language that allows people enacting the identity of being a *Yu-Gi-Oh* player and to do the things they need to do to play *Yu-Gi-Oh*." Social languages, as in-group recognition devices, purportedly disguise meanings from out-groupers. They are what Halliday (1976: 570) refers to as "antilanguages." Of course, these styles all use some resources from the vernacular version of the language, but they are enriched with special words, phrasings, and grammatical patterns to mark out their identity and to engage in their distinctive work or play (Gee 2015). Thus, players seeking to establish in-group power and belonging engage in discourses in order to perform "quasi-meritocratically as powerful in-group members, who know about games and gaming, and who are aware of in-group behavioral norms" (Ensslin 2012: 107–08).

3 Bad language in videogames

Politeness or impoliteness can be perceived as good or bad language behavior based on a variety of factors. These include the context, the relationship of the interlocutors, the subject matter, the communicative event, and the medium (spoken or written). Taking into consideration that politeness is wedded to context, place, and time, participants have to consider whether what they are saying will maintain, enhance, or damage their own face. Thus, in most contexts, sociolinguistic interactions are typically oriented toward maintaining saving face, and the participants usually engage in mutually face-saving acts by creating rapport with one another. This implies a culturally and socially contingent balance between personal involvement and independence (Scollon and Scollon 2001). Goffman's notion of face, however, both implies social constraints as well as allowing for changes in behavior between contexts (1967: 6–7). Taking, for example, the debates that arise between politicians in parliaments, mutual criticism, ridiculing, and challenging are not only acceptable communicative strategies; they are also "part of the discourse expectations of a good parliamentary speaker" (Paltridge 2006: 76).

Allan and Burridge (2006) discuss politeness and impoliteness with the notions of *euphemism* (sweet talking), *dysphemism* (speaking offensively), and *orthophemism* (straight talking). Orthophemisms are linguistic patterns employed by speakers in order to avoid possible loss of face in a conversation with other interlocutors. They are either direct or not sweet-sounding expressions (e.g., "vagina" instead of "pussy"). Euphemisms are words or phrases used as an alternative to dispreferred expressions (e.g., "I am going to the bathroom" instead of "I am going to pee"). Finally, dysphemisms are words or phrases with connotations that are offensive either about the subject matter and/or to audience addressed, or overhearing the utterance (e.g., "You cunt!").

However, people also use linguistic patterns at odds with the intentions that lurk behind them in order to cause less face-loss or offense. These are the linguistic mechanisms of euphemistic-dysphemisms and dysphemistic-euphemisms. Let me describe an example. The expressive exclamation "fuck!" is typically considered a dysphemism. Thus, a person may feel the inner urge to swear, but at the same time, in order to prevent face-loss or offense, may prefer to use conventionalized euphemistic-dysphemisms like "Gosh!" instead. Yet, given another context with a different set of interlocutors (e.g., among gamers during gameplay), the same expression could just as well be described as cheerfully euphemistic: here, "fuck!" could be used and understood as an expression of support and/or admiration (see Ensslin and Finnegan, this volume).

Speakers resort to dysphemisms to talk about people and things that frustrate and annoy them, that they disapprove of, and that they wish to disparage, humiliate, or degrade. By and large, the use of words or phrases with connotations that are offensive for the interlocutors are tabooed. Dysphemisms are therefore often used privately among cliques when talking about their opponents. Some examples of dysphemistic expressions include curses ("Die!"), name-calling ("You asshole"), and any sort of derogatory comment directed toward others in order to insult or offend them ("You are fat as fuck!"). If we focus on the concept of dysphemism, and more precisely on swearing, we see that it exists in most people's repertoire accompanied by high-level emotional and linguistic patterns, but it is also considered in most contexts as taboo language. Swearing as an identity marker functions in various ways, such as revealing strong emotions and indicating social distance or social solidarity, stress, anger, or disappointment (Allan and Burridge 2006). In these situations, though, it has actually been shown that swearing can have a stress-reducing, and even pain-reducing, effect (Crystal 2003: 173). In other cases, swear words can function as purely stylistic expressions (Ljung 2011) or as in-group solidarity markers within a shared colloquial style (see Allan and Burridge 2006: 88). The management of social status (power and social distance relations) involves the management of face, and it is important to consider that relative status, relative power, and social distance of the interlocutors plays a silent but salient role and affects the perceptions of profane swearing (Allan and Burridge 2006).

Gamers use language in subversive ways in order to function effectively and negotiate their social identity in their in-group relationships (Ensslin 2012; Wright et al. 2002), and they often lift the rules of politeness as do people in many other areas of everyday life. For instance, male adolescents use swearwords during storytelling as linguistic strategies that enable the narrators to construct for themselves the identity of the powerful members of a group who share strong friendship bonds (Karachaliou and Archakis 2015).

4 Unified discourse and frame analysis in videogames

In this chapter, a combination of unified discourse analysis (Gee 2015) and frame analysis (Goffman 1974) will be employed. Discourse in videogames focuses on action taking (Gee 2004, 2015), and this is reflected in (digital) literacy practices (Kiourti 2018; Steinkuehler 2007) and linguistic practices (Ensslin 2012) used in the gaming context. Gamers become agents during gameplay and through conversational interactions and other semiotic tools they perform in order to accomplish their goals (Gee 2015). Through participation in a discourse community, gamers come to understand the world (and themselves) from the perspective of their in-community of practice (Wenger 1998) and they share and creatively operate upon a shared set of social practices, of which language is an important tool.

Gamers, via their avatars, are allowed, for instance, to steal, harm, or kill other beings in the gameworld. Similarly, it is fully acceptable to use bad language and express emotions freely and loudly, which, in many actual-world situations, would be quite or even completely inappropriate. For the investigation and analysis of these behaviors, I prefer the concept of frame analysis (Goffman 1974) over the notion of the "magic circle" (Huizinga 1950; Salen and Zimmerman 2004). A frame is "what" the participants are allowed to do or say in a specific situational context. The "what" are the rules, the norms, the expectations, the possible roles, and so forth, which are available to social actors to make sense of any given situation or encounter. For Goffman, social actors are similar to players in a card game, drawing from an already set and ordered deck of options (Goffman 1961: 25). Unlike the magic circle, in frame analysis, gameplay is not considered to be removed from other aspects of social life (Chayko 1993) but is embedded in one frame within a social order that is saturated with other, often multilayered, frames. Goffman (1961) describes the boundaries of each frame as a "membrane" or a "screen." The fact that social frames permeate human life helps us locate gameplay within a wider social context and understand gameplay as just one form of social encounter without dichotomizing gamers and gameplaying from life, and online from offline. Furthermore, adopting the concept of frame analysis allows us to better understand the gaming context and the functionalities of dysphemisms from a different perspective. While in other contexts, impolite language behavior is perceived as dysphemistic, in gaming contexts, the same dysphemisms are considered as acceptable linguistic behaviors in the community of gamers. As I will analyze in Section 7, gamers consciously and frequently use dysphemisms such as swearing and face-threatening utterances (e.g., insults) in order to negotiate their in-group power relationships, to bond with their teammates, and to provide fast feedback on their co-players' gameplay.

5 Methodology

5.1 Philosophy and methods

This study follows the philosophy and methods of ethnography (Hammersley and Atkinson 2007) and virtual ethnography (Hine 2000). The research focuses on the practices of dysphemisms as linguistic strategies performed by a group of four male gamers (age 16–17) during gameplay in the first-person shooter game *CS:GO*. To protect research participants' identity, their names are replaced by the pseudonyms Burnt-Bread, Hookah, Kutch_me, and Purplezz. The research data were collected in the geographical context of Cyprus with a diglossic Cypriot Greek speech community, and they consist of multimodal data (e.g., screen gameplay recordings with oral conversations among the participants in Greek-Cypriot dialect).

In terms of the linguistic context in Cyprus, Standard Modern Greek (SMG) and Greek-Cypriot dialect (GCD), the native language acquired at home, are used interchangeably by the same speakers of all socioeconomic backgrounds (Karyolemou and Pavlou 2001). GCD, though, does not have a standard official orthography and is generally used in informal oral communication, while SMG enjoys more prestige as it is used in formal schooling and is associated with professionalism, prosperity, and modernity (Papapavlou 1998). However, with the rise of computer-mediated communication (CMC), young Greek Cypriots seem to have positive attitudes toward GCD nowadays, for example, in online environments such as Facebook (Sophocleous and Themistocleous 2014). This might be because GCD is now more evident in local TV series and also because young Greek Cypriots use forms of GCD as part of youth slang, which indicates youth identity (Tsiplakou 2004).

The research has a twofold aim. First, it seeks to investigate how the players' "noncooperative linguistic" communicative strategies and, more precisely, bad language are employed in order to manage and maintain their identities as players during gameplay. Second, it examines how breaches of politeness are correlated to gaming literacy.

5.2 Research process and data analysis

The research was conducted for nine months (May 2015–January 2016), and the participants were systematically observed in steady intervals with 46 observations (195 hours in total). Data collection was systematically performed through video recordings of the participants via Go Pro HERO3 + action camera, video-screen recordings of their gameplay using Open Broadcaster Software (OBS), rich field notes, postfield diary notes, and semistructured interviews. The video-screen recordings of the gameplays and most of the observations of the participants were recorded at the gaming cafe they were visiting almost on a daily basis in order to play videogames, have fun with their friends and other gamers, and discuss general and gaming-related topics. The observations were not structured because they were dependent on the participants' decisions regarding the place and time they wanted play. During observations, especially at Burnt-Bread's house, I was present in the same room observing, following, and sometimes participating (especially at the beginning of the research) in their discussions and taking rich field notes on the way they played, and the linguistic practices they were using. As I started being accepted into their gaming community, the observations were less participatory. The rationale was to maximize their naturally occurring talk and practices and to be less intrusive in their space. The video recordings and the interviews were transcribed, and data analysis was conducted following a two-way model. First, for the content analysis, the main themes and bottom-up categories were identified and categorized using the qualitative data software MAXQDA. Since the data were highly multimodal, the software helped me link and code different types of data (e.g., conversations during gameplay, screen recordings, field notes, and/ or interviews). Additionally, to re-examine or reconfirm specific categories that arose from the first analysis, I focused on a simultaneous analysis of specific episodes of all the participant's screen recordings. This was a type of coding with more detailed categories which enhanced access to the sizeable dataset and allowed me to organize the observations in different thematic categories linked to dysphemistic words or phrases. Finally, critical incidents were selected from the data corpus which shed light on the main research objectives of the study.

6 The context of *CS:GO*

CS:GO is a multiplayer first-person shooter game developed by Valve Corporation and Hidden Path Entertainment. As part of the *Counter Strike* series, *CS:GO* has a few adjustments to the original video game, such as rebalancing of the weapon damage models, improvisation of bullet penetration through walls, and graphical updates on crosshair and on Valve's Graphical User Interface (VGUI) buy menu. The game allows cross-platform multiplayer play between Microsoft Windows and Mac OS, and it consists of five different offline or online game modes: Casual, Arms Race, Demolition, Deathmatch, and Competitive.

The participants were observed in Competitive mode, because this mode was the only one used by their team. Players are eligible to play in Competitive mode after they reach level two by playing any of the other four modes. The aim of this constraint is for the players to gain experience and knowledge of the game in order to compete online with other more experienced gamers. In Competitive mode, players are encouraged to act more strategically in *CS:GO* than in most other multiplayer games due to the inability to respawn once killed. This context affects gamers in such a way that they need to communicate effectively through linguistic means in order to increase their chances for a win.

In Competitive mode, the game features a Competitive matchmaking system and a ranking system which includes 18 skill groups, with Silver I being the lowest rank and Global Elite the highest. Ranks are based on rounds won. The formula calculations generate a rank and add it to a handicap. This handicap is implemented based on how the player performs in a game against either higher or lower ranks. The ranking system takes into account a game consisting of 30 rounds and 10 players. The final outcome is then matched against that of the ranges. Thus, it gives the players their matchmaking ranks.

6.1 Gameplay

Two opponent teams consisting of five players each are divided into Terrorists and Counter-Terrorists in a 30-round match. Each roundtime lasts 1 minute and 55 seconds. During this time, a bomb must be planted by Terrorists, while Counter-Terrorists need to defuse it. Once the bomb is planted by Terrorists, it takes 40 seconds to explode. The aim of both sides is to manage eliminating the other while also completing separate goals (e.g., exploring the map locations, shooting exercises, and organizing their strategies). Players are not allowed to switch sides during the game except at halftime. After the first 15 rounds, the game reaches halftime and the two teams switch sides. The first team to score 16 points wins the game. If both teams score a total of 15 points by the end of the 30th round, the match will end in a tie. Within this context, time management is crucial for the effectiveness of the gameplay and affects players' linguistic practices (e.g., short phrases, long pauses).

7 Data and findings

7.1 Violation of the rules of communication

As mentioned in the theoretical preamble, videogames are highly multimodal spaces (see Figure 7.1), and they have a subversive frame in terms of both performative actions and language use on the players' part. In *CS:GO*, tactics, strategies, time management, and effective communication are vivid factors that players need to take into consideration during gameplay. The

FIGURE 7.1 CS:GO *as a multimodal environment.*

following episode focuses on the use of dysphemisms (swearing, expletives) and dysphemistic-euphemisms (irony) by players when significant violations of the communicative patterns occur, for example, when talking about other topics rather than focusing on communicative exchanges for their gameplay. Two of the players (Burnt-Bread and Hookah) are using dysphemisms, and more specifically swearing, in an attempt to get voiced in their team when they realize they confront performative face-loss.

(1) **Burnt-Bread:** e maˈlaka⁵ laˈŋgaɾo ˈaɣɾia..... maˈlaces laˈŋgaɾo ˈpaɾa poˈlːa!

[E dude, I am lagging wildly.....Dudes, I am lagging so much!]

(2) **Purplezz:** maˈlaka eˈɣo aˈkuo traˈuθca. eˈpelːana toˈɾa. (Dancing while sitting).

[Dude I am listening to songs. I am going crazy now.]

(3) **Hookah:** ɾe ˈkamete ˈekʰːo t͡ʃe ˈpcanːumen do eˈpomenon. ɾe Kutch_me ˈkame ˈekʰːo. ˈmen aˈɣoɾasis.

[Re go eco and we win the next (round). Re Kutch_me go eco. Don't buy (any equipment)].

(4) **Purplezz:** ɾe i ˈpomba ˈpɾepi na ˈpai ˈmbi ɾe maˈlaka, na ˈpcasumen do pʰːlant ɾe. ˈðoste mu tin ˈpomba nːa ˈpao ˈmbi.

[Re, the bomb needs to be planted in B (site) in order to take the plant re. Give me the bomb to go to B (site)].

(5) **Kutch_me:** enːa se kaɾteˈɾuːsin

[They will be waiting for you.]

(6) **Purplezz:** ˈjolo ɾe, ˈjolo. suˈag ɾe.

[Yolo, re . Yolo. Swag, re.]

(7) **Burnt-Bread:** ɾe maˈlaces ˈfkaɾte ˈponiman. ɣaˈmo ti ˈrːat͡sʰːa mːu ɣaˈmo.

[RE ASSHOLES SHUT THE FUCK UP. Fuck my life, fuck]

(8) **Purplezz:** ˈt͡ʃil braː (high pitch).

[Chill brah!]

(9) **Kutch_me:** aː, eˈpeksan sːe epiˈði emiˈlusamen. (Turns his body, seeing Burnt-Bread who is sitting behind him.)

[Oh, so they shot you, because we were talking.]

(10) **Hookah:** ˈfausa nːa ˈfkalete ɾe! ˈkame pʰːlant ˈmbam mbam! ... oˈɾeoː!

[Shut the fuck up re! Plant the bomb fast! Nicee!!]

(11) **Burnt-Bread:** ˈmen lːaˈlite tin istoˈɾian dis zːoˈi sːas ɾe.[...] ˈθelis na miˈlo ˈulːon do ˈɲʲeim? enːa su aˈɾesi?

[Don't tell the story of your life re [...] Do you want me to keep talking during the whole game? Would you like that?]

While the roundtime of the game initiates, Burnt-Bread (the player with the lowest rank (Silver I)) informs his co-players about the lagging⁶ problem he faces in the game (1). High frame rate in *CS:GO* is vital for player performance, because it means less input latency. The higher frames per second (fps) a game has, the more quickly the player's input is registered and played back via his avatar. Burnt-Bread does not get any response from his co-players about the lagging problem he faces in the game. Thus, after five seconds, he decides to reiterate his utterance by repeating (laˈŋgaɾo *[I am lagging]*) and altering the adverbs (ˈaɣɾia, ˈpaɾa poˈlːa! *[wildly, so much!]*). Articulating repeatedly the object of his agony (laˈŋgaɾo *[I am lagging]*) signals his concern that his avatar movements will be processing with latency in the game vis-à-vis the movements of his co-players and/or his opponents. Lagging, along with the fact that Burnt-Bread is not an experienced player in *CS:GO*, increases the chance of him losing (e.g., being eliminated by an opponent's gun) compared to the other players. In other words, Burnt-Bread confronts a high-level chance of having performative face-loss among his co-players and more generally in the gaming community.

Purplezz (one of the most experienced players in the team) shifts footing⁷ as he changes the topic of the conversation from the lagging problem that his co-player Burnt-Bread is concerned about to his own emotional condition, which is far more relaxed and playful (2). Purplezz is listening to songs, shaking his head and body with dancing moves, while sitting and playing in front of his computer. Hookah then shifts the conversation again, requesting that his co-players apply the strategy he proposes (ɾe ˈkamete ˈekʰːo *[re go eco]*) (3). He demands that his co-players employ the economic (eco) mode strategy. The aim of this strategy is twofold: (a) saving money for the next rounds and (b) reading and analyzing the opponents' tactics and strategies. In this round, the team do not have a high budget, and spending any money on gear will negatively affect their performance in the following rounds of the game. Hookah, as experienced player of *CS:GO*, is processing all the information he receives about his team's financial status and the opponents' gameplay and decides that the best solution to increase the percentage of winning in the next rounds is to employ eco mode. This means insufficient equipment (e.g., guns, molotovs, flashbangs) will increase the chances of losing this round (t͡ʃe ˈpcanːumen do eˈpomenon *[and we win the next (round)]*).

Purplezz, while still dancing on his chair, brings into the conversation the main aim of the game, the bomb-plant (4). Planting the bomb successfully will lead to the team's budget increase. He immediately requests his co-players to give him the bomb in order to take it and plant it in B site: ðoste mu tin ˈpomba nːa ˈpao ˈmbi. *[Give me the bomb to go to B (site)]*. The player Kutch_me replies to Purplezz' request with a warning that the decision to plant the bomb in this moment is wrong, because he hypothesizes that the opponents have already calculated the Terrorists' strategies and will be hiding at B site, waiting to eliminate them before they plant the bomb

(5). Purplezz is footing once again the frame of gaming, a rather serious activity in which gamers invest time, knowledge, and performance that also include moments of fun, into a clearly monolithic enjoyable practice where players should take risks just for the fun (6). With the repetition of the acronym *Yolo*[8] ('jolo re, 'jolo *[Yolo, re . Yolo]*) accompanied by the acronym *Swag*,[9] Purplezz starts running to B site. By this performative action, he takes control of the gameplay, with an increased likelihood of losing. In addition, he imposes on his co-players that they should take risks and have fun during gameplay, without being concerned about the consequences (e.g., of losing the round of the game).

As was mentioned in Section 6.1, performative actions and conversations among players are highly dependent on the short duration of roundtime in *CS:GO*. Taking this into consideration, Purplezz violates (2, 6) the communicative practices and performative actions that are considered effective for his team's gameplay (e.g., talking about tactics, strategies, giving information about locations or opponents) by taking control of the round time. While Purplezz thus violates the communicative rules (2, 6) along with the fact that he is also approaching B site, at the same time, Burnt-Bread's avatar gets eliminated by an opponent. Immediately after his elimination, Burnt-Bread raises his voice and uses dysphemism (7) in a form of a deictic expression (re ma'laces *[RE ASSHOLES]*), followed by more dysphemisms ('fkarte 'poniman. ɣa'mo ti 'r:atsʰ:a m:u ɣa'mo *[SHUT THE FUCK UP. Fuck my life, fuck]*), urging his co-players to stop talking.

Purplezz requests Burnt-Bread to calm down (8), a linguistic strategy which aims to cool the situation by showing sympathy for his bro's failed gaming performance (being eliminated by the opponents) in this playful context. Kutch_me, a co-player who sits behind Burnt-Bread, turns around, looks directly at Burnt-Bread, and responds to him with a response cry (a:; *[Oh]*). The response cry functions as an indicator of hesitation, and it is followed by a dysphemistic-euphemistic utterance (9), indicating irony. Kutch_me does not accept that there was a violation of the communicative rules and indirectly criticizes Burnt-Bread's abilities, skills, knowledge, tactics, and strategies as a player.

Hookah interrupts the discussion (10) with a dysphemism in the form of swearing ('fausa n:a 'fkalete re! *[Shut the fuck up re]*), asking Kutch_ me and Burnt_Bread to stop violating the communicative rules. He then shifts the conversation to the performative actions they should take in the game. Specifically, he observes that Purplezz has already arrived in B site, and he requests him to plant the bomb fast (kame pʰ:lant 'mbam mbam! *[Plant the bomb fast!]*). After three seconds, when Purplezz finally plants the bomb successfully, Hookah replies to him with a euphemism (o'reo:! *[Nicee!]*). The use of the euphemism o'reo:! functions as positive feedback on Purplezz' performative action of planting the bomb.

Burnt-Bread, at last, requests his co-players to stop having long and inefficient conversations during gameplay (11). Right afterwards, he sees Purplezz posing a question to him as to how he would feel if Burnt-Bread were to keep talking during the whole game (11). Burnt-Bread points out the communicative rules that are acceptable and unacceptable in the frame of gaming.

7.2 Cooling the tension and bond with teammates

The following short episode was one of the most common linguistic subversive strategies employed by the players during gameplay. Expressions like expletives were often employed by the players in order to ease tension, agony, and stress arising during gameplay due to the expectations of each player individually and those of all players as team to win the game. The use of these linguistic strategies among players have a twofold aim: first, to reconfirm that dysphemisms such as swearing and expletives are not considered as taboo language in the gaming frame, and second, that dysphemisms function as linguistic in-group strategies to reduce tension, agony, and stress that players feel during gameplay, thus creating an in-group entertainment space and a bond with teammates.

(12) **Hookah:** ′ela re ′putʰ:oː!!
      [COME ON re pussy!!]
(13) **Purplezz:** sta′mata na iv′riz:is.
      [Stop swearing.]
(14) **Hookah:** {hahahaha!}
(15) **Purplezz:** {hahahaha!}

In Episode 2, one of the players in the team informs his co-players that an opponent has been seen in Long A Site. The player Hookah, whose his avatar is near Long A site, analyses the given information and decides to take advantage of it based on the assumption that the opponent will pass from A site. The analysis of the given information about the opponents' location leads Hookah to decide to hide behind a wall in A site waiting for the opponent to approach and eliminate him. The opponent, however, does not meet Hookah's expectations as he does not pass from the site. The opponent's unexpected action causes stress and agony to Hookah because the chances of implementing a successful performative action are getting lower and lower with every second that passes in the game. Hookah expresses his anxiety (12) with a dysphemism in the form of expletive deixis referring to the opponent (′ela re ′putʰ:oː!! *[COME ON re pussy!!]*). Purplezz then requests Hookah to stop swearing (12). In the framework of gaming, of course, the use of bad language is fully acceptable, and any attempts to censor it will likely result in laughter (14, 15). Players have an in-group knowledge of what is and isn't acceptable in this context. Consequently, the request functions as a rather symbolic linguistic strategy to ease the tension of any negative feelings and bond with teammates during gameplay, especially when player expectations of a successful performative action are not employed.

7.3 Dysphemisms during gameplay as linguistic strategy for quick feedback

In a ranked context like *CS:GO*, gamers have several aims, such as to win the game, level up, implement knowledge in a situated environment, develop tactics and strategies, and have fun. In *CS:GO*, a team-based online game, the performative actions of each player are vital for the game because they impact collective gameplay positively or negatively. Bearing this in mind, the following episode, extracted from a semistructured interview, describes the use of dysphemisms by the players when their co-players do not employ effective performative actions during gameplay. Due to the fact that time is limited in each round, the implementation of these dysphemisms functions as quick feedback in order to force co-players to reconsider their performative actions during gameplay as an attempt to prevent future performative face-loss of the team.

(16) **Purplezz:** o <Burnt-Bread> piˈstefko ˈe ˈvːalːi to nːun du ˈpano sto peˈxniði. enːoˈo su aˈpla ˈpezːi. ˈama ˈθːeli na ˈkat͡sʰːi na sceˈfti, ˈsceftete, ˈopos eˈxtes ˈepezːeŋ gaˈla. ˈtroi poˈlːes ˈoɾes ˈpano sto peˈxniðin, aˈlːa ˈθeli ˈlːiim baɾaˈpano proˈspaθia, nːa ˈvali to nːun du na sceˈfti. ˈeʃi foˈɾes ˈi enːa to aˈkusi t͡ʃe enːa su pi efcharistˈo pu mu to ˈipes ˈi enːa su pi ˈindambu mu to laˈlis, eˈɣo ˈiθela ˈtuto nːa ˈkamo.

[I believe that Burnt-Bread is not thinking seriously about the gameplay. What I mean, he just plays. When he really wants to sit and think seriously he will do so, for example like yesterday. He spends a lot of time on gameplay, but he needs to try more, to think more. He will either listen to you and he will tell you "thank you for telling me this" or he will tell you "why are you telling me this, it's what i wanted to do"]

(17) **Elisavet:** epiɾeˈazːi se ˈtuton do ˈpraman eˈsena?

[Does this affect you?]

(18) **Purplezz:** nːe. nevɾiˈazːi me, epiˈði o ˈiðios ˈkseɾi oti ˈimasteŋ gaˈlːitʰːeɾi tu t͡ʃe laˈli mas to. o <Kutch_me> ja paˈɾaðiɣman enːa su

pi ˈindaˈmbu kames ɾe ˈmbastaɾte. t͡ʃini pu to andˈexun mboˈɾu nːa ˈpeksu mːaˈzːin du. ˈt͡ʃini pu ˈen aˈndexun ˈe mboˈɾun. e ˈfleim.

[Yes. He makes me feel angry, because he does know we are better than him and he is also telling this to us (that we are better than him). Kutch_me for example will tell you "what have you done re bastard". Those who can handle it can play with him. Those who can't handle it cannot. It's flaming].

In the above extract, Purplezz evaluates (1) Burnt-Bread's performances and explains how he gets frustrated and angry by his co-players' nonsituated performative actions (2). Purplezz emphasizes the fact that even though Burnt-Bread spends valuable time playing *CS:GO*, he nonetheless needs to make an effort to apply his scaffolding[10] package of knowledge to the different situational problems that arise during gameplay and implement effective performative actions. Although Burnt-Bread takes into account the feedback given by his co-players, he also wants to challenge himself as a player by experimenting with his own decisions for performative actions. This frustrates his co-players, because he adopts and implements performative actions that he reads or observes via gaming websites, YouTube, and Twitch channels (professional player tournaments) without adjusting them to the problems that need to be solved during gameplay. In an attempt to change his low identity status as a player among the team, Burnt-Bread risks and experiments with nonsituated performative actions during their gameplay. These actions reveal his doubts about the higher identity status his co-players have. This causes high frustration levels among his co-players. When frustration levels are high, Kutch_ me is flaming with expletives (*inda pu kames re ˈmbastarte*). Consequently, for players, effective performance is vital in *CS:GO*. Effective performance is not just random implementation of actions. It is the cognitive procedure of the player to interconnect (a) the knowledge he/she has on various aspects of the game (e.g., mechanisms, software, gameplay, rules), (b) the variety of problem/s that arise during gameplay, and (c) the context of the game. These factors lead players to make decisions for effective performative actions (e.g., tactics, strategies) that benefit their identity as individual players and as teams in the community. In this context, players perform dysphemistic feedback strategies in order to force co-players to reconsider their performative actions during gameplay as an attempt to prevent future performative face-loss of the team.

In the following episode, all player characters have already been executed by the opponents; only one player character (Hookah's) is still alive in the game. In an attempt to help Hookah to successfully carry out the mission single-handedly, his co-players provide him with information about the location of the opponents.

(19) **Burnt-Bread:** oˈɾeːo!

[Niːce!!!]

(20) **Purplezz:** kaˈloː!

[Gooːd!!!]

(21) **Hookah:**ˈeʃi t͡ʃ ˈalːon? (He requests information to decide on his next moves in the game)

[Is there anyone else?]

(22) **Purplezz:** θco, ˈmbi

[Two (opponents in) B (site)]

(23) **Kutch_me:** θco, ˈmbi

[Two (opponents in) B (site)]

(24) **Purplezz:**ˈpienːe ˈlouer! ˈpienːe ˈlouer, ˈerkunde patiˈmeni.

[Go at Lower! Go at Lower! They are coming extremely fast.]

(25) **Purplezz:**ˈercete ˈpano su!

[He is coming over to you]

(26) **Hookah:** noːːː (he gets killed by an opponent)

[Noooo!]

(27) **Kutch_me:** oˈreon do maˈʃeri sːu re Hookah.

[Your knife's nice, Hookah!]

(28) **Hookah:** a?

[Huh?]

(29) **Kutch_me:** poˈlːa oˈreon do maˈʃeri sːu.

[Your knife is very nice!]

Hookah kills one of the opponents, and his co-players directly provide him with euphemistic feedback in high pitch (19, 20) for his successful performance in the game with positive adjectives (oˈreːo! *[Niːce!!!]*), (kaˈloː! *[Gooːd!!!]*). Hookah then asks his teammates to give him more information about the location of the opponents. The request is aimed at facilitating his decision-making on his next actions in the game (21). Purplezz and Kutch_ me immediately provide him with information about the number of the opponents and their location (22, 23). Purplezz observes on his screen that the two opponents are running very fast (ˈerkunde patiˈmeni *[They are coming extremely fast]*) approaching Lower B site, so he repetitively commands Hookah to go to Lower B site (ˈpienːe ˈlouer! ˈpienːe ˈlouer, *[Go at Lower! Go at Lower!]*) (24). Hookah, at this specific moment, evaluates within seconds the information and request given by his co-player Purplezz, and he decides to change his gun to a knife (Figure 7.2) while approaching Lower B site. Knives in CS:GO weigh less than guns, thus enabling player characters to walk faster. While he is running to the exit of Lower B site, the opponent suddenly appears in front of him. Hookah, with very fast moves, changes the

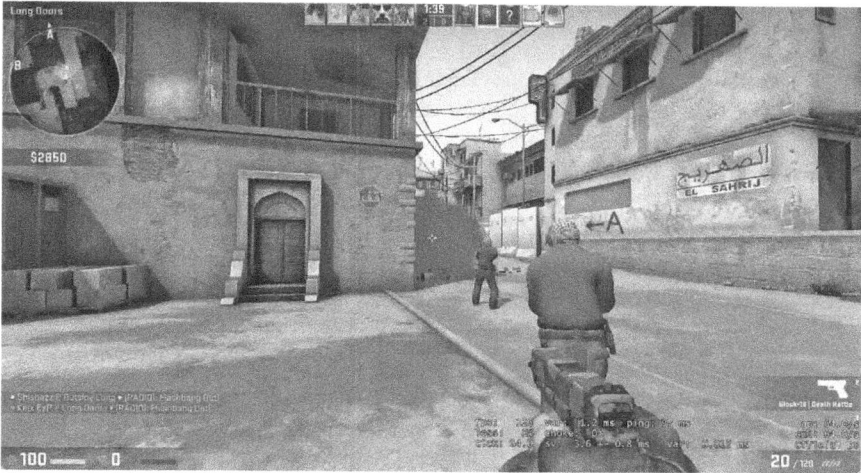

FIGURE 7.2 *Hookah is changing his gun to a knife. Screenshot by Elisavet Kiourti,* © *Valve Corporation.*

knife to a gun. Simultaneously, Purplezz warns Hookah that the opponent is preparing to kill him (25). Hookah attempts to change his knife back to a gun but his willing performative action fails as the opponent, who was already holding a gun, eliminates him. Hookah's response cries (26) (no:::*[Noooo!]*) indicate his disappointment at his failed performative action. Right after the end of the game, Kutch_me uses an irony (27) (oˈreon do maˈʃeɾi sːu ɾe Hookah *[Your knife's nice, Hookah!]*) as a frustrated but fast linguistic strategy to provide feedback on the unsuccessful decision of his co-player Hookah to carry a knife rather than a gun. Hookah first doesn't understand (28) the hidden meaning of Kutch_me's use of euphemism-dysphemism. Kutch_me responds again (29) with the same euphemistic-dysphemism (oˈreon do maˈʃeɾi sːu *[Your knife's nice]*), adding emphasis via the adverb "poˈlːa" *[very]*. Kutch_me's euphemistic-dysphemism has a twofold aim: First, to express his frustration over losing the game and, secondly, to provide quick feedback on the wrong decision that his co-player Hookah made.

8 Conclusions

This chapter presented the findings of an ethnographic study examining cooperative gameplay in the FPS game *CS:GO* in Cyprus, aiming to investigate the linguistic strategies that underlie the gaming discourse and, more specifically, players' bad language practices and their interconnection with the notions of identity and literacy in videogames. From the analysis of the oral discussions occurring during gameplay, it became evident that videogames function as frame social spaces in which players use dysphemisms in various ways (swearing, expletives, irony) and for effective purposes. More specifically, players use swearing as a linguistic strategy, with the aim to prevent individual or team-performative face-loss when communicative violations occur during gameplay. Violation of the "rules" of gameplay often leads to expletives and swearing as a fast linguistic strategy to get the attention of the co-players and negotiate their demands and reconstruct the balance of the team. Secondly, the use of swearing and expletives functions as linguistic strategy to ease stress and to ensure in-group bonding. Due to the fact that time and effective language practices are essential in the context of *CS:GO*, players use swearing, expletives, and irony as the shortest linguistic strategies in order to provide feedback to their co-players for their low performative actions in the game. They actually trigger negative emotions in their listeners in order to force them to reconsider the effectiveness of the performative actions during gameplay. The paradox that happens between gamers is that language practices function in opposition to those we tend to employ in other social environments (e.g., workplaces and school). They choose to linguistically face-threaten their co-players in order to achieve performative face-saving for themselves as players and for their team. The dysphemistic linguistics strategies they use reflect a hidden battle of identity relations among players which are indirectly interconnected with their social status as players in the broader community of gamers. Players break the rules of politeness for the sake of successful performance during gameplay and for the sake of maintaining a positive identity for themselves as gamers.

Notes

1 *Pε [re]* is a term-pragmatic marker used in informal conversations in Greek language and Cypriot dialect. It indexes speech organization and shows the stances of the speaker toward his/her utterance of/and the interlocutors (Taavitsainen and Jucker 2003). *Pε* indicates that the speaker perceives the communication situation and the relationship to the interlocutor as relaxed and quite familiar. For instance, in Karachaliou's research (2015: iv), the use of *ρε* in storytelling functions as term-pragmatic marker to highlight unexpected events and to add evaluative comments on speaker's stories. The recipients of the story use *ρε* to show their alignment toward the unprecedented events and aspects of the stories.

2 Note: laboratory experiments take place in controlled rather than physical environments.

3 "A condition, episode, person, or group of persons emerges to become defined as a threat to societal values and interests; its nature is presented in a stylized

and stereotypical fashion by the mass media; the moral barricades are manned by editors, bishops, politicians and other right-thinking people; socially accredited experts pronounce their diagnoses and solutions; ways of coping are evolved or (more often) resorted to; the condition then disappears, submerges or deteriorates and becomes more visible" (Cohen 1980: 1).

4 *Yu-Gi-Oh* is an adapted strategy card video game from manga card game.

5 *Malakas* is a name-dysphemism which signifies the word *asshole*. In this context, though, it means *dude*.

6 Lagging happens in online video gaming when there is low frame rate in the game because of the slow process of the computer system (frames per second = fps 92 [see Figure 7.1]).

7 Within the participation in framework, footing is "the alignment we take up to ourselves and the others present as expressed in the way we manage the production or reception of an utterance" (Goffman 1981: 128). Within a conversation, interlocutors shift footing as they change their alignments with each other from moment to moment based on contextual and linguistic utterances.

8 *Yolo* is an acronym of the phrase *You Only Live Once.*

9 *Swag* has several acronymic explanations, for example, *Secretly We Are Gay* or *Stuff We All Get*, and it is used as slang term between young people to describe anything thought to be cool.

10 Burnt-Bread was known in the gaming community for searching and reading numerous articles on various websites on a daily basis, and for watching professional gamers' gameplay on YouTube.

References

Activision (2018), "Leaderboard." Available online: https://my.callofduty.com/bo3/stats/leaderboard (accessed June 12, 2018).

Allan, K. and K. Burridge (2006), *Forbidden Words: Taboo and the Censoring of Language*, Cambridge: Cambridge University Press.

American Psychological Association (2015), "APA review confirms link between playing violent videogames and aggression." Available online: http://www.apa.org/news/press/releases/2015/08/violent-video-games.aspx (accessed June 12, 2018).

Anderson, C.A., et al. (2010), "Violent video game effects on aggression, empathy, and prosocial behavior in Eastern and Western countries: A meta-analytic review," *Psychological Bulletin*, 136 (2): 151–73.

Bourdieu, P. (1994), "Structures, habitus, power: Basis for a theory for symbolic power," in N.B. Dirks, et al. (eds.), *Culture/Power/History: A Reader in Contemporary Social Theory*, 155–99, Princeton, NJ: Princeton University Press.

Chayko, M. (1993), "What is real in the age of virtual reality? 'Reframing' frame analysis for a technological world," *Symbolic Interaction*, 16 (2): 171–81.

Christopher, D. (2017), "The negative effects of video game addiction." Available online: https://www.livestrong.com/article/278074-negative-effects-of-video-game-addiction/ (accessed June 12, 2018).

Cohen, S. (1980), *Folk Devils and Moral Panics*, New York: St. Martin's Press.

Crystal, D. (2003), *The Cambridge Encyclopedia of the English Language*, Cambridge: Cambridge University Press.

Ensslin, A. (2012), *The Language of Gaming*, Basingstoke: Palgrave Macmillan.

Fairclough, N. (1995), *Critical Discourse Analysis: The Critical Study of Language*, London: Longman.

Fairclough, N. (2001), *Language and Power*, Harlow: Longman.

Gee, J.P. (2003), *What Videogames Have to Teach Us about Learning and Literacy*, New York: Palgrave Macmillan.

Gee, J.P. (2004), *Situated Language and Learning: A Critique of Traditional Schooling*, New York: Routledge.

Gee, J.P. (2005a), *Why Videogames Are Good for Your Soul*, S.I.: Common Ground.

Gee, J.P. (2005b), *An Introduction to Discourse Analysis: Theory and Method*, New York: Routledge.

Gee, J.P. (2015), *Unified Discourse Analysis: Language, Reality, Virtual Worlds, and Video Games*, London: Routledge/Taylor & Francis Group.

Goffman, E. (1961), *Encounters: Two Studies in the Sociology of Interaction*, Oxford, England: Bobbs-Merrill.

Goffman, E. (1967), *Interaction Ritual: Essays on Face-to-Face Behavior*, Garden City, NY: Doubleday.

Goffman, E. (1974), *Frame Analysis: An Essay on the Organization of Experience*, Cambridge, MA: Harvard University Press.

Goffman, E. (1981), *Forms of Talk*, Philadelphia: University of Pennsylvania Press.

Halliday, M.A.K. (1976), "Anti-languages," *American Anthropologist*, 78 (3): 570–84.

Halliday, M.A.K. (1985), *An Introduction to Functional Grammar*, London: Edward Arnold.

Hammersley, M. and P. Atkinson (2007), *Ethnography: Principles in Practice*, London: Routledge.

Hine, C. (2000), *Virtual Ethnography*, London: Sage.

Huizinga, J. (1950), *Homo Ludens*, Boston, MA: Beacon Press.

Karachaliou, R. (2015), "Προσφωνήσεις ως πραγματολογικοί δείκτες σε συνομιλιακές αφηγήσεις: Η περίπτωση του ρε και των συνδυασμών του," Thesis, University of Patras. Available online: http://thesis.ekt.gr/thesisBookReader/id/36877#page/1/mode/2up.

Karachaliou, R. and A. Archakis (2015), "Identity construction patterns via swearing: Evidence from Greek teenage storytelling," *Pragmatics and Society*, 6 (3): 421–43.

Karyolemou, M. and P. Pavlou (2001), "Language attitudes and assessment of salient variables in a bi-dialectal speech community," in *Proceedings of the First International Conference on Language Variation in Europe*, Universitat Pompey Fabra, 110–20.

Kiourti, E. (2018), "(Απο)συνδέοντας γραμματισμούς: Από το διαδικτυακό βιντεοπαιχνίδι Counter Strike Global Offensive στη σχολική τάξη," (Disconnecting Literacies: From Online Videogame Counter Strike Global Offensive into the Classroom), Thesis, Nicosia: University of Cyprus.

Ljung, M. (2011), *Swearing: A Cross-Cultural Linguistic Study*, Basingstoke and New York: Palgrave Macmillan.

Mahardy, M. (2014), "Infographic details 10 years in World of Warcraft." Available online: http://www.ign.com/articles/2014/01/28/infographic-details-10-years-in -world-of-warcraft (accessed June 12, 2018).

Paltridge, B. (2006), *Discourse Analysis: An Introduction*, London: Continuum.

Papapavlou, A.N. (1998), "Attitudes toward the Greek Cypriot dialect: Sociocultural implications," *International Journal of the Sociology of Language*, 134 (1): 15–28.

Salam, M. and L. Stack (2018), "Do videogames lead to mass shootings? Researchers say no." Available online: https://www.nytimes.com/2018/02/23/us/ politics/trump-video-games-shootings.html (accessed June 12, 2018).

Salen, K. and E. Zimmerman (2003), *Rules of Play: Game Design Fundamentals*, Cambridge, MA: The MIT Press.

Scollon, R. and S.B.K. Scollon (2001), *Intercultural Communication: A Discourse Approach*, Malden, MA: Blackwell Publishers.

Shaffer, D.W., et al. (2005), "Videogames and the future of learning," *Phi Delta Kappan* [Online], 87 (2): 105–11.

Sophocleous, A. and C. Themistocleous (2014), "Projecting social and discursive identities through code-switching on Facebook: The case of Greek Cypriots," *Language at Internet*, 11 (5). Available online: http://www.languageatinternet. org/articles/2014/sophocleous (accessed August 24, 2018).

Squire, K.D. (2008), "Videogames and education: Designing learning systems for an interactive age," *Educational Technology*, 48 (2): 17–26.

Statista (2016), "League of Legends MAU 2016 | Statistic." Available online: https:// www.statista.com/statistics/317099/number-lol-registered-users-worldwide/. (accessed June 12, 2018).

Steinkuehler, C. (2007), "Massively multiplayer online gaming as a constellation of literacy practices," *E-Learning and Digital Media* [Online], 4 (3): 297–318.

Taavitsainen, I. and A.H. Jucker (2003), *Diachronic Perspectives on Address Term Systems*, Amsterdam: John Benjamins Publishing Company.

Tsiplakou, S. (2004), "Στάσεις απέναντι στη γλώσσα και γλωσσική αλλαγή: Μια αμφίδρομη σχέση (Attitudes towards language and language change: A two-way relation?)". In *Proceedings of the 6th International Conference on Greek Linguistics*, Edited by: Catsimali, G., Kalokairinos, A., Anagnostopoulou, E. and Kappa, I. Rethymno: Linguistics Lab. CD-Rom.

Valve (2018), "Unique players last month." Available online: http://blog.counter-strike.net/ (accessed June 12, 2018).

Wenger, E. (1998), *Communities of Practice: Learning, Meaning, and Identity*, Cambridge: Cambridge University Press.

Wright, T., E. Boria, and P. Breidenbach (2002), "Creative player actions in FPS online videogames: Playing counter-strike," *Game Studies*, 2 (2). Available online: http://www.gamestudies.org/0202/wright/ (accessed August 24, 2018).

8

"I Cut It and I …
Well Now What?"

(Un)Collaborative Language
in Timed Puzzle Games

Luke A. Rudge

1 Introduction

Language, in its production and reception, allows us to perform collaborative tasks. These tasks may involve language solely, such as the sentencing of a criminal by a judge, or using language alongside physical actions. In the world of video gaming, particularly in collaborative multiplayer situations wherein a task needs to be completed within a certain amount of time, effective collaborative communication between players is critical.

This chapter presents a small-scale, preliminary study on the use of language between players cooperating to achieve a goal in a time-limited situation. Specifically, this work analyses the use of (un)collaborative language employed between players in the game *Keep Talking and Nobody Explodes* (2015). Linguistic analysis is performed from two different yet complementary perspectives: via Systemic Functional Linguistics (SFL) and Conversation Analysis (CA). Both approaches have strong links regarding the use of language in context and can analyze interactive elements of communication, but they do so from different angles. This approach is taken to show, both quantitatively and qualitatively, a sample of linguistic factors that may contribute to the (un)successful completion of a collaborative task.

If it is assumed that collaborative efforts and communication are precursors to greater task success (i.e., the more people work together to complete a task, the more likely it will be performed successfully, correctly, or on time; see e.g., Orasanu and Salas 1993; Sexton and Helmreich 2000; and Krifka, Martens and Schwarz 2004), then it may be argued that certain factors in communication exist that correlate with instances of task success or failure. In other words, certain features in communication may be understood to be "collaborative" (e.g., permitting appropriate time for turn-taking) and "uncollaborative" (e.g., making deliberate and frequent interruptions). Nonetheless, given the preliminary nature of such an investigation in video gaming, further study is encouraged.

2 Literature review

2.1 Language, collaboration, and the impact of stress

Communication between two or more language users can be viewed as a collaborative effort and as a means toward a goal (see, for example, Sacks, Schegloff and Jefferson 1974; Halliday 1978; Halliday and Hasan 1989; Clark and Wilkes-Gibbs 1990; Halliday and Matthiessen 2014). Language users employ different linguistic techniques to achieve whatever this goal may be, regardless of whether said goal is accomplished primarily via language itself (e.g., building interpersonal solidarity between interlocutors when telling a joke; see Fiksdal 2001), or when language accompanies other actions that are "outside" of language (e.g., when pilots and air-traffic control towers communicate; see Garcia 2013). Arguably, this latter type of collaboration incorporates the former: collaborative tasks that are accompanied by language also require collaboration to occur *within* communication. In other words, to get anything done between two or more people, it is important to "get on" linguistically.

Halliday and Matthiessen (2014: 33) refer to communicative instances as "socio-semiotic activities," which span "from contexts where language does all the semiotic work to contexts where all the semiotic work is done by some semiotic system or systems other than language" (2014: 38). For instance, when two employees of a removal company are working to maneuver a sofa through a small stairwell, they will communicate with one another to be aware of aspects such as weight distribution and potential obstacles. Depending on the size of the sofa and the complexity of the stairwell, other people may also be involved in communication to ensure, for example, that no damage is done to property, or that the movers are approaching a step or a large potted plant. As such, the successful completion of this task requires both physical and linguistic collaboration and is classed as the sociosemiotic activity of "doing." Considering further sociosemiotic activities such as "narrating" and "advising" (see Halliday and Matthiessen 2014: 37), it may be argued that all communication can be viewed as some form of collaborative task. While the example of moving a sofa presents a more physical and "overt" task, the small talk used between two employees in a taxi can be viewed as the completion of a more social task, albeit still an instance of collaboration (i.e., maintaining cordial relationships; see Coupland 2003). Nonetheless, it is the area of "doing" that is focused on in this chapter.

A further point to note is that each of these examples can be viewed as relatively stress-free, if *stress* is understood as a negative emotional response from a stimulus, which may manifest physically or verbally (see the "stressor-strain" approach exemplified in Beehr and Franz 1987). While there are aspects that may increase stress levels in the examples (e.g., the possibility of job loss if the furniture or walls are damaged), they may be viewed as less stressful than other situations. If the examples are compared to a surgeon in an operating theater who has minutes to successfully revive a patient and must rely on (among other things) successful communication with her team to complete the task, factors including the increased pressure due to timing, the heightened risk to life, and the potential complications that may arise would heighten stress levels considerably. Given that stress can manifest verbally (Jaffe and Feldstein 1970), communication will likely be affected in such situations.

There is a small but informative body of literature concerning the intersection of language, collaboration, and stress. Sexton and Helmreich (2000) observe collaborative language within interactions between flight-crew members during simulations consisting of low and high workload tasks. While Sexton and Helmreich identify that language use is one of many factors contributing to successful flights, they note that there are "links between pilot language use and flight outcome" (2000: 66). Statistical analyses performed on the language data allow the authors to posit various conclusions, such as the correlation between the use of larger words (defined by Sexton and Helmreich as words containing more than six letters) and reduced task performance: "those individuals who expend the cognitive resources necessary to speak more elaborately (using bigger words) do so at the expense of decreased situational awareness" (Sexton and Helmreich 2000: 66).

Khawaja, Chen, and Marcus (2012) also observe linguistic variation in collaborative tasks, namely the language of an incident-management team working together to solve simulated bushfires. The authors analyze relationships between cognitive load and language, presenting comparisons between language when cognitive load is low—when "participants were involved in communication not related to their task, for example, conversation about personal life"—and when cognitive load is high—when "participants were involved in challenging tasks, for example, handling unexpected events, producing information reports, and completing tasks within time constraints" (Khawaja, Chen and Marcus 2012: 523). Khawaja, Chen, and Marcus present numerous conclusions, including that more speech is used during tasks with a high cognitive load (calculated by the difference in the number of words used between tasks); that complex tasks result in greater collaboration and coordination (derived from the more frequent use of the plural pronoun "we" in high cognitive load tasks); and that disagreements are more common during stressful tasks (derived from the number of "disagreement words" employed between the different tasks, although such words are not overtly reported in the findings). However, the analysis provided by Khawaja, Chen, and Marcus (2012) requires caution. The utterances appear to be subjectively coded depending on task load, and the authors note that these codings had an initial interrater reliability score of 72%, which later increased to 83% after coders "discussed further the points of difference" (2012: 523). Although the statistical test regarding interrater reliability is not mentioned (e.g., Cohen's Kappa or Krippendorff's Alpha), both percentages imply that between one-quarter and one-fifth of codings were not agreed on. While issues persist surrounding the arbitrariness of "acceptable" percentages (see e.g., McHugh 2012: 279), the calculated values suggest that further investigation into this data could be useful.

More recently, in McKendrick et al.'s (2014) study into collaboration in simulated environments with unmanned aerial vehicles, similar results to those studies discussed here are found. For example, "an increase in words used per message was associated negatively with task performance" (2014: 472). However, McKendrick et al.'s approach to this area of study, alongside the approaches of Sexton and Helmreich (2000) and Khawaja, Chen, and Marcus (2012), contain a notable omission: while each study claims to focus on "linguistic cues," the relationship to linguistic theory is often disregarded. Each of these investigations rely to some extent on counting words, calculating word lengths, or assigning words to subjective categories, without considering deeper linguistic principles. For instance, McKendrick et al. (2014: 465) associate the number of words produced as a measure "of communication frequency and complexity," while overlooking other linguistic features that may also play a part in task collaboration.

Although such a critique is not intended to dismiss the results presented in these works, it does suggest that approaches to the analysis of language employed during collaborative interactions can be enhanced. Krifka,

Martens, and Schwarz (2004) address this point in their work, applying a more in-depth linguistic analysis to a subset of data from Sexton and Helmreich (2000). By applying and adapting Searle's (1975) speech act theory, Krifka, Martens, and Schwarz (2004) note that successful simulations correlate with, among other things, a heightened use of speech acts that are positive in their nature and that seek support from other team members. Conversely, speech acts related to opposition and re-establishing known facts correlate with poor outcomes. Moreover, Nevile's (2001) study on the collaborative language of air-traffic controllers and pilots (commented and expanded on by Garcia 2013) analyses language via Conversation Analysis. In doing so, Nevile identifies further linguistic and interactional elements that are conducive to collaborative communication, such as the use of short, succinct turns.

The studies discussed here analyze collaborative language to varying extents, and (with the exception of Nevile's work) focus on simulated environments with varying levels of stress. Although simulations may have parallels to video games (e.g., the requirement for successful task completion to advance, preprogrammed events, etc.), they are not video games per se. Interaction in collaborative gaming environments has been researched to varying extents, yet there remains the opportunity for the language used in these environments to be explored and analyzed in greater detail. For instance, Egenfeldt-Nielsen, Smith, and Tosca (2016) identify that the language employed in gameplay has the ability to enhance social cohesion, but they do not present a detailed linguistic analysis to support this statement. Similarly, Ducheneaut and Moore (2004) observe the social side of gaming and interaction patterns in online social settings, but no in-depth analysis of language occurs.

Interestingly, Taylor (2009: 38) notes that "the importance of linking design with the social life of a game cannot be overemphasized," with this linkage being facilitated via language. It is here that a divide may be drawn between studies that observe communication to build and sustain social elements (i.e., Halliday's sociosemiotic activities of "sharing," "reporting," etc.), and communication within games that *works alongside* the completion of another, "extralinguistic" task (i.e., Halliday's socio semiotic activity of "doing"). Prior to investigating this point further, however, a short review of how linguistic analysis could be performed in these environments is presented.

2.2 Analyzing collaboration in communication

As evidenced by the range of subdisciplines in linguistics and the many convergent and divergent theories found therein, the potential for linguistic analysis is vast. Nonetheless, certain theories are more "suitable" in their applications to linguistic analysis than others, including the analysis of collaborative gameplay. The approaches presented here are Systemic Functional Linguistics (SFL) and Conversation Analysis (CA), chosen for their applicability to analyzing language in action and their recognition of the importance of context in interaction.

The theoretical groundings of SFL and CA are such that there are similarities in their approaches and epistemic positioning. SFL understands language as a "social semiotic," with roots in the Firthian concept of context of situation (Firth 1935) and how language use varies according to environment. Language and context are also understood to act in a dialogic manner (see, for example, Hasan 2014), influencing and "constructing" one another. Likewise, CA developed with strong influence from Garfinkel's (1967: vii) ethnomethodology, which noted a complementary phenomenon: communicators understand a context and then employ language that reinforces that context. Both SFL and CA therefore promote the importance of context in linguistic analysis and the fact that any communications have analyzable elements that can be explored in further detail. There are similarities in their philosophical positioning, yet they are distinct in their approaches, which are explained briefly here.

2.2.1 Systemic Functional Linguistics (SFL)

SFL is a broad, functional approach to the description and analysis of language (Butler 2003). The theories it presents are in-depth and numerous, such that a full account of SFL cannot be provided here (see, for example, Halliday and Matthiessen 2014; Thompson 2014, for more information). Nonetheless, a fundamental idea of SFL is that language produces several strands of meaning simultaneously, known as *metafunctions*. These are *ideational*, or how experience is represented and logically organized in language; *textual*, or how a text develops over time; and *interpersonal*, or how social relationships are enacted and maintained through language. It is this final metafunction that is focused on in this chapter.

The observation of the three Hallidayan metafunctions is performed at clause level: the occurrence and order of certain functional elements within a clause go toward explaining the meanings that are expressed. Focusing on the interpersonal area of meaning in English, the key functional elements are the Subject—"the entity […] that the speaker wants to make responsible for the validity of the proposition being advanced in the clause"—and the Finite—the element that "makes it possible to argue about the validity of the proposition" (Thompson 2014: 55). These elements allow for an interpersonal "move" to be made by the speaker. For instance, when the speaker wishes to provide information, such as "he is young," the Subject (he) is followed by the Finite (is). Conversely, in requesting clarification, the Subject and Finite are inverted, thereby creating "is he young?" Other configurations are possible, such as the removal of the Subject to create a command: "(you) look at this!"

While interpersonal elements present far more complexity in English than what has been discussed here (see, for example, Chapter 4 of Halliday and Matthiessen 2014), with similar levels of complexity across languages (Caffarel, Martin and Matthiessen 2004), even the identification of the Subject and the Finite alone allows for linguistic analysis from the perspective of social interaction to occur. The configuration of Subject and Finite result in the use of different clause types in communication (and therefore the different kinds of interpersonal moves made between communicators), thereby presenting insights into social elements of language, including, but not limited to, areas such as collaboration. This has been observed in previous work (e.g., Jacobs and Ward 2000) and is explored in greater depth later in this chapter.

2.2.2 Conversation Analysis (CA)

CA (Sacks, Schegloff and Jefferson 1974) is an approach that understands the use of language as action within sociocultural contexts, due in part to influence from ethnomethodology (Garfinkel 1967). Similarly to SFL, the interaction between context and language is imbued with high importance: "CA offers an alternative to the view [...] that our conduct automatically reflects the context in which it occurs" (Woofit 2005: 69). However, rather than focusing on the clause as a unit of analysis, CA observes "the properties of the ways in which interaction proceeds through activities produced through successive turns" (2005: 8). CA therefore permits the analysis of data that may be "omitted" in SFL by observing the "ostensibly 'minor' contributions and non-lexical items [that] may be interactionally significant" (2005: 12).

In Ten Have's (2007) words, there is a distinction between *pure* CA and *applied* CA, primarily defined by their scope: pure CA is concerned with the elements of interaction when "interaction" is understood as an intrinsic phenomenon, whereas applied CA extends toward the observation of interaction within specific contexts. Put another way, pure CA is used to understand the general strategies employed in interaction across contexts, but in applied CA, "the scope of one's findings will often be intentionally limited to a specific setting or interaction type" (2007: 147). As such, using CA to analyze interaction in the context of collaboration in multiplayer gaming may be understood as "applied."

Whether pure or applied in scope, the same underlying principles of CA are generally observed. The primary principle is that any conversation is split into sequences of turns, and as Sacks, Schegloff, and Jefferson (1974: 730) note, "one party talks at a time." Furthermore, while no two conversations are completely identical, turn-taking in conversation has an overall systematic development and usage. As such, there are similar elements in any conversation that can be analyzed, including overlaps in communication, interruptions, repair, and pausing (see, for example, Liddicoat 2007; Ten Have 2007).

As seen in some of the studies mentioned in the literature review, the application of CA to various communicative situations can produce noteworthy results. When combined with the applicability of SFL in observing the interpersonal strategies realized in language, alongside the opportunity to explore collaborative language in video games in greater detail, the following questions may be proposed: when considering the outcome (i.e., successful vs. unsuccessful) of collaborative tasks that involve a certain level of stress and that require collaborative communication, are there specific linguistic patterns that can be found? If so, what may be observed at clause level (i.e., via SFL) and at the level of the turn (i.e., via CA)?

3 Methodology

To investigate the language of collaboration, an experiment was set up to record the vocal interactions of a group of participants playing the video game *Keep Talking and Nobody Explodes* (2015). According to the developers, this game began as a rough contribution to the 2014 Global Game Jam—an event wherein game developers work together around a theme. However, its popularity during this event would be the precursor to its success, eventually being developed for Windows, OS X, Linux, and various VR platforms.

The game requires a minimum of two players to work cooperatively using spoken communication to defuse a timed bomb by successfully disarming different modules (i.e., completing short tasks, such as cutting a specific wire or pressing buttons in a certain order). To successfully defuse the bomb, one player—the "defuser"—listens to the instructions given from another player or players—the "expert(s)." However, the defuser may only see the bomb, and the expert(s) may only see the instructions. As there is also a variable time limit assigned to each bomb, each scenario comprises numerous stressors. In order to enhance the chances of successful task completion, effective and collaborative communication must be used.

Five participants were recruited to take part in a series of rounds, and these participants were chosen based on several factors. First, each participant confirmed that their production and comprehension of spoken and written English was suitable for the task (i.e., they were native English speakers or had at least a C1 level of English according to the CEFR scale; see Council of Europe 2001). Secondly, the participants were briefly asked about their previous experiences with video games and technology to ensure that they could easily understand the game mechanics and how to interact both with the bomb (as defuser). Finally, it was confirmed that each participant had little to no previous exposure to *Keep Talking and Nobody Explodes* (2015) to ensure a similar ability level across the sample.

The five participants were introduced to each other prior to gameplay to become acquainted. A demonstration of the first level was also provided, allowing players to understand how to interact with the interface (i.e., point and click via a mouse) and for any queries to be answered. Furthermore, the demonstration confirmed how players were not permitted to look at what the other player could see, which was reinforced by the configuration of the players during gameplay: players sat at opposite ends of a table with an opaque screen in the middle. This configuration allowed for easy verbal communication while ensuring both that players saw only what they were permitted to see, and that nonverbal signals were removed from communication. A schematization of this setup can be seen in Figure 8.1.

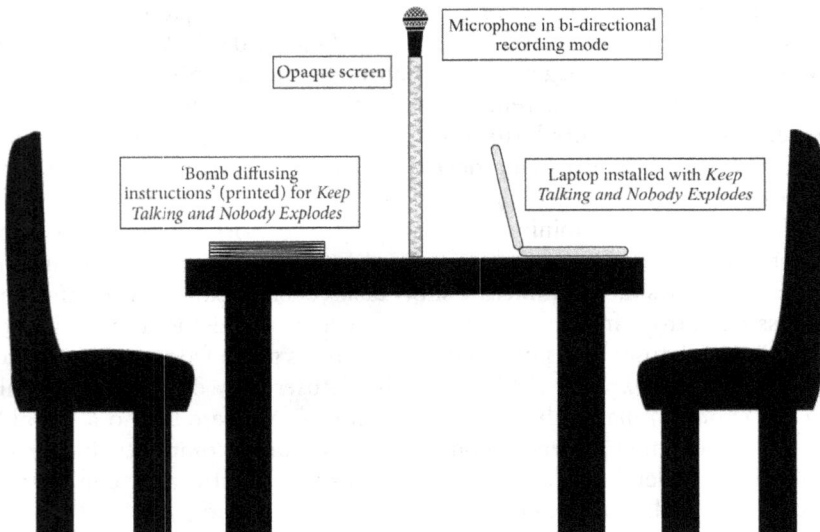

FIGURE 8.1 *A side view of the experimental setup.*

Rounds were organized so that each player worked with all other players once. This resulted in ten rounds, with each player having two rounds as the defuser and two rounds as the expert. The organization of participants per round is shown in Table 8.1, with the final column identifying if the round was successful (i.e., the bomb was defused before time elapsed) or unsuccessful (i.e., the bomb exploded due to either three incorrect moves or because time elapsed). In total, six rounds were successful, and each participant was part of a successful and unsuccessful round at least once.

Each round had the same level of difficulty, requiring four modules to be defused in three minutes. However, each bomb was unique in its composition: no two configurations of modules were the same, and noninteractive parts of the bomb (e.g., the serial number and the number of LEDs and batteries, which all contribute to correct defusing) were randomized. As such, each round was unique, allowing for a moderate level of challenge and ensuring that any "previous answers" could not be used in to subsequent rounds.

In each round, screen-capture software recorded the display showing the bomb, which was time-aligned with an audio recording of the two players verbally interacting. The audio was recorded using a microphone placed

TABLE 8.1 *The organization of participants in the ten rounds played*

| Round | Players | | Round result |
	Defuser	Expert	
1	1	2	Exploded
2	3	1	Defused
3	1	4	Defused
4	5	1	Defused
5	2	3	Defused
6	4	2	Exploded
7	2	5	Defused
8	3	4	Defused
9	5	3	Exploded
10	4	5	Exploded

on top of the opaque screen separating the players, set in a bidirectional recording mode (i.e., configured to focus on the voices of the participants sitting opposite from one another, rather than other noises). Although only the audio was transcribed, the screen capture allowed for greater clarity in cases where the defuser used various deictic words. For instance, if the defuser said "I don't know what this is" while moving the cursor over a button, the screen capture clarified the intended referent.

Data were transcribed and analyzed via SFL and CA. The former required the identification of clause types (declarative, interrogative, etc.) and clause composition, whereas the latter looked at conversational elements including turns, pauses, and interruption.

4 Findings and discussion

4.1 The systemic functional perspective

This study analyzed language in a similar method to that of Eggins (2004): clauses were identified and counted based on their function and composition. For this study, these counts were then further split based on whether the round was successful or unsuccessful. These totals are tabulated in Table 8.2 and also represented in a chart format in Figure 8.2:

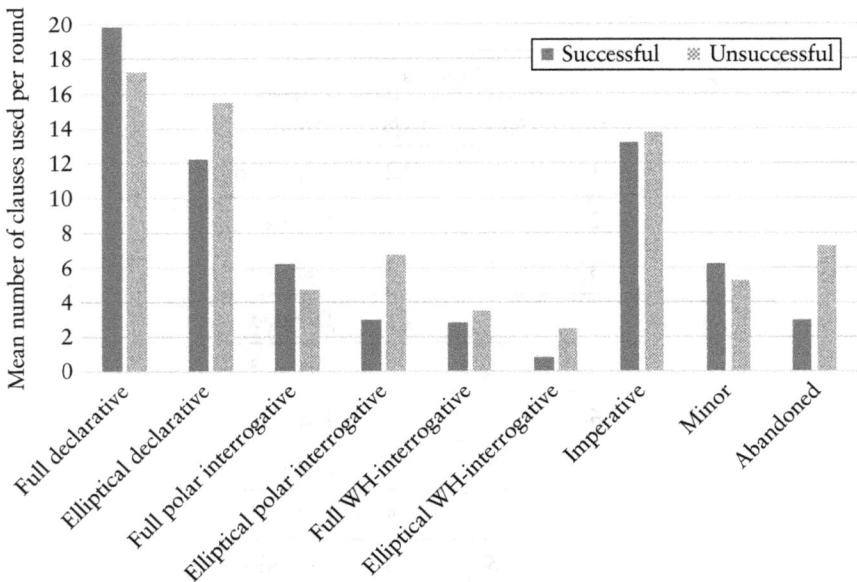

FIGURE 8.2 *Mean values of clauses used per round, split for task success.*

The "clause types" in Table 8.2 are defined as follows. Full clauses contain all the mandatory clausal elements to be deemed "complete":

(1) A full declarative contains the Subject and Finite in that order (e.g., "<u>I have</u> cut it").

(2) A full polar interrogative contains the Subject and Finite in reverse order (e.g., "<u>Have you</u> cut it?").

(3) A full WH-interrogative contains a wh- question particle with a Finite element (e.g., "<u>Who cut</u> it?").

Conversely, elliptical clauses omit one or more of these mandatory elements but are nonetheless understood to be an interpersonal move in the dialogue:

(1) An elliptical declarative may be a short response to a question (e.g., responding to "Which one did you cut?" with "<u>Red</u>").

(2) An elliptical polar interrogative may use intonation to differentiate it from an imperative clause (e.g., "Cut it?" with rising intonation).

(3) An elliptical WH-interrogative may only contain the wh- question particle (e.g., "Which?").

TABLE 8.2 *Number of clauses used (with mean values per round in parentheses), split by task success*

	Task success			
Clause type	*Successful*		*Unsuccessful*	
Full declarative	119	(19.83)	69	(17.25)
Elliptical declarative	73	(12.17)	62	(15.50)
Full polar interrogative	37	(6.17)	19	(4.75)
Elliptical polar interrogative	18	(3.00)	27	(6.75)
Full WH-interrogative	17	(2.83)	14	(3.50)
Elliptical WH-interrogative	5	(0.83)	10	(2.50)
Imperative	79	(13.17)	55	(13.75)
Minor	37	(6.17)	21	(5.25)
Abandoned	18	(3.00)	29	(7.25)
Total	*484*		*469*	

Imperative clauses omit the Subject and have the force of "commanding" the recipient of the message (e.g., "Cut it" with falling intonation, to differentiate it from an elliptical polar clause). Minor clauses contain neither Subject nor Finite elements but are still interpreted as interpersonal moves (e.g., exclamations and alarms such as "Oh!"), and abandoned clauses are those that are started but not completed due to interruption from another source or the speaker themselves (e.g., "You should probably…").

Despite the small sample size of this study, inferential statistical analyses (two-sample t-tests) were performed to identify statistically significant trends with regards to clause occurrence. While differences are already apparent in the "Task success" subcolumns of Table 8.2 and in the chart area of Figure 8.2, three statistically significant differences were calculated. First, the number of elliptical polar clauses used in successful rounds ($M = 3.00$, $SD = 1.41$) compared to the number found in unsuccessful rounds ($M = 6.75$, $SD = 2.63$) was significantly different ($t = 2.96$, $p < 0.02$). Secondly, the difference in number of elliptical WH-interrogative clauses observed in successful rounds ($M = 0.83$, $SD = 0.98$) compared to those used in unsuccessful rounds ($M = 2.50$, $SD = 0.58$) was calculated at a similar level of significance ($t = 3.02$, $p < 0.02$). Finally, the difference in instances of abandoned clauses in successful rounds ($M = 3.00$, $SD = 0.63$) in comparison with those observed in unsuccessful rounds ($M = 7.25$, $SD = 1.26$) was calculated to be highly significant ($t = 7.17$, $p < 0.001$). In all three cases, these clause types occurred more in unsuccessful rounds.

Some initial suggestions may be posited for this patterning, with the difference between these suggestions being a matter of "direction": the use of particular clause types (i.e., elliptical polar, elliptical WH-interrogative, and abandoned clauses) contributed toward the round being unsuccessful; the unsuccessfulness of the round resulted in the occurrence of these clause types; or a "downward spiral" effect occurred wherein the use of these structures increased the likelihood of being unsuccessful, thereby creating more opportunities for these clause types to occur. Therefore, it is necessary to more closely observe these clause types in context to understand whether they were a contribution toward, a result of, or a self-fulfilling consequence of unsuccessful rounds.

Concerning abandoned clauses, while not unexpected in spontaneous spoken language (see Eggins 2004), there was a prominent pattern in their usage during unsuccessful rounds. Of the total 29 instances, 24— roughly 83% of the total—were used once there was at least one "strike" on the bomb (i.e., when at least one wrong move had been made). Each bomb had a tolerance for two incorrect moves, with a third leading to explosion. Furthermore, the countdown rate increased by roughly 1.25% for one strike and 1.50% for two strikes, arguably increasing stress levels as the task needed to be completed in a shorter time. This may therefore manifest as abandoned structures, as seen in the following example from Round 1 (note that the use of ellipses [...] indicate a pause of 0.6 seconds or longer):

Round 1—Player 1 (defuser) and Player 2 (expert)

(Player 1 clicks an incorrect button and receives the first "strike" on the bomb.)

Player 1	That was wrong and it's counting down quicker.
Player 2	Okay so you need to... Okay oh god. Do you have a... Wait.
Player 1	What do I do?

This extract is from the first game, suggesting that it was the first time that either player had played the game. In this instance, the module required players to press colored buttons in a certain sequence, but the sequence altered depending on the number of strikes obtained. This confused Player 2, leading to two abandoned clauses and the requirement for Player 1 to then use an interrogative clause to clarify the next steps. Following this exchange, there was a silence of roughly 10 seconds while Player 2 tried to advise on the correct sequence, but subsequent attempts to defuse the module were performed incorrectly, causing the bomb to explode and classifying the round as unsuccessful.

Another instance of abandoned clauses can be seen from Round 10:

Round 10—Player 4 (defuser) and Player 5 (instructor)

(Player 4 selects the wrong wire and receives their second "strike" on the bomb.)

Player 4	No that wasn't it.
Player 5	Really? Did you cut the...
Player 4	I cut it and I... well now what?
Player 5	Cut the other one and move to...
Player 4	Which other one?

Unlike the extract from Round 1, this was the final round of the session and therefore represented each players' fourth attempt at the game. While both players were more aware of how to complete the module in question (cutting colored wires), prior miscommunication had occurred: Player 4 incorrectly stated the number and color of wires in the module, causing Player 5 to provide the wrong answer, and for Player 5 to assume that only one wire was left to cut. As such, Player 5 asked for clarification, followed by a partial response from Player 4, and then a further partial command from Player 5. Importantly, these instances were not abandoned due to interruption from the other player, and this effect is discussed from a CA perspective in the following section.

Abandoned structures also appeared in successful rounds, albeit infrequently, and they mostly appeared as faults due to the expert misreading or misinterpreting the instructions. Often, the expert would realize the error, stop, and reformulate the utterance, as exemplified in Round 2:

Round 2—Player 3 (defuser) and Player 1 (expert)

Player 3   I can see a bunch of wires.

(Player 1 searches through the instructions for roughly 7 seconds)

Player 1   Right, do you see all... sorry. How many wires do you see?

Regarding elliptical polar and WH-interrogative clauses, these appeared to be used in various stages of unsuccessful rounds, unlike abandoned clauses that commonly appeared after strikes were made. For these elliptical clauses, the trend was for further clarification to be required after their use, therefore delaying progress. This can be seen in the following extracts from Round 6 and Round 9:

Round 6—Player 4 (defuser) and Player 2 (expert)

Player 2   Hold the button and tell me the color.
Player 4   Which?
Player 2   The button that says "hold."
Player 4   I mean which color do you want: the button or the light?

Round 9—Player 5 (defuser) and Player 3 (expert)

Player 3   Select the wire to cut it.
Player 5   What?
Player 3   Click on the correct wire to cut it.
Player 5   Yeah I get that but which color is the correct wire?

In both instances, the use of a single interrogative particle by the defuser results in ambiguity. The expert interprets each response but fails to respond in the way that the defuser expects, requiring further clarification to rectify confusion.

It may also be suggested that the use of elliptical interrogative clauses is self-fulfilling: if a player wishes to save time by assuming shared understanding of certain elements, they may reduce the elements of an utterance to a minimum by use of ellipsis (Halliday and Hasan 1976). This, however, increases the potential level of ambiguity. As such, additional utterances may have to be used, thus requiring more time. It is suspected that given more practice and exposure to the tasks, the use of elliptical structures would produce less ambiguity as players would know "key elements" to save time. However, until a higher level of proficiency with the tasks is reached, full clauses appear necessary, even though they take marginally longer to produce. The use of elliptical and abandoned clauses, then, seems to show a downwards spiral, at least at a novice level.

4.2 The conversation analysis perspective

As noted in various CA works (e.g., Ten Have 2007), there is no "correct" method to analyzing texts from the CA perspective. However, a trend among most CA analyses is that the starting point seems to be without a specific or deliberate point of linguistic interest. In other words, these analyses are "not prompted by prespecified analytic goals […] but by 'noticings' of initially unremarkable features of talk" (Schegloff 1996: 172). Reading through the transcribed data allowed for these "noticings" and patterns of occurrences in the data, which appear to correlate within and between (un)successful interactions.

The first area of interest concerns effects within a basic unit of discourse: the adjacency pair. At its most simplistic, Schegloff (2007) notes that an adjacency pair forms two parts, the first of which initiates (first pair parts [FPPs]), and the second of which responds (second pair parts [SPPs]). In everyday speech, there may be items leading up to, separating, or following the FPP and the SPP (i.e., expansions), but the notion of initiation and appropriate response allows for communication to progress in a collaborative manner.

In Rounds 6 and 9 (both unsuccessful), several instances were observed where an FPP was presented along with expansions, yet an SPP did not occur. An extract from Round 6 is presented here, wherein the defuser recommends a module to begin with but then changes their mind (notation conventions are explained at the end of this chapter):

Round 6—Player 4 (defuser) and Player 2 (expert)

1	Player 4	I have six wires s let's do wires	[there]	is
2	Player 2		[oka:y]	
3	Player 4	i:[s six w]ires		
4	Player 2	[ju just] wires?		
5	Player 4	y:es=		
6	Player 2	=okay blues?		

Player 4 looks at other modules (1.2)

7	Player 2	Do you have a[ny blues]
8	Player 4	[yeah yea]h just gimme a [momen]
9	Player 2	[b u t y]ou
10		need to (0.4) t tell m[e how m]--
11	Player 4	[should]we <u>change</u> mods

In line 4, Player 2 asks a question and immediately receives a response from Player 4 (i.e., a minimal adjacency pair is formed). However, when another question is asked by Player 2 in line 6, no response is provided despite the use of "yeah yeah" by Player 4 in line 8, which appears to be used in a dismissive manner. As such, the FPP is not paired with an SPP. In fact, line 11 shows Player 4 invoking another FPP to steer the conversation, and the task focus, into a different area.

Pauses and silences were also noted to show interesting effects. For some modules, the information required by the expert would only appear intermittently. In the case of the following extract from Round 1, the module in question flashes a color or sequence of colors every few seconds:

Round 1—Player 1 (defuser) and Player 2 (expert)

1	Player 1	The first is red.
2	Player 2	Okay okay (0.8) so: red i:s blue=
3	Player 1	=blue okay?

Player 1 clicks button and waits for next sequence (2.6)

4	Player 2	okay?=
5	Player 1	=okay the color is blue=
6	Player 2	=blue is red.
7	Player 1	°oka.

Player 1 clicks button and waits for next sequence (3.5)

8	Player 2	.hh hello the <u>next</u> o[ne]
9	Player 1	[gr]een gr green sorry I ha
10		°to °t-- it's green (0.3) .hh <u>green</u>

In this instance, Player 2 was not aware that Player 1 was waiting for the next color to appear, resulting in short but perceptible periods of silence. Player 2, however, is conscious of the limited time, and so in lines 4 and 8 tries to re-engage Player 1 (despite Player 1 already being fully engaged). Previous studies have noted that situations evoking higher levels of stress usually result in shorter silences between turns (see, for example, Jaffe and Feldstein 1970), and it appears that Player 2 follows this pattern, viewing extended silences as halts in communication and a threat to successful task completion. Nonetheless, the silences were necessary to successfully complete this task.

Conversely, in Round 8, silence is viewed in a different way between the players when completing a module consisting of four buttons with symbols that had to be pressed in a specific order:

Round 8—Player 3 (defuser) and Player 4 (expert)

1	Player 4	you should ha:ve symbols in front o[f you]
2	Player 3	[yep I] can
3		see four symbols=
4	Player 4	=great. what do they look like.
5	Player 3	er:m .hh okay there's a backwards L (0.4) a: W
6		thing (0.5) copyright symbol and then star.

Player 4 looks over possible combinations (5.2)

7	Player 4	o::kay got it (0.4) click on copyright the:n W
8		thing then L thing then star

Player 3 selects the buttons in the order given (1.6)

9	Player 3	okay done brill[iant]
10	Player 4	[yes:] <u>nice</u> one

The number of silences seen in this extract reflects a common pattern seen throughout successful rounds. Unlike the extract from Round 1, the silences between turns (notably from line 5 to line 9) allowed the players to perform the necessary steps to successfully complete the module and, eventually, defuse the bomb. Although the bomb was counting down at the same rate as that of the extract from Round 1, a "calmer" approach that allowed for silences was common in successful rounds. This echoes what was noted by Sexton and Helmreich (2000): superfluous speech is understood to impede task success, so understanding these "longer" pauses as necessary for tasks to be completed, rather than instances to be filled with extra linguistic information, appears to increase the likelihood of successful task completion.

In addition, this extract from Round 8 shows little cross talk and interruption. A further distinction between successful rounds and unsuccessful rounds may be observed when considering interruption, as exemplified in Round 10:

Round 10—Player 4 (defuser) and Player 5 (expert)		
1	Player 4	so there's a number with four numbers below it.
2	Player 5	ri:ght so what do[es the fir]--
3	Player 4	[and the bi]g one is a two=
4	Player 5	=I I
5		was about to a:sk y[ou that so y]--
6	Player 4	[yeah it's a]tw a tw[o]
7	Player 5	[a] two
8		right?
	(4.6)	
9	Player 4	so: what now.
10	Player 5	well is it a two o[r not]
11	Player 4	[yes y]es it's a two=
12	Player 5	=okay er:
13		so you nee:d to pr[ess]--
14	Player 4	[yea]h press which one time's
15		running out

In this extract, multiple overlaps and interruptions are observed, creating an impasse after line 8: Player 5 wishes to know the response to their question, which is asked (partially or fully) in lines 2, 7, and 8, and despite Player 4 giving this information in lines 3 and 6, Player 2 still desires clarification. As such, after line 8, Player 4 is expecting the next instruction, while Player 5 awaits clarification. After nearly 5 seconds, Player 4 breaks the silence and tries to resolve the issue. However, the issue persists in the remainder of the extract, including Player 4's interruption and more forceful expression of "which one" in line 14.

This brief CA analysis suggests that there are conversational patterns in (un)successful rounds. First, there was a higher likelihood of success when adjacency pairs were completed appropriately (see Schegloff 2007), regardless of any expansions that were added before, in between, or after the pair. Secondly, allowing for periods of silence between turns correlated with more successful rounds. If these were interrupted by a repeated request, this resulted in wasted time and a generally more "panicked" approach. Finally, when players allowed each other to complete their turns with little or no interruption, there was greater task success. Of course, no interaction was free of interruption or cross talk—as is the nature of spoken language (Heldner and Edlund 2010)—but there were far fewer occurrences of these in successful rounds.

5 Conclusions and further study

In videogames such as *Keep Talking and Nobody Explodes* (2015), players have no choice but to collaborate effectively if they wish to be successful. Given the added constraint of being able to view only half of the overall information and the addition of a variable time limit, the reliance on effective verbal communication increases dramatically. Such environments are not unlike other "real world" environments (e.g., air-traffic control; see Nevile 2001), although the consequences of miscommunication in each differ markedly.

From this preliminary (albeit limited) study, several linguistic features may be suggested as markers of collaborative language, if collaboration is understood to correlate with successful task completion. From an SFL perspective, the use of "full" clauses (as opposed to elliptical and abandoned clauses) has a stronger association with successful task completion. From a CA perspective, the completion of adjacency pairs, allowing for pauses between turns, and fewer interruptions were all observed more frequently in successful rounds. Overall, despite a mixture of a short time limit, penalties for incorrect responses, and deliberate difficulty in collaboration, success occurred when time was taken over communication and the relative "stress" of the situation was ignored (cf. Jaffe and Feldstein 1970). The language of collaboration therefore appears to be at its most effective when ambiguity is low and when turns are taken in a logical and nonoverlapping manner. When clauses were elided and/or abandoned, and periods of silence were viewed as detrimental rather than necessary, the likelihood of success dropped, suggesting that these features are uncollaborative.

Nonetheless, there are likely other elements within communication that "fly under the radar" of both SFL and CA analyses. For instance, questions may be posed regarding the balance of power in these interactions, such as in Round 1, wherein Player 1 uses quieter speech and apologies while Player 2 appears to "dominate" with louder speech and commands.[1] Observations of other collaborative games with specific short tasks and time limits, or games with longer tasks wherein in-depth strategies are required, would also be beneficial to observe and compare with the findings presented in this chapter. However, it will need to be borne in mind that collaborative games may use text chat rather than vocal chat, adding another level of complexity to turn-taking in these environments.

Furthermore, recorded play throughs of modified versions of *Keep Talking and Nobody Explodes* are accessible on various websites, wherein multiple experts assist one defuser to disarm bombs consisting of numerous high-difficulty modules in 10 minutes. Analyzing the complexity of such multiplayer communication would undoubtedly prove interesting, and more extensive studies will be needed to further corroborate, qualify, and/or finesse the observations made in this study.

Notation conventions

[and]	- points where speech overlap begin and end
=	- speech between participants without a gap
.	- silence of less than 0.3 seconds
(0.0)	- amount of silence in seconds
?	- rising intonation (not necessarily a question)
:	- extension of preceding phoneme
--	- location of abandoned clause
.hh	- audible exhalation
underlined	- word(s) pronounced noticeably more forcefully
°	- following word pronounced noticeably less forcefully

Note

1 It is noted (e.g., Fairclough, 1995: 23) that CA is not suited to or indeed "resistant to linking properties of talk with higher-level features of society and culture [including] relations of power." However, Hutchby (1999) contends this fact in his various works. As such, observing power relations via CA should not be completely dismissed.

References

Beehr, T.A. and T.M. Franz (1987), "The current debate about the meaning of job stress," in J.M. Ivancevich and D.C. Ganster (eds.), *Job Stress from Theory to Suggestion*, 5–18, New York: Haworth Press.

Butler, C.S. (2003), *Structure and Function: A Guide to Three Major Structural-Functional Theories*, Vol. 1, Amsterdam: John Benjamins.

Clark, H.H. and D. Wilkes-Gibbs (1990), "Referring as a collaborative process," in P.R. Cohen, J. Morgan, and M. E. Pollack (eds.), *Intentions in Communication*, 463–93, Cambridge, MA: Massachusetts Institute of Technology.

Council of Europe (2001), *Common European Framework of Reference for Languages: Learning, Teaching, Assessment*, Cambridge: Cambridge University Press.

Coupland, J. (2003), "Small talk: Social functions," *Research on Language and Social Interaction*, 36 (1): 1–6.

Ducheneaut, N. and R.J. Moore (2004), "The social side of gaming: A study of interaction patterns in a massively multiplayer online game," in *Proceedings of the 2004 ACM Conference on Computer Supported Cooperative Work*, 360–69, New York: ACM Press.

Egenfeldt-Nielsen, S., J.H. Smith, and S.P. Tosca (2016), *Understanding Video Games: The Essential Introduction*, 3rd edn., Oxford: Routledge.

Eggins, S. (2004), *An Introduction to Systemic Functional Linguistics*, 2nd edn., London: Continuum.

Fairclough, N. (1995), *Media Discourse*, London: Edward Arnold.

Fiksdal, S. (2001), "Prosody in the study of conversation," in A. Wennerstrom (ed.), *The Music of Everyday Speech: Prosody and Discourse Analysis*, 166–99, Oxford: Oxford University Press.

Firth, J.R. (1935), "The technique of semantics," *Transactions of the Philological Society*, 1935: 36–72.

Garcia, A.C. (2013), *An Introduction to Interaction: Understanding Talk in Formal and Informal Settings*, London: Bloomsbury.

Garfinkel, H. (1967), *Studies in Ethnomethodology*, Englewood Cliffs: Prentice-Hall.

Halliday, M.A.K. (1978), *Language as a Social Semiotic: The Social Interpretation of Language and Meaning*, London: Edward Arnold.

Halliday, M.A.K. and R. Hasan (1976), *Cohesion in English*, London: Longman.

Halliday, M.A.K. and C.M.I.M. Matthiessen (2014), *Halliday's Introduction to Functional Grammar*, 4th edn., Oxford: Routledge.

Hasan, R. (2014), "Towards a paradigmatic description of context: Systems, metafunctions, and semantics," *Functional Linguistics*, 2 (9): 1–54.

Heldner, M. and J. Edlund (2010), "Pauses, gaps and overlaps in conversations," *Journal of Phonetics*, 38 (4): 555–68.

Hutchby, I. (1999), "Beyond agnosticism? Conversation analysis and the sociological agenda," *Research on Language and Social Interaction*, 32 (1–2): 85–93.

Jacobs, G.M. and C.S. Ward (2000), "Analysing student–student interaction from cooperative learning and systemic functional perspectives," *Electronic Journal of Science Education*, 4 (4): 1–28.

Jaffe, J. and S. Feldstein (1970), *Rhythms of Dialogue*, New York: Academic Press.

Keep Talking and Nobody Explodes (2015), *Videogame*, Ottawa, ON: Steel Crate Games. Available online: www.keeptalkinggame.com (accessed February 20, 2018).

Khawaja, M.A., F. Chen, and N. Marcus (2012), "Analysis of collaborative communication for linguistic cues of cognitive load," *Human Factors*, 54 (4): 518–29.

Krifka, M., S. Martens, and F. Schwarz (2004), "Linguistic factors," in R. Dietrich and T. M. Childress (eds.), *Group Interaction in High Risk Environments*, 75–86, Farnham: Ashgate Publishing.

McHugh, M.L. (2012), "Interrater reliability: The kappa statistic," *Biochemia Medica*, 22 (3): 276–82.

McKendrick, R., T. Shaw, E. de Visser, H. Saqer, B. Kidwell, and R. Parasuraman (2014), "Team performance in networked supervisory control of unmanned air vehicles: Effects of automation, working memory, and communication content," *Human Factors: The Journal of the Human Factors and Ergonomics Society*, 56 (3): 463–75.

Nevile, M. (2001), "Understanding who's who in the airline cockpit: Pilots' pronominal choices and cockpit roles," in A. McHoul and M. Rapley (eds.), *How to Analyse Talk in Institutional Settings*, 57–71, New York: Continuum Press.

Orasanu, J.M. and E. Salas (1993), "Team decision making in complex environments," in G.A. Klein and J.M. Orasanu (eds.), *Decision Making in Action: Models and Methods*, 327–45, Westport: Ablex.

Sacks, H., E. Schegloff, and G. Jefferson (1974), "A simplest systematics for the organization of turn-taking for conversation," *Language*, 50 (4): 696–735.

Schegloff, E. (2007), *Sequence Organization in Interaction: A Primer in Conversation Analysis*, Cambridge: Cambridge University Press.

Schegloff, E.A. (1996), "Confirming allusions: Towards an empirical account of action," *American Journal of Sociology*, 104 (1): 161–216.

Sexton, J.B. and R.L. Helmreich (2000), "Analyzing cockpit communications: The links between language, performance, error, and workload," *Journal of Human Performance in Extreme Environments*, 5 (1): 63–68.

Ten Have, P. (2007), *Doing Conversation Analysis*, 2nd edn., London: Sage.

9

"Watch the Potty Mouth"

Negotiating Impoliteness in Online Gaming

Sage L. Graham and Scott Dutt

1 Introduction

Gaming is a highly prevalent and multicultural mode of interaction in first-world countries and is growing exponentially as more young people turn to gaming as a primary setting for interaction and community. As Brenden Maher (2016) puts it, "By the age of 21, the average young gamer will have logged thousands of hours playing time. That fact alone makes dichotomies such as 'digital world' and 'real world' ring false—for many, game-playing is the real world." And while many associate gaming with recreation, it is also increasingly used in other ways—as a teaching tool in the classroom, an assessment tool in business, and an evaluation/diagnostic tool in certain medical contexts. As gaming takes a more prominent place in a wider variety of situations, examination of this ready-made environment for establishing social skills and connections is warranted. As Newon argues, "[Since gaming] may no longer be considered a niche interest held by a small minority, but rather a popular and pervasive activity undertaken by people of all ages, genders and geographies, it is important to understand how gaming and digital media intersect with people's everyday lives" (Newon 2016: 289).

Despite (or perhaps because of) the widespread popularity of online games, it is not uncommon for people to associate these games (particularly those played for entertainment or recreation) with violence and inappropriate behavior. Impoliteness in the form of flaming, spamming, trolling, and cyber-bullying is, in fact, a "hot topic" of discussion, particularly among parents, who are concerned that online gaming promotes violent behavior among their children. For others, however, impoliteness is simply a normal component of online interaction and is something that should be ignored. Flaming and trolling,[1] for example, are commonly understood as negative behaviors, as evidenced by the frequency of discussion on "how to deal with trolls and flaming" in many digital environments, and yet the definitions of what constitutes these behaviors (and tolerance for them) can vary widely.

In multicultural environments like online games, navigating the norms and expectations for appropriate/polite behavior is particularly tricky, since differing expectations for what counts as (im)polite, appropriate/ inappropriate, and/or a violation of the "rules" may differ across groups. It is therefore not surprising that many online spaces create FAQs or "Codes of Conduct" that are meant to serve as guides for appropriate behavior. In digital interaction, this codification of how one should behave includes an appeal to a "moral order" (Kádár and Haugh 2013) that establishes a set of ideal behaviors and places them as a "yardstick" by which behavior in that community will be measured. Creating codes of conduct that will be (a) understood universally by people with widely different cultural backgrounds and (b) enforceable is no small enterprise, however. It often results in breakdowns of communication and/or debate about what kinds of behavior will or will not be tolerated. As our experience in more (and more complicated) digital venues grows, it will be increasingly important to examine rules imposed in digital environments and communities that are meant to identify and mediate negative behaviors, while at the same time making them accessible to people with different cultural and social backgrounds.

2 Previous research

2.1 Online gaming

While originally text-based, online games have evolved into complex multimodal platforms for interaction that often include visual elements, audio, and simultaneous text-based synchronous chat. Perhaps because of this complexity, coupled with their rise in popularity, online gaming has become an increasingly frequent research focus across disciplines as far ranging as computer science, communication studies, sociology and anthropology, art and visual design, and business. Early research on gaming explored it as a potential tool for establishing cultural and cognitive literacy (e.g., de Freitas and Maharg 2011; Gee 2003, Gee 2015; Guimaraes 2005; Jenkins 2006; Prensky 2001), especially in language-learning environments (e.g., Thorne 2008; Thorne and Fischer 2012). While this focus on gaming as a cognitive and teaching tool is certainly valuable, the undeniably large presence of gaming outside the classroom should not be neglected. Some researchers have therefore turned their attention to the social aspects of online gaming, particularly with regard to constructing digital identities and communities (e.g., Nardi 2010; Newon 2011, 2016; Pearce 2011; Taylor 2006), reflecting cultural and literary narratives (e.g., Ensslin 2012, 2014), and negotiating game conflict talk (e.g., Wright et al. 2002).

In particular, team-oriented games such as MMOGs have been a common area of investigation, largely because they are (a) among the best-selling and most popular games, and (b) because they create an environment that brings large numbers of people with different backgrounds and cultural expectations together around a common enterprise.[2] The second feature is of particular interest here, since successfully working with others to achieve goals within the game is dependent upon rules for interaction and collaboration that a gaming community develops in order to coherently accomplish objectives.

Notions of (im)politeness are a critical part of this type of collaborative interaction. Graham (2015), for example, argues that impoliteness is an important (and possibly required) element in achieving status within an online community as a "core" group member. There is, however, little research examining (im)politeness in gaming contexts (although see Ensslin [2012] for a discussion of subverting politeness in gaming), and no research to our knowledge on rules of conduct and overt impoliteness as manifested in online gaming practice.

2.2 Digital (im)politeness and rules for behavior

As noted here, while the developing complexity of online gaming can provide a richer experience for participants, it can also make adhering to social expectations more difficult, since users must juggle not only multiple technical constraints, but often multiple audiences (each with their own expectations of what counts as appropriate behavior). In many of the most popular games, such as *League of Legends* (Riot Games 2009) and *Overwatch* (Blizzard Entertainment 2016), players control avatars that attempt to achieve specific goals in-game, either alone or through collaboration with other players. This collaboration, however, relies on some common understanding of what constitutes effective and appropriate interaction so that the members of the team can achieve the goals at hand.

Virginia Shea's 1994 book *Netiquette* outlines some basic "polite" behaviors for communicating online that include respecting others, being forgiving, and "remembering the human," which are echoed in one form or another in most guidelines for online behavior (1994: 35). These rules seem to be simple common-sense and would presumably facilitate effective interaction, but negative, impolite, threatening, and sometimes harassing behavior are common in digital contexts. As digital interaction (including gaming) has become more popular, we have been flooded by rules for digital interaction (e.g., Terms of Acceptable Use, Codes of Conduct, FAQs, etc.). Creating these types of behavioral rules is no easy task, however. Locher and Watts' (2005) model of (im)politeness as "Relational Work" is helpful here, since it identifies different levels of problematic behaviors as either polite, politic (i.e., appropriate), nonpolitic (i.e., inappropriate for the given context), and impolite. According to this approach, most interactions are made up of politic utterances—those that are appropriate to the context and are therefore unmarked. Nonpolitic utterances are inappropriate to the context but do not indicate action that goes "over-and-above" what we might expect to occur in the given setting. Polite and impolite acts, on the other hand, fall outside the norms of appropriate (unmarked) behavior and reflect overt behaviors (either positive or negative). For example, if a driver sees a large object in the roadway, it is appropriate and unmarked to change lanes to avoid hitting it. However, if the driver instead stops and moves the object out of the roadway to remove the hazard for others, it is an overt act that goes beyond what is necessarily expected and is therefore classified as polite. The same structure applies on the negative side of the spectrum, with impoliteness requiring overt negative action.

This distinction is important for any exploration of digital (im)politeness because norms of behavior can vary so greatly. Behavior that is seen as normal and unmarked in online gaming might be seen as highly problematic and impolite in a different digital community/setting. And as Graham and Hardaker (2017) observe, there are many intertwined elements that affect which behaviors will be viewed as nonpolitic/impolite. Yet despite the complexity, there are still common facets of digital interaction that affect how people enact and interpret acceptable behavior and (im)politeness— many of which align with Shea's (1994) initial model.

We see these elements reflected in FAQs and codes of conduct, but given the variability and subjectivity in classifying online negative behaviors like trolling, flaming, and online bullying, we might question how effective these rules are (or even whether they can be effective). As Graham (2008: 302–03) notes, for example, "While the FAQ brings guidelines to the table, the guidelines themselves are contradictory, placing the newbie in a difficult position—one where s/he must learn by trial and error what the norms of the community are."

Specifically within a gaming context, Graham (2017) examines the posted rules of online game streamers, noting that while rule-breaking is often punished, it is also in many cases encouraged by the streamers themselves. It is often inequitably enforced; some (in-group) participants can break the posted rules without consequences, while others are punished via timeouts and bans. Using a corpus of open chat postings and corresponding video from live streams hosted online at Twitch.tv (n.d.), this case study takes an interactional sociolinguistic approach in expanding this previous research by exploring the ways that rules are defined, communicated, and enforced. We will explore the interrelationship between rules and actual behavior to determine (1) how problematic behavior is identified in streamer rules, and (2) how it is addressed when it occurs.

3 Data and methods

In examining online gaming, we use data collected from Twitch, a live game-streaming platform where gamers broadcast themselves playing an array of games (single player or multiplayer). Twitch streams are highly multimodal—most allow viewers to (1) watch the gameplay as it unfolds, (2) hear audio of the gamer while s/he plays, (3) see a camera feed of the streamer while s/he plays, and (4) participate in an open chat forum (see Figure 9.1).

For the purposes of this study, we will focus on two multiplayer online team games. *Overwatch* (Blizzard Entertainment 2016) is a team-based first-person shooter which, in 2017, was the third most popular game streamed on Twitch (Twitchstats 2017). *League of Legends* (Riot Games 2009) is a multiplayer online battle arena (MOBA) that, at the time of this study, was the most popular game streamed. In each game, teams of five or six players are compiled (either through random server matchmaking or through player selection) and play together to defeat an opposing team of the same size. Each player chooses an avatar with a predetermined set of abilities, and teams are most often constructed so that these abilities will complement those of the other team members. By working cooperatively, teams can take advantage of the distribution of skills/abilities to complete the required tasks of the game.

In each game, three sets of rules are at play: those of Twitch, those of the games distributed by the game designers, and those that individual streamers create for chat interaction on their channels. Although there is some variation across the rules from these three sources, they tend to share some common denominators (i.e., racism, sexism, homophobia, hate speech, and spam are prohibited). These basic guidelines are frequently

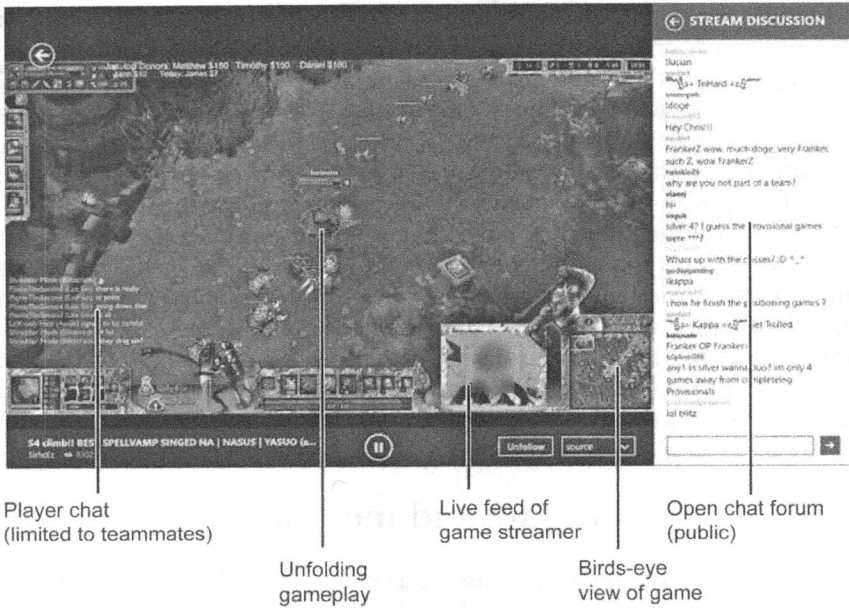

Player chat
(limited to teammates)

Unfolding
gameplay

Live feed of
game streamer

Birds-eye
view of game

Open chat forum
(public)

FIGURE 9.1 *Twitch stream screen layout using* League of Legends *as an example,* © *by Riot Games.*

supplemented with rules established by each individual streamer for his/her own channel. Streamers are responsible for enforcing both their rules and the rules of Twitch. To assist in this, streamers often appoint moderators and use automated programs (or *botmods*) to identify violations and mete out the appropriate punishments. These might include being blocked from participating for a specified amount of time, banned (i.e., blocked from participating until "unbanned" by a moderator), or "permabanned" (i.e., permanently banned) from Twitch.

Using screen-capture and text-logging software, chat postings were collected from the open chats of five streamers—1,000 postings each. All are professional streamers[3] and are recognized within the community as specializing in either *Overwatch* or *League of Legends*. Rules were collected from the profiles of all five streamers (two male and three female), compared to the concurrent rules on Twitch, and categorized based on ethnographic knowledge of the communities of practice to identify common themes. Punishments for problematic behavior (which were flagged automatically in the chat-log software) were tracked, analyzed, and coded based on the type of rule violation that triggered the punishment based on ethnographic knowledge of the communities of practice.

4 Analysis

4.1 Rules as multipurpose, multifaceted guidelines

As noted here, there are multilayered rules that govern behavior on Twitch—some established by the company itself, some suggested by the game developers, and some defined by the individual streamers. The Twitch rules for chat participants, which will be the focus of our investigation here, prohibit behavior that includes "Targeted harassment, threats, and violence against others" and "Hate speech or other harassment."

In May 2015, Twitch guidelines also specified the following:

> We're not going to tell you to watch the potty mouth, that's between you and your mother who doesn't approve anyway. What we will tell you, however, is that if you choose to use language or produce content that is racist, sexist, homophobic or falls under generally accepted guidelines for hate speech you will disappear from Twitch. (https://www.twitch.tv/p/rules-of-conduct)

These guidelines were amended in November 2015 with the statement, "Any content that promotes or encourages discrimination, harassment, or violence based on race, ethnicity, gender identity, sexual orientation, age, religion, or nationality is prohibited." The guidelines also prohibit "spam, scams and other malicious content." Streamers write rules that reflect Twitch policies and supplement them with rules specific to their stream's norms and expectations. Rules are most frequently posted on the streamer's Twitch profile, which appears below the video feed (although it is increasingly the case that visitors must check a box explicitly accepting the rules before being allowed to participate). It is also important to note that, while many streamers post rules for their streams, the methods for interpreting, enforcing, and/or ignoring the rules can differ widely and are influenced by factors such numbers of viewers and subscription rates. All of the streamers in the current study (Fenn3r [m], Ceweina [f], LoserFruit [f], VesperVonDoom [f], and Wingsofdeath [m]) posted rules on their Twitch profile pages.

Each rule set encourages creating an environment that is instructive and/or friendly for participants. In each case, the rules also all address what counts as inappropriate behavior. The data analysis revealed five (sometimes overlapping) categories of rules (see Table 9.1). Each category addresses different types of problematic behaviors.

It is important to note that these categories are nonexclusive; the same post could include multiple rule violations. Since the categories are based on subjective constructs (e.g., what constitutes aggressive language), a

TABLE 9.1 *Types of rules*

Category	Description	Features of violations	Examples
(1) Legalistic	Rules that reference legally actionable behavior	— Aggressive — Harassing	— Prejudicial language — Threats — Racist, Sexist, Homophobic language
(2) Adversarial	Rules that address antagonistic behavior	— Overtly impolite — Critical	— Name-calling — Insults — Overt sexual objectification — Obscenity — Disregarding others
(3) Authoritative	Rules that address challenges to the streamer's power or expertise	— Antagonistic — Oppositional	— Backseat gaming — Orders/commands to streamer — Unsolicited advice
(4) Disruptive	Rules prohibiting behavior that is deemed disruptive	— (Non-)politic — (In)appropriate — Distracting — Overly long	— Excessive emojis — All caps — Copypasta — Links — Foreign languages
(5) Collaborative	Rules that encourage collaboration	— Politic	— Unmarked participation

particular utterance might be classified differently depending on the context. Our focus in the next sections, however, is the rules themselves and how they index particular behaviors.

4.1.1 Category 1: Re-enforcing legality

In the first category, all of the streamers included instructions about discrimination and hate speech as seen in the following examples:

- "No racism" (Wingsofdeath)
- "Please refrain from insults or sexual harassment" (Ceweina)
- "Incendiary comments about race, gender or sexual preference, and bullying in general, are unwelcome here. You could be banned if we feel you've gone too far" (Fenn3r)

The rules in Category 1 attempt to address the prejudicial, threatening, harassing, and aggressive behavior that was described in the Twitch policy discussed here, which prohibits "any content that promotes or encourages discrimination, harassment, or violence based on race, ethnicity, gender identity, sexual orientation, age, religion, or nationality." The language used in the streamers' individual rules mirrors this phrasing and lexicon by using the terms "racism" (Wingsofdeath) and "sexual harassment" (Ceweina), and using strings of words that index demographic categories ("race, gender, sexual preference" [Fenn3r]). These patterns are often seen in nondiscrimination and antiharassment policies in political and corporate discourse (particularly in the United States) as well. The inclusion of terms found in legal documents, such as *racism, sexism,* and *harassment,* give these rules an impression of severity.

This is heightened by the fact that violation of antiharassment/ discrimination policies can have grave "real world" ramifications that include criminal consequences, loss of employment, or financial penalty. Combining linguistic elements from both Twitch policy and broader legal discourse, these streamers establish an undercurrent of "real life" ramifications. In these instances, the rules that reference racism and sexism must be evaluated in the context of the stream (i.e., what counts as racism/sexism/homophobia on Twitch), but they are also anchored in social consciousness away from the keyboard and carry a severity that violating other types of rules (like typing in ALL CAPS) does not have.

4.1.2 Category 2: Preventing an adversarial environment

The second category of rules is meant to prevent an adversarial environment. While Category 1 rules index demographic groups of people, Category 2 rules, in contrast, address behaviors that target individuals. In these cases, streamers present problematic behavior in terms that are more specific to their streams:

- "Don't be a dick" (Wingsofdeath)
- "If you don't have anything nice to say then don't say it at all" (VesperVonDoom)
- "Keep it chill" (Fenn3r)
- "Respect the mods and each other" (LoserFruit)
- "No weird trolling" (Ceweina)

These rules are less "legalistic" than those in Category 1 in the sense that the Category 1 rules might have legal consequences, whereas there is no *law* against being "a dick." Category 2 rules also have no benchmarks (even imperfect ones as in the first set of rules) for how problematic behaviors might be identified. The notion of "keeping it chill," for example, requires a great deal of subjective judgment on the part of the participant and/or moderator to determine what counts as chill (or *not* chill). Similarly, participants who want to strictly follow the rules in Ceweina's stream would have to evaluate what constitutes *weird* trolling (as opposed to nonweird trolling—or trolling at all for that matter), or what constitutes saying something nice (as opposed to not-nice) in Fenn3r's stream. These rules reflect an attempt to protect the stream environment and make it a space where there is cooperative interaction—as Fenn3r puts it, to "promote mutual respect among the viewers, and a pleasant and relaxed environment for everyone," but their vagueness has the potential to create misunderstanding.

4.1.3 Category 3: Maintaining authority

A third type of rules is also prevalent: a prohibition against challenging the streamers' gaming abilities or power to regulate his/her stream. Examples include the following:

- "No backseat gaming of any kind" (Wingsofdeath)
- "Don't tell me how to run my stream" (Wingsofdeath)
- "Don't suggest games, unless I ask" (LoserFruit)
- "NO BACKSEAT GAMING!" (VesperVonDoom)
- "Don't tell me to play the game" (VesperVonDoom)

Rather than address conduct that is discriminatory (Category 1) or insulting/impolite (Category 2), these types of rules relate to the autonomy and power of the streamer to regulate his/her stream, including how (or whether) that authority can be challenged. Unlike the second category, which addressed how participants collaboratively communicate with one another (e.g., "respect the mods and each other"), rules in this group focus on the streamer's ultimate control in crafting the overall content and tone of the stream. This is consistent with Grimes and Feenberg (2009) and Ensslin (2012), who note that challenging social order and discursive power is a norm in gaming. The existence of these rules, in fact, reaffirms this norm, since if the challenges were not present, there would be no need to have rules to control them.

4.1.4 Category 4: Regulating disruption

Fostering the "relaxed environment" outlined in Category 2 rules, streamers also include guidelines that minimize disruptions to the flow of the chat. Rules that fall into this category are often predictive in nature and can therefore be preprogrammed and enforced by automated programs (botmods) as well as human moderators. Using the botmods as a resource requires identifying rule-breaks in an objective, concrete way; examples include repetitive messages,

messages with excessive emojis, and messages that include hyperlinks. Each streamer has the option to activate a botmod (some of which are provided preprogrammed by Twitch) to identify specific disruptive characteristics of posts and respond.

On Twitch, some default regulations programmed into botmods are:

- No hyperlinks—hyperlinks in chat can be misleading or destructive (e.g., phishing[4]) and put participants at risk or lead them away from the chat interaction.

- No messages with more than eight emojis—long sequences of emojis take up multiple lines of the chat stream and can therefore be distracting or take up enough lines of chat to disrupt the flow of conversation.

- No messages in all caps—while fine for occasional emphasis, all caps are perceived as shouting and are therefore too extreme for normal chat posts.

- No copypasta[5]—as with excessive use of emojis, copypasta can take up visually available space in the chat window and, if posted in quick succession, could interfere with the visibility of other messages.

- No posts in foreign languages—while there are streams conducted in all languages around the world, most streamers use only one or two languages in their streams. It is frequent practice to limit discussion to those languages since (a) if mods do not speak the language in which posts are written, they cannot screen them for problematic content, and (b) posts in nonstream languages exclude any member of the conversation who does not speak that language.

By nature, these botmod rules are the most objective, but they also offer limitations in enforcement due to the difficulty in distinguishing behavior that is disruptive (which is often highly context-dependent) from that which is merely community-specific or part of the "discourses of fun" that exist in online gaming (Ensslin 2012: 110).

Spam is one manifestation of disruption that is frequently referenced in rules (by both Twitch and many streamers), but, like the vague notions of creating a "positive environment" outlined here, spam is equally difficult to define. Heyd (2013), for example, examines spam as one type of email hoax in which unsolicited email is distributed for profit—a practice with quite negative associations. In contrast, in online gaming, spam is treated more like a collaborative game than as an aggressive move. Fenn3r's rules, for example, say

No spam. Don't spam links and don't spam questions/statements. Copypasta is welcome unless you're purposefully spamming without any real occasion to do so. You will be timed out if your spam is considered disruptive.

In this case, Fenn3r "welcomes copypasta" (which is often explicitly identified in rules as spam) and uses *intentionality* as the measure of what is acceptable. Wingsofdeath, whose rules prohibit "one-man spam," bases his definition on whether initial spam posts are taken up and repeated by other participants (in which case reposting becomes a collaborative game). As is the case with the notion of "keeping it chill," spam must be assessed in context and as part of an emergent interaction so that it can be identified when it is disruptive and appropriately controlled.

4.1.5 Category 5: Bolstering the in-group

Finally, rules in a Category 5 index emotion, whimsy, or insider knowledge of the stream community.

- "Love the alpaca and be one with it" (Ceweina)
- "Don't fall in love with me" (VesperVonDoom)
- "Be awesome" (LoserFruit)

While the other sets of rules govern the content of the chat postings, this type of rule refers to the participants' experience beyond the scope of the chat room. In saying, "Don't fall in love with me," for example, Vesper is giving instructions not about what chatters say within the chat, but how they feel in a more global sense. Similarly, LoserFruit's instruction to "Be awesome" references a state that potentially extends beyond her stream. In her instruction to "Love the alpaca and be one with it," Ceweina refers to in-group information by referencing the mascot/logo of her stream (which she has nicknamed the "Alpaca Kingdom"). By including this insider reference in her rules, she provides a playful reference that taps into and recognizes those who have the knowledge and longevity in the community to understand the "inside joke."

These instructions prescribe an emotional state that bridges the online community with life "away from the keyboard" and give guidelines for behavior that is appropriate or politic rather than attempting to control behavior that is impolite or aggressive. While "packaged" as serious rules, these instructions define an atmosphere that is more consistent with the play that is inherent in games through humor ("be one with the alpaca") and support ("be awesome"). In this sense, they are consistent with Ensslin's discourses of "cool" and "fun" that index in-group knowledge and align with "an important pragmatic presupposition underlying metaludic discourse" which is that "gameplay is meant primarily to entertain and please" (2012: 110).

We see in these sets of rules, then, a breadth of purpose. The rules that mirror those provided by Twitch allow streamers to bridge the corporate requirements of the streaming platform (and the larger expectations of a "politically correct society") with their individual interpretations of what constitutes problematic conduct (e.g., as manifested in their instructions to "keep it chill"). They also establish their authority in their streams (Category 3) while providing rules that bridge the world away from the keyboard with what happens in the chat. Overall, these categories encapsulate rules that exist on a spectrum from the most severe (Category 1) to the least (Category 5), and, when combined, allow a comprehensive set of guidelines for behavior.

4.2 Sanctions for violations: Enforcing the rules

Despite the extensive guidelines laid out by the streamers, rule-breaking is an accepted and frequent facet of online gaming, and there are an array of possible consequences for rule violations. It is certainly the case that problematic behavior can be ignored or addressed in situ by members of a chat community, or streamers can warn participants verbally. Twitch also provides tools to prevent participants from posting to the chat, although they can still see the stream and the messages posted by others. As is the case with the categories of rules, punishments exist on a scale of severity. In the least severe, participants receive a predetermined length of timeout (usually ranging from 1 second to 10 minutes or more) that will automatically reinstate posting privileges after the specified amount of time has elapsed. In other cases, however, the suspension of privileges does not include a specified time limit, and the participant will remain banned until explicitly unbanned by the streamer or a (human) moderator. In extreme cases, Twitch may "permaban" users; in these cases, the users have their accounts suspended and are permanently banned from participating in all Twitch streams barring a successful appeal to the Twitch organization. In many cases, punishments are also progressive, beginning with a warning for the first violation, progressing to timeouts in increasing increments for subsequent violations, and possibly culminating in a ban (until reinstated by a human moderator).

In the data that is the focus of this study, streamers do not usually make a ready distinction between the terms for timeouts and banning. Fenn3r is the only streamer, in fact, to make a distinction in his rules, in which he specifies that disruptive spam will result in a timeout while "incendiary comments about race, gender, or sexual preference and bullying in general" will result in a ban. For the purposes of this paper, we use the term *ban* to indicate being barred from participating for an undetermined length of time (i.e., until further notice). We separate punishments that block

TABLE 9.2 *Total sanctions by rule category*

	Warnings (1–10s)	Timeouts (10 + min)	Bans (Indefinite)	Total sanctions
1. Legalistic	0	2	0	2
2. Adversarial	0	9	3	12
3. Authoritative	0	0	0	0
4. Disruptive	12	4	3	19
5. Collaborative	0	0	0	0
Total	12	15	6	33

users from posting using a time limit into two categories: *warnings* have a duration of 10 seconds or less, and *timeouts* have a duration of 10 minutes or more.[6]

Although there were hundreds of posts that could have conceivably been classified as rule violations, in the 5,000 chat posts that comprise this data set, there were only 33 cases where punishments were levied. The relatively small number of punishments compared to the number of violations, therefore, does not necessarily reflect a congenial environment, but instead indicates that procedural punishment for offenses is relatively rare. Four of the five streamers imposed at least one sanction, although the majority of punishments (22/33) were issued by one streamer (LoserFruit). Distribution of the sanctions across the rule categories is shown in Table 9.2.

While each category addresses a different facet of what it takes to create an ideal stream, there is uneven distribution of occurrence across the categories. In our data, for example, there were no punishments issued for violations of rule Categories 3 or 5—all punishments were issued for violations of rules in Categories 1, 2, and 4, which will be discussed in turn.

4.2.1 Category 1 violations—Harassment and hate speech

In our data, there were only two violations of this rule category. As noted here, Category 1 rules address the most objectionable behaviors and index legalistic policies based on prejudice, aggression, and hate speech. It is no surprise, therefore, that violations of these rules do not receive warnings; moderators instead go straight to timeouts and bans (minimum 10 minutes) when these rules are broken. The first Category 1 violation occurred when a participant said "fk ur gay" after multiple participants said that they play the game *HotS* (*Heroes of the Storm*) (Blizzard Entertainment 2015). The comment resulted in a 10-minute timeout. While a 10-minute block may seem minor for an aggressive and impolite act of hate speech, this length provides benefits (to the punishers) that a longer timeout or ban would not. Participants receiving a longer sanction would likely leave the chat altogether, while participants who receive a shorter block are more likely to remain as a viewer until the timeout expires and they can participate again. In our data, there were cases where the streamer and participants used this time to make fun of the offender—reinforcing their rapport and adding a layer of punishment (public ridicule) that makes this sanction more severe than the 10-minute timeframe might imply.

The more severe punishment in this category occurred when a moderator issued an 8-hour and 4-minute timeout after a participant criticized a streamer's ethnicity.

Example 1:

217	wlsgus6355)	fucking mix blood chainise brodcasting go to hell got it?
218	Erika_Miss_America	been here D:
219	ElMexicanRanger	cmonBruh
220		BAN: wlsgus6355 (28800s)
221		BAN: wlsgus6355 (242s)

Given the offensiveness of the racial comment, we might ask why the punishment was not worse. There are multiple factors at play, but we speculate that the moderator's (self-declared) inexperience may have meant that s/he didn't know how to issue a full ban (and so, instead, issued a lengthy timeout). This interpretation is based in part on the unusual duration of 8 hours and 4 minutes. Regardless, while this punishment is not a *ban*, the 8-hour timeout was enough to significantly curtail behavior because the stream ended well before the 8-hour timeout expired. Therefore, wlsgus6355 was unable to participate again until the next streaming session (the next day), which counts as a fairly severe punishment in this community of practice (CofP).

4.2.2 *Category 2 violations—If you can't say something nice...*

Category 2 violations most often took the form of personal attacks, but they didn't necessarily represent what would be classified as hate speech since they didn't index a particular demographic group as a whole. The messages that broke these rules were overtly impolite by insulting or showing a lack of consideration for others. Examples in our data include name-calling, overt sexual objectification, and insults related to either physical appearance or gaming skill. Just as in the case of Category 1 rules, violations in this category did not receive any warnings—only timeouts and bans. This again is consistent with the expectations that more severe violations would elicit more severe punishments.

Most violations of Category 2 rules received timeouts of 10 minutes, as seen in Examples 2 and 3:

Example 2:

<theragingbrit420gamer>	WTF SHE HAS A BOYFRIEND????
	UNSUBBED/UNFOLLOWED FUCK U

Example 3:

<tightcomecloser>	looking for tramps like you

In each of these cases, there was a personal attack aimed at the streamer—telling the streamer "FUCK U" in Example 2 and calling her a tramp in Example 3—which resulted in a timeout of 10 minutes. While both of these posts refer in some way to gender or sexual behavior, they do not fall under Category 1 violations for two reasons. First, Category 1 rules attempt to control aggressive or discriminatory behavior toward a demographic category of people (Muslims, women, people over 65, etc.). "Tramps" or "women with boyfriends" do not constitute such a category and therefore are better classified as personal insults directed at an individual (which is consistent with Category 2). Secondly, with regard to Example 2, we would argue that it is not inherently sexist to comment on this streamer having a boyfriend. The statement does not imply that the streamer is inferior because she has a boyfriend; it indicates that <theragingbrit420gamer> has chosen to withdraw his/her subscription to her stream (which entails withdrawing a financial contribution). The implication is that the poster is upset because the streamer is not available for a romantic relationship and therefore does not want to invest in her by donating to her stream. This would be equivalent to a person in a pub choosing not to buy a drink for someone who s/he initially perceived as a potential partner if it were revealed that that person was not available.

We would argue that refusing to give someone else a gift (e.g., a cocktail) in such a situation would not be considered as discriminatory or harassing, since the person in the pub is not entitled to receive free drinks. While the obscenity in Example 2 is potentially offensive, since there is no entitlement and the poster is not in a position of authority over the streamer, we cannot claim that these statements "encourage discrimination, harassment or violence" against the streamer as the Twitch rules and the rules in Category 1 indicate.

While the "letter-of-the-law" dictates classifying these specific examples as Category 2 violations, we would also concede that Examples 2 and 3 reflect a larger systemic bias against females within Twitch and the gaming community in general. In these contexts, female power and authority are compromised by pressure to accommodate sexualized roles and accommodate inequitable gendered practices to be successful. This systemic gender bias (for further discussion, see Graham 2018 and Salter 2017) should not in any way be dismissed, but in-depth examination of this is beyond the scope of this chapter.

Within the Category 2 violations, there were two instances where participants were issued full bans (i.e., of an undetermined duration) after insulting the streamer's gameplay, saying, for example, "you can't win because you *be lame* your team" (emphasis added). This post contains a double insult—indicating that the streamer is lame, *and* that he blames his team for it. In another instance, a viewer received an 8-hour timeout after insulting the whole chat community by saying, "this chat is so aids lol." It is no surprise that violations of Categories 1 and 2 (those that are meant to address the most objectionable behaviors) result in severe punishments rather than warnings. What is more surprising is the range of punishments of Category 4 violations, as will now be discussed.

4.2.3 Category 4 violations—Rules are made to be broken?

As noted in Section 4.1.1, Category 4 rules are intended to address behaviors that are seen as (in)appropriate or (non)politic and are frequently identified by participants as annoying or irritating rather than aggressive or impolite. Since these behaviors are less severe than those addressed in Categories 1 and 2, we would expect the punishments to be less severe, but these violations fall less neatly into a clear pattern. While all of the warnings (10 seconds or less) in our data were issued for Category 4 violations, timeouts and bans were also given in this category. In these cases, the punishment does not seem to fit the crime—reposting the same message multiple times or posting more than eight emojis in a single message seems minor compared to aggressive hate speech or overtly impolite insults (as seen in Examples 1–3). We might ask, then, "Why do these relatively minor infractions receive the whole spectrum of punishments rather than being limited to the least severe ones?" Our data show two possible explanations for this disconnect: (1) what we call *progressive punishments* and (2) the community-building function of spam.

The first case is related to the mechanisms by which disruptive behaviors are punished. These rules are often predictive in nature, so disruptive behaviors can be identified ahead of time and preprogrammed into a botmod that will handle violations automatically. It is simple to program a botmod to automatically delete any message that contains more than eight emojis, for example, and streamers frequently rely on botmods to help ensure that blocky messages like this do not take up all the space in their chat streams (which, as noted here, could interfere with other users' ability to participate).

Botmods are often also programmed to increase punishments for each subsequent infraction. So while a user might initially receive a 1-second warning for posting a message in all capital letters, a subsequent message with a violation by the same user might receive an increased timeout of 10 minutes—not because of severity, but because of *frequency*. In the data for this study, all examples of Category 4 violations that received more severe punishments received less severe punishments first. One illustrative case can be seen in Example 4:

Example 4:

<ELENA7676516321>	BEST SEX SITE ▶ ▶ ▶ http: wbt.link/KpPsZ
	BAN: ELENA7676516321 (5s)
	ELENA7676516321 ⟶ No Links! [warning]
	BAN: ELENA7676516321 (10s)
	[1 message]
	BAN: ELENA7676516321

What is noteworthy here is the progression and sequence of punishments. This message elicits three simultaneous actions: a 5-second prohibition accompanied by a warning not to post hyperlinks, a 10-second warning, and a ban. The most likely explanation is that it contains multiple violations—posting links, posting messages in all caps, and restricting explicit content (in this case, a reference to sex)—each of which imposes a predetermined length of timeout. In this case, it appears that a 5-second timeout was given for one of these violations, and a separate automated process gave a second time out of 10 seconds for the second violation. The (full) ban could have been the result of the third violation or could have been because a specified maximum number of violations occurred within a given time period. It is likely, then, that this ban was caused by the sequence and close time proximity of the messages (rather than increasing severity of the offenses). The user here, in posting one message that contained three violations, received two progressive timeouts and a resultant ban for repeat offenses without ever having the opportunity to adjust the problematic behavior. Although this is an extreme case (and the poster was likely an automated bot), the pattern exists for other participants in our data as well. The automated enforcement of the progressive punishments negates the warning (since there is no intervening time to adjust behavior to fit the rules) and calls into question the purpose of having warnings at all. In this case, although the ultimate result is a ban, users were only banned due to a flaw in the automated process where *frequency* trumps *gravity*.

The second factor influencing the wide distribution of punishments in Category 4 is the potentially positive role of "disruptive" behaviors in the Twitch community. While hate speech is seen as objectionable and against the rules across the board, tolerance for what is identified as disruptive behavior is more varied. Spam (which falls under Category 4) is a normal and expected part of interaction on Twitch. We see this in the rules about spam posted by Fenn3r, who says, "Copypasta is welcome unless you're purposefully spamming without any real occasion to do so." This statement includes an inherent contradiction. On Twitch, the semantic meaning of spam is secondary or even unimportant. It serves a metapragmatic/phatic function as a community-builder through an invitation to play. There will therefore always be "a real occasion" to post it. Fenn3r's rules, by both welcoming and restricting spam, reflect the balance streamers must strike between encouraging viewers to participate (by allowing them to post normal and unmarked messages), while also preventing disruptive messages that would interfere with that participation.

To chatters, spam is an avenue through which they can bond together through shared rule violation—buying into the playfulness of breaking the rules together and seeing how long they can get away with it. One such case is seen in Example 5:

Example 5: Copypasta

01	jubarhd	is this the same loserfruit in top 500??
15	daddymoonmoon	[13 messages] is this the same loserfruit in top 500?
16		BAN: daddymoonmoon (1s)
20	Bebzii	[3 messages] is this the same loser fruit? oh crap deleted
24	Bebzii	[3 messages] too laaaate
29	Bebzii	[4 messages] im just copying my daddy

While daddymoonmoon's 1-second warning (line 16) is likely a response to a previous (unrelated) violation, Bebzii is interrupted in the process of recopying daddymoonmoon's message and interprets the 1-second sanction as a response to the copypasta. Benzii's next messages "oh crap deleted" and "too late" acknowledge that s/he didn't act fast enough in participating in the "spam game" before the spam was sanctioned by mods. S/he then explains that s/he was "just copying my daddy" (a pun on the username "daddymoonmoon"), indicating that participating in the "spam game" is a trivial offense and shouldn't be punished.

Over the next four minutes, daddymoonmoon posts another copypasta message repeatedly in a bid to get others to play. Finally, after only one participant reposts the second attempt, daddymoonmoon says, "I will now have meaningful conversation and stop spamming." We can see here that spam can be used as an engaging game that allows collaboration among participants and thereby strengthens community engagement. This only works, however, if others participate. Not having buy-in from other chatters results in "one-man spam" (which is explicitly prohibited in Wingsofdeath's rules) and is seen as annoying or even somewhat pathetic. As these examples show, Category 4 rule violations (e.g., spamming) are a valued part of Twitch conversation but are only effective if there is widespread participation. The presence of botmods that impose inconsistent and sometimes over-the-top sanctions increases the stakes for users. In these cases, a choice must be made whether to adopt lighthearted (and unmarked behavior), but with the realization that there is a risk of disproportionate punishment.

5 Discussion and conclusions

Building on Graham 2008 and 2017, this study has brought additional patterns to light by illuminating the different types of rules at play in online gaming and how they address behaviors that range on a spectrum from overtly aggressive and impolite to unmarked and appropriate. All streamers in this study constructed rules that prohibited forms of hate speech and insults (both aggressive and overt forms of impoliteness) and, when sanctions were levied for these violations, they were given without warnings rather than requiring multiple violations before sanctions were imposed.

All of these streamers also addressed disruptive behaviors like spam and copypasta, which required balancing two, often competing, interests that involve potentially disruptive behaviors:

- maintaining order in the chat by enforcing the rules and fostering a safe and welcoming environment, while
- not ostracizing participants and viewers by enforcing the rules so harshly that their viewership and donations decrease through reduced participation (i.e., if you ban everyone, there won't be anyone left to give you money).

This process was heavily influenced by the medium and automation, however, which may have resulted in punishments being out of sync with the actions that instigated them. While the data set for this study is relatively small, it shows a sliding scale of behavior management that is affected by the nature of the rules themselves and the goals of the streamer, which may have a large impact on how rules are understood and followed (or broken).

This exploration, while certainly valuable, has only scratched the surface in understanding how codes of behavior are enacted and/or challenged in online gaming. Since streamers must operate within the constraints of the Twitch platform, future research should continue to untangle complexities of streamer practice such as how they see their authority to manage their own streams in relation to Twitch's authority to enforce its rules, how streamers use the available tools to enforce the rules (how they program their botmods, what instructions they give to human mods, etc.), and what concessions they allow for different types of participants (e.g., subscribers vs. nonsubscribers). Also beyond the scope of this chapter is a more detailed examination of the frequency of violations compared to the frequency of punishments. There are often cases where users break the rules but are *not* punished while others are, and continued exploration of this complex process would be worthwhile.

Finally, the role of gender in how rules are chosen, presented, and enforced merits attention. Gender discrimination in gaming has been a highly visible topic of discussion of late. This has particularly been discussed in relation to the strategies that female streamers must adopt to be financially successful (e.g. Graham 2018, 2019). These often require catering to participant's sometimes sexist expectations, which female streamers often do by adopting objectified or highly sexualized personas. Such cases make identifying and sanctioning sexist behavior highly problematic, since these "booby-streamers" are both enabling and prohibiting sexist talk. In this setting, then, the balance of authority and hospitality that female streamers must adopt in their rules—both how they are written and how they are enforced—cannot be separated from gendered norms and expectations, and it would be beneficial to examine more closely the ways that streamers navigate this balance.

Gaming is an increasingly more prevalent mode of interaction and community involvement. It is also one where significant aggressive and impolite (i.e., toxic) behavior is prevalent, despite the existence of rules meant to curb it. While many argue that there should be action to address/control it, this is not a simple proposition, and greater understanding and research will facilitate an informed and appropriate response to destructive behavior and the means available to control it.

Notes

1 Definitions of *flaming* and *trolling* can vary greatly from community to community. While a full discussion of flaming and trolling is far beyond the scope of this chapter, both are commonly associated with negative behavior in which one person attacks or criticizes another. The two are often distinguished

from one another on the basis of intent—flaming is frequently associated with expressing genuine negative sentiments, while trolling is more often associated with making vitriolic statements that are designed to inflame others and cause discord rather than expressing sincere feelings.

2   This is consistent with Lave and Wenger's (1991) notion of Communities of Practice (hereafter CofP).

3   Professional streamers use subscriptions to their streams and donations from viewers as a source of income. This group therefore is dependent on the viewership of the stream. This is in contrast to professional gamers, who derive income from tournament winnings and sponsorships (and therefore derive income from their expertise in the game).

4   *Phishing* is an attempt to harvest personal information from individuals for profit.

5   *Copypasta* are large preformed messages that are copied and pasted into chat streams. They consist of blocky images or tongue-in-cheek story-like texts and are often labeled as rule violations because, if posted in quick succession, they can obscure visual access to other chat posts.

6   In our data, there were no timed punishments for the range between 10 seconds and 10 minutes.

References

Blizzard Entertainment (2015), *Heroes of the Storm* [Videogame], Irvine, CA: Blizzard Entertainment.
Blizzard Entertainment (2016), *Overwatch* [Videogame], Irvine, CA: Blizzard Entertainment.
De Freitas, S. and P. Maharg (2011), *Digital Games and Learning*, London: Continuum Press.
Ensslin, A. (2012), *The Language of Gaming*, Basingstoke: Palgrave Macmillan.
Ensslin, A. (2014), *Literary Gaming*, Cambridge: MIT Press.
Gee, J.P. (2003), *What Videogames Have to Teach Us about Learning and Literacy*, New York: Palgrave Macmillan.
Gee, J.P. (2015), *Unified Discourse Analysis: Language, Reality, Virtual Worlds and Video Games*, New York: Routledge.
Graham, S.L. (2008), "A manual for (im)politeness? The impact of the FAQ in an electronic community of practice," in D. Bousfield and M. Locher (eds.), *Impoliteness in Language: Studies on its Interplay with Power in Theory and Practice*, 281–304, Berlin/Boston: de Gruyter.
Graham, S.L. (2015), "Relationality, friendship and identity in digital communication," in A. Georgakopoulou and T. Spilioti (eds.), *The Routledge Handbook of Language and Digital Communication*, 305–20, New York: Routledge.
Graham, S.L. (2017), "Politeness and impoliteness," in C. Hoffmann and W. Bublitz (eds.), *Handbook of the Pragmatics of Social Media*, 459–91, Berlin: de Gruyter.

Graham, S.L. (2018), "Impoliteness and the Moral Order in Online Gaming," *Internet Pragmatics*, 1 (2): 302–326. https://doi.org/10.1075/ip00014.lam

Graham, S.L. (forthcoming May 2019), "Interaction and conflict in digital communication," in L. Jeffries, J. O'Driscoll, and M. Evans (eds.), *The Routledge Handbook of Language in Conflict*, Ch 16, New York: Routledge.

Graham, S.L. and C. Hardaker (2017), "(Im)politeness in digital communication," in J. Culpeper, M. Haugh, and D. Kádár (eds.), *Palgrave Handbook of Linguistic (Im)politeness*, 785–814, London: Palgrave Macmillan.

Guimaraes, M. (2005), "Doing anthropology in cyberspace: Fieldwork boundaries and social environments," in C. Hine (ed.), *Virtual Methods: Issues in Social Research on the Internet*, 141–56, New York: Berg.

Heyd, T. (2013), "Email hoaxes," in S. Herring, D. Stein, and T. Virtanen (eds.), *The Pragmatics of Computer-Mediated Communication*, 387–410, Berlin: Mouton de Gruyter.

Jenkins, H. (2006), *Fans, Bloggers and Gamers: Exploring Participatory Culture*, New York: New York University Press.

Kádár, D. and M. Haugh (2013), *Understanding Politeness*, Cambridge: Cambridge University Press.

Lavé, J. and E. Wenger (1991), *Situated Learning: Legitimate Peripheral Participation*, Cambridge: Cambridge University Press.

Locher, M.A. and R.J. Watts (2005), "Politeness theory and relational work," *Journal of Politeness Research*, 1: 9–33.

Maher, B. (2016), "Can a video game company tame toxic behavior?" *Nature*, March 30. Available online: http://www.nature.com/news/can-a-video-game-company-tame-toxic-behavior-1.19647 (accessed April 25, 2018).

Nardi, B. (2010), *My Life as a Night Elf Priest: An Anthropological Account of World of Warcraft*, Ann Arbor, MI: University of Michigan Press.

Newon, L. (2011), "Multimodal creativity and identities of expertise in the digital ecology of a *World of Warcraft* guild," in C. Thurlow and K. Mroczek (eds.), *Digital Discourse: Language in the New Media*, 203–31, Oxford: Oxford University Press.

Newon, L. (2016), "Online multiplayer games," in A. Georgakoupoulou and T. Spilioti (eds.), *The Routledge Handbook of Language and Digital Communication*, 289–304, New York: Routledge.

Pearce, C. (2011), *Communities of Play: Emergent Cultures in Multiplayer Games and Virtual Worlds*, Cambridge: MIT Press.

Prensky, M. (2001), *Digital Game-Based Learning*, New York: McGraw-Hill.

Riot Games (2009), *League of Legends* [Videogame], Los Angeles, CA: Riot Games, Inc.

Shea, V. (1994), *Netiquette*, San Francisco: Albion Books.

Taylor, T. (2006), *Play between Worlds: Exploring Online Game Culture*, Cambridge: MIT Press.

Thorne, S.L. (2008), "Transcultural communication in open internet environments and massively multiplayer online games," in S. Magnan (ed.), *Mediating Discourse Online*, 305–30, Amsterdam: John Benjamins.

Thorne, S.L. and I. Fischer (2012), "Online gaming as sociable media," *ALSIC: Apprentissage des Langues et Systèmes d'Information et de Communication*, 15 (1), Available online: http://alsic.revues.org/2450 (accessed August 24, 2018).

Twitch (n.d.). Available online: http://www.twitch.tv (accessed August 24, 2018).

Twitchstats (2017), "Most popular Twitch games by average viewers," June 6. Available online: http://www.twitchstats.net/most-popular-games (accessed August 24, 2018).

Wright, T., E. Boria, and P. Breidenbach (2002), "Creative player actions in FPS online video games," *Game Studies*, 2 (2). Available online: http://www. gamestudies.org/0202/wright/ (accessed August 24, 2018).

PART THREE

Beyond the "Text": Multimodality, Paratextuality, Transmediality

10

On the Procedural Mode[1]

Jason Hawreliak

1 Introduction

During a mission of the open-world action-adventure game *Mafia III* (Hangar 13 2016), the playable character, Lincoln Clay, enters a pawn shop and is immediately confronted by the shop's owner. The shop owner tells Lincoln that if he does not leave immediately, the police will be called and Lincoln will be arrested. Sure enough, if Lincoln does not exit the premises promptly, the police will be called and will attempt to arrest or even shoot Lincoln. The player cannot help but feel this is unfair, as Lincoln has not stolen anything, broken any merchandise, or done anything else which would seem to warrant such a response. What Lincoln has done, however, is enter a white-owned shop as a black man in 1960s New Bordeaux (ostensibly New Orleans). Lincoln has ignored the signs of "No Colored Allowed" posted outside of the shop and broken a strictly enforced rule which segregates white people from black people in places of business, worship, and education.

Throughout the game, Lincoln must drive to various parts of the city to complete missions or purchase items. If Lincoln commits a crime, such as hitting something with his car or firing a weapon—a large portion of the gameplay—witnesses may call the police. The police response time, however, in part depends upon the district in which the offense is committed. As the game's lead writer, Bill Harms, explains in an interview with *Waypoint's* Austin Walker (2016): "the police respond quickly in the highly populated, wealthy and white downtown district. 'But if you're in the Hollows—one of the poorer, much blacker districts in New Bordeaux...' He [Harms] shrugs." In Walker's words, "It's the first time I've seen this element of structural racism systematized in a big-budget game" (para. 10). *Mafia III*'s rules and systems provide an interactive expression of structural racism and white supremacy that are ontologically and semiotically distinct from textual, filmic (audiovisual), or performative expressions of the same systems. While interacting with racist systems in a videogame of course does not capture the brutal reality of racism in the "real world," and videogames themselves have a long history of racist and stereotypical representation (Gray 2015), such interactions nevertheless point to an avenue of expression that is distinct from other forms of representation.

The mode of representation illustrated in the aforementioned examples of in-game interaction (i.e., committing crimes, driving, etc.) is what I will call the *procedural mode*, which is well established in game studies but notably absent in the realm of semiotics. The primary goal of this chapter is to demonstrate the usefulness of conceiving procedurality as a semiotic mode in the context of multimodal discourse analysis. It argues that procedurality—in this context, the rules, systems, and parameters of a game—should be viewed as a semiotic equal alongside established modes such as text, image, and music. At the core of this argument is the somewhat innocuous premise that videogames in particular, and computers in general, allow for modes of communication which are not readily available to other communications media (Murray 1997; Galloway 2006; Bogost 2007). Simply put, we can communicate certain kinds of information through videogames and procedurality differently—though not necessarily better— than we can in other media. Moreover, accepting procedurality into the catalogue of acknowledged semiotic modes will fulfill the secondary aim of this chapter, which is to bring multimodal studies and game studies into a deeper conversation with one another. Such a conversation can only result in a mutually beneficial relationship. Videogames are highly multimodal artifacts which have the potential to communicate via most of the known semiotic modes (Gee 2013). A multimodal approach to videogame analysis may therefore lead to rich, highly nuanced semiotic readings of videogames (Ensslin 2012). On the other hand, multimodal studies can benefit from game studies, as games and other interactive texts can communicate meaning in ways not yet fully incorporated into interactive semiotics.

2 Procedurality and procedural rhetoric

Procedurality is an established, core concept in game studies. The idea that meaning can be communicated via rule-based models predates digital game studies by decades with examinations of games as cultural and ritualistic practices (Huizinga 1955; Caillois 1961). However, the term *procedurality* in its current usage gained prominence in Janet Murray's essential 1997 book, *Hamlet on the Holodeck: The Future of Narrative in Cyberspace* (also see Manovich 2001; Bogost 2006; Flanagan 2009). In this text, Murray argues that procedurality is one of four affordances of computational media and computational narratives. For Murray, procedural authorship "means writing the rules for the interactor's involvement, that is, the conditions under which things will happen in response to the participant's actions. It means establishing the properties of the objects and potential objects in the virtual world and the formulas for how they will relate to one another" (1997: 152–53). In this conception, the procedural mode is expression through interactive, rule-based systems enacted by a player.

Similarly, Miguel Sicart describes procedurality as "the ways arguments are embedded in the rules of the game, and how the rules are expressed, communicated to, and understood by a player" (2011: para. 12). For Michael Mateas, "procedural literacy" is an essential aspect of new media practitioners and scholars. He defines procedural literacy as "the ability to read and write processes, to engage procedural representation and aesthetics, to understand the interplay between the culturally-embedded practices of human meaning-making and technically-mediated processes" (2005: para. 2). Mateas argues that just as literary scholars should have an understanding of the written word, scholars of computational media should have an understanding of "how code operates as an expressive medium" and that in fact, "Code is a kind of writing" (para. 3). Procedures, of course, are not confined to videogames nor to communications media in general (Bogost 2007). The procedure for walking is placing one foot in front of the other, for instance. The procedure for framing a room involves measuring, cutting, and fastening pieces of wood together in a particular order. If I cut a piece of wood before measuring it, for example, I have not followed the proper procedure and will likely have an uneven or unstable wall frame. Had this carpentry task been set in the context of a game, I would have lost. Procedurality has gained currency within game studies because computer-based media run via sets of procedures—executable code which is enacted by the user. Therefore, the argument goes, computers are the perfect form for procedural representation since they simulate processes with processes (Murray 1997; Bogost 2006). Furthermore, since all games contain rules of some sort, it is appropriate to adopt an analytical system which takes rule-based behavior as its foundation.

Perhaps the clearest discussion of procedurality comes from the work of Ian Bogost in his discussions of procedural rhetoric.[2] "Procedurality," explains Bogost, "refers to a way of creating, explaining, or understanding processes. And processes define the way things work: the methods, techniques, and logics that drive the operation of systems" (2007: 2–3). In the context of videogames, a procedure is typically a series of tasks undertaken to execute an in-game action or to fulfill a particular goal. In *Persuasive Games*, Bogost explains how procedural expression is distinct from other modes of expression and why it is a potentially powerful means of communication:

> Procedural representation is significantly different from textual, visual, and plastic representation. Even though other inscription techniques may be partly or wholly driven by a desire to represent human or material processes, only procedural systems like computer software actually represent process with process. This is where the particular power of procedural authorship lies, in its native ability to depict processes. (2007: 14)

As this passage illustrates, games scholars treat procedurality as a semiotic mode even if they do not typically adopt the language of multimodal semiotics.

The central point of *Persuasive Games* goes beyond discussions of procedurality generally and demonstrates an application of procedural authorship which Bogost calls "procedural rhetoric" (also see Voorhees 2009; Harper 2011; Layne 2015). Bogost defines procedural rhetoric as "a technique for making arguments with computational systems and for unpacking computational arguments others have created" (2007: 3). Procedural rhetoric can be thought of as the means by which the game's rules and parameters guide action in the game world and can be expressed as a series of questions: Which actions do the game's rules require or allow? Which do they forbid? Which do they reward or punish? We can thus identify the procedural rhetoric of a given game by examining its rules, parameters, and reward/punishment structures. A simple illustration of procedural rhetoric is found in many military-themed games.

For instance, in the single-player campaign of the first-person shooter *Call of Duty: Modern Warfare* (Activision 2007), the player takes the role of an American or British special operations soldier attempting to stop a terrorist organization from obtaining nuclear weapons. It is a typical military shooter that requires the player to virtually kill hundreds of enemies throughout the game. There is no other way to deal with enemies. You cannot negotiate with them; they will not surrender. There is no diplomatic option available at any time (Gagnon 2010). The logic of the game is kill-or-be-killed. The procedural argument here—even if unintentional—is that terrorists are illogical, violent beings who must be destroyed, full stop (Stahl 2006). Through playing the game, the player is exposed to the argument that terrorists are either incapable or unworthy of discourse. If the player could engage in conversation or even use nonlethal means for dispatching enemies, the procedural rhetoric of the game would be much different. This same propagandistic argument has been made in essentially every medium and mode after 9/11, but the way it is communicated in an interactive, procedural manner is distinct (see Payne 2016).[3] Put another way, the player encounters this argument in the game in a way that is different from how they encounter it in text (e.g., a Tom Clancy novel) or through audiovisual media (e.g., a Tom Clancy film). This does not mean it is more convincing, but simply that it is communicated differently.[4]

While procedurality has appeared in thousands of academic articles, chapters, and books—as of June, 2018, Google Scholar lists 2,548 citations for Bogost's *Persuasive Games* alone—it has not gained much traction outside of game studies in general and in semiotics in particular. One reason may be that it is perhaps not immediately clear how, if at all, procedurality can be classified as a mode of expression. Unlike text or speech, there is a certain abstraction to rules and processes. However, although "the procedural mode" is absent from multimodal scholarship, there may be analogs. For instance, within multimodal studies, there has been a considerable amount of work on *layout*, or the ordering of information in graphic design (e.g., Kress and van Leeuwen 2001; Hiippala 2015). As Kress and van Leeuwen remark in their discussion of a biology textbook, "[t]he organization of material through layout produces specific social and ontological arrangements," and furthermore, "positions semiotic elements and their relations; [layout] 'orients' viewers/readers to classifications of knowledge, to categories such as 'centrality' or 'marginality,' 'given' or 'new,' 'prior' and 'later,' 'real' and 'ideal'" (2001: 92).

The order and arrangement of information on a page is itself a way to communicate meaning. How this ordering and arrangement of text/image in a book impacts meaning is not unlike the procedures—that is, the sequential processes—players must undertake when playing a game. In both cases, the arrangement and rules of interpretation contribute to the meaning potential of the artefact. Likewise, the arrangement and ordering of still images to create the illusion of a moving image is crucial to conveying meaning in film (Bateman and Schmidt 2012). For instance, a series of sequential shots shown in order means something very different than if placed out of order into a montage (Eisenstein [1949] 1977). Even if the shots are exactly the same, the order in which we view them changes their meaning. The difference with these examples, of course, is that they lack the same type of interactivity found in computational media. Nevertheless, they demonstrate the way that rules, sequences, and processes impact meaning in their own ways.

Returning to games, the procedures of *Tic-Tac-Toe* look something like this:

(1) A 3 × 3 grid is drawn for two players;

(2) Player 1 places an *X* in one of the (empty) squares;

(3) Player 2 then places an *O* in any empty square;

(4) Steps 2 and 3 are repeated until

    a. There are no empty squares left, which results in a tie, or

    b. One player wins by successfully placing three *X*s or *O*s in a row.

Based on these simple rules, it is possible to play the game and devise strategies for winning. But what if we change just one of the steps in the procedure? For instance, instead of alternating turns of placing one X or O at a time, let us say that players get to place two Xs or Os at a time. Even if the win condition and symbols remain the same, the change in procedure has changed something fundamental about the meaning of the game. Players will adopt different strategies and will understand what they are supposed to do to win in a different way than they would with the original rules in place. Like rearranging text and image in a biology textbook, changing the rules in *Tic-Tac-Toe* changes its meaning. This would not be the case if procedurality was not a semiotic mode; it is the ordering itself which contains meaning.

Perhaps the strongest objection against designating procedurality as a mode is that, unlike other modes, it does not really stand by itself but almost always requires other modes to make it intelligible. In the context of videogames, procedurality is, in essence, the mode of computer programming, or software authorship (Bogost 2007). It is communication through mathematical formulae, conditional (if) statements, algorithms, and so on. Of course, rules and systems can be—and often are—developed before any coding occurs via paper prototyping and other methods, and contemporary videogame engines often take care of much of the mathematical work behind the scenes; however, these rules and systems must ultimately be translated in a programming language to create a videogame. At this point, a clear challenge arises. Apart from computer programmers and computers themselves, this mode of expression is esoteric and ultimately unintelligible until it is tied to other forms of representation, such as auditory and visual modes.

For instance, the procedure for shooting an enemy in a first-person shooter can roughly be expressed as

(1)   aim at enemy;

(2)   pull trigger to fire at enemy;

(3)   if enemyHit = True, then Killcount = Killcount + 1;

(4)   else if, return to step 1.

Here the player tries to shoot an enemy, and if they succeed, they get a "kill," and if they miss, they try again. This procedure, however, is only intelligible if the player can see the enemy, the gun, the feedback which indicates a successful hit (or not), and so on. Procedurality, then, requires the assistance of other semiotic modes for reification, unless one is literate in the game's programming language and somehow has access to the code. Nevertheless, what is expressed in the example of firing a weapon is a process, designed and constructed through software authorship. Just as layout requires material text to lay out in the first place, procedurality requires material signs (images, an input device, etc.) to be enacted and made intelligible. The experience of firing a simulated weapon in a simulated battle conveys something about combat, war, weapons, and so on that is distinct from how it is conveyed through watching it happen in a film or reading about it in a book.

3 Semiotic modes and multimodality

To make the case that procedurality is a legitimate semiotic mode, it is necessary to set out some criteria for what constitutes a mode in the first place. On the face of it, this appears to be more or less straightforward. Established modes, such as text, gesture, moving image, and music, all create meaning in their own ways and so are semiotically distinct. But as we move away from specific instances, it becomes increasingly difficult to provide a satisfactory definition which encompasses all modes, or to find that singular essence of modality. As Charles Forceville plainly puts it, "there is no generally accepted definition of what counts as a mode" (2016: 20). There are some very useful ways to approach defining *mode*, however. In *Introducing Social Semiotics*, Theo van Leeuwen borrows the term "semiotic resource" from M.A.K. Halliday to describe "the actions and artefacts we use to communicate, whether they are produced physiologically—with our vocal apparatus; with the muscles we use to create facial expressions and gestures, etc.—or by means of technologies—with pen, ink, and paper; with computer hardware and software" (2005: 3). The advantage of this definition is that it draws attention to the materiality of semiosis: communication is never ahistorical, but happens in particular contexts—social, cultural, economic, technological, and so on—that are indelibly linked to its usage by actual human beings in time and space.

Keeping the materiality of communication in mind, perhaps the most concise (and best) definition of *mode* comes from Carey Jewitt, who defines it as "a means for making meaning" (2009a: 2). This definition is both sufficiently informative and flexible enough to account for the myriad ways human beings materially communicate ideas, and it will certainly do for our purposes. When it comes to modes, then, what we are really talking about are different types or genres of signifiers. Each mode has its own way of communicating information, and each mode has certain affordances that the others do not (Kress 1993). In Kress and van Leeuwen's words, "the question, 'What mode for what purpose?'" is central when crafting any communication (2001: 46). A text message on a mobile phone, for example, is often more useful for relaying short bits of information like "running late, be there in 5," than speaking to someone on the phone, and it is certainly more efficient than drawing a picture or composing a song to convey the same message.

To determine if something is its own semiotic mode, therefore, it should satisfy the following criteria:

(1) Is it a material "means of making meaning?"
(2) Does it have its own affordances or communicative advantages relative to other modes?

If these are satisfactory criteria for counting something as a mode, then procedurality certainly gets a pass. As illustrated in the example of *Mafia III*, interacting with and confronting systemic racism through the game's procedures communicates systemic racism in a way that is both meaningful and different than representations of systemic racism in other modes such as text.

Another test is how a mode behaves when combined with other modes, or within a multimodal ensemble (Jewitt 2009b). Kress and van Leeuwen define *multimodality* "as the use of several semiotic modes in the design of a semiotic product or event, together with the particular ways in which these modes are combined" (2001: 20). Multimodality is ultimately concerned with the inherent complexity of a semiotic event, from face-to-face conversations—which include nonverbal signs like facial expressions—to navigating online spaces, which may include image, speech, and text (Sindoni 2013). Multimodality has been applied in a wide variety of contexts, such as education (Rowsell 2013; Crawford-Camiciotti and Fortanet-Gomez 2015), film studies (Bateman and Schmidt 2012), literature (Gibbons 2012), and discourse analysis (e.g., Kress and van Leeuwen 2001; O'Halloran 2004). As I will discuss in greater detail here, multimodality has also been applied to the analysis of videogames (e.g., Kromhout and Forceville 2006; Machin and van Leeuwen 2007; Ensslin 2012; Toh 2015), but not as much as one might think, given the sheer number of modes available to game developers.

In all cases, multimodality "proceeds on the assumption that representation and communication always draw on a multiplicity of modes, all of which have the potential to contribute equally to meaning" (Jewitt 2009b: 14). A multimodal analysis does not simply catalogue or list all of the modes utilized in an artefact but also, crucially, examines how the modes are configured and work together. Like individual instruments in an orchestra, each mode in an ensemble contributes to the overall meaning of a semiotic event in its own way (Kress and van Leeuwen 2001). Each mode has the potential to radically influence the other modes in an ensemble, and therefore, the meaning of the message overall. As Jewitt puts it, "[t]he meanings in any mode are always interwoven with the meanings made with those of all other modes co-present and 'co-operating' in the communicative event. The interaction between modes is itself a part of the production of meaning" (2009b: 15). Multimodality is a dynamic system wherein changing one component (e.g., the music in a film) impacts the meaning of another (e.g., the images on screen) and, therefore, the artefact as a whole.

A multimodal approach requires a form of complex analysis which factors the meaning potential of multiple semiotic resources simultaneously. As the number of available modes increase, so too does the potential for complexity (Lemke 1998). This can make it challenging for semioticians to conduct a full multimodal analysis of richly multimodal artifacts since it requires i) a broad understanding of signification practices across many different modes and ii) a potentially overwhelming matrix of dynamic meaning potential (Ensslin 2012). Untangling the web of semiotic interactions between multiple modes can be a Herculean effort, especially when dealing with highly multimodal media such as videogames. Therefore, it is important that analyses focus on only the most salient modal interactions for their intended purpose.[5] To the two criteria listed above then, we might add

(3)   Does adding or subtracting the potential mode change the overall meaning of a multimodal artefact?

As I discuss here, adding, subtracting, or altering the procedure(s) in a game can radically alter the game's potential meanings and how players experience it.

4 Multimodality and game studies

It is certainly true that all communication is multimodal (Kress and van Leeuwen 2001). A seemingly monomodal medium, such as print, still communicates information in ways apart from text. The feel of the paper on one's hands or the smell of the binding, for instance, contribute to the overall meaning of the experience of reading a book. However, although all communication is multimodal, there are varying degrees of multimodality and therefore of semiotic complexity. As James Paul Gee observes, there "is no other more multimodal media today than video games" (2013: 49). Videogames are particularly well suited to multimodal analysis as they rely upon the communicative resources of a wide array of modes all at once. To use Sigrid Norris' terminology, videogames possess a high "modal density" (2004: 102). Videogames "remediate" and adopt the representational practices of other multimodal media such as film (Bolter and Grusin 2000) but employ haptic, and—as I am arguing here—procedural forms of expression as well (Bogost 2007; Ensslin 2012). As such, "videogames call out to be analyzed multimodally in the sense of how multiple representational modes displayed on screen create complex layers of meaning, which are decoded and interacted with by players" (Ensslin 2012: 118).

Somewhat surprisingly, until now, little work has directly addressed the potential intersections between multimodal analysis and game studies. James Paul Gee (2003) has written about the "multimodal principle" of videogames; however, multimodality in videogames is not the focus of

Gee's work. Burn and Schott (2004) have explicitly linked multimodality and videogames in their analysis of player-avatar relations, and the book *Computer Games: Text, Narrative and Play* (Carr et al. 2006) also makes this link. Furthermore, there is scholarship on multimodality in the field of human-computer interaction (e.g., Jaimes and Sebe 2005), and Kromhout and Forceville (2013) have examined how the concept of multimodal metaphor and cognitive metaphor theory (Lakoff and Johnson 1980) can be applied to videogame analysis. Toh (2015) has written a dissertation on multimodal discourse analysis in the context of videogames, but the focus is on the interaction between story and gameplay and gathering empirical data. Perhaps most notably, Astrid Ensslin's chapter on multimodality in *The Language of Gaming* (2012) provides a fantastic introduction to multimodal videogame analysis, analyzing interface, player-to-player communication, haptics, and narrative. In a later work, Ensslin (2017) examines often ignored modes in interactive media like the olfactory and gustatory modes, which at once pose daunting technical challenges and rich semiotic potential. Ensslin's work is really the first to provide a detailed argument outlining the value of multimodality in game studies. Again, this chapter hopes to build upon the existing discourse by bringing videogame multimodality into sharper focus.

A multimodal approach should be viewed as one component in an analytical toolkit and cannot possibly account for all the myriad, unpredictable ways meaning is ultimately negotiated between game and player. The advantage of a multimodal approach is that it allows us to interrogate how videogames signify in all their complexity. It lets us examine what each mode is "saying," and, more importantly, how each mode interacts with the others to construct meaning potentials. Jewitt's point that at any given time all modes in a multimodal ensemble "have the potential to contribute equally to meaning" (2009b: 14) parallels existing work within game studies. For example, as Ian Bogost writes in a 2009 blog post, "all aspects of a game's existence have the same potential to matter." In other words, no mode— including the procedural mode—is inherently "superior" to any other, and given the sheer number of videogame modes, we are always bound to focus on some and omit others according to our analytical goals. Echoing Kress and van Leeuwen's point about multimodality generally, the question we can then pose is, for a particular game in a particular circumstance, which aspects, or modes, "matter" in a given situation? Which bits of the meaning equation are we interested in examining, and which will we leave aside? Such questions are essential for thinking through a semiotics of videogames.

5 Multimodal (con)figurations and case studies

If we accept that procedurality is a mode that can exist within a multimodal ensemble alongside other videogame modes, such as moving image, music, sound effect, and haptics, then we can think about ways that procedurality might fit within particular configurations. A multimodal ensemble can be configured in any number of ways, but one useful way to think about arrangement is in terms of multimodal agreement and disagreement (Lemke 1998), or what I will call *consonance* and *dissonance* or *irony*. Multimodal consonance occurs when two or more modes reinforce one another either thematically, aesthetically, or rhetorically. In this configuration, modal components align to create a coherent, holistic message. For example, in horror films, multimodal consonance often occurs when ominous or off-putting music accompanies a tense or fear-inducing set of visual images on screen (Chion 1994). In this instance, both music and moving image combine to (ideally) produce a sense of fear or anxiety within the audience.

In the context of videogames, modal consonance is the default configuration. Unless there is a specific purpose for doing otherwise, it is usually advantageous to use all available modes to achieve an expressive purpose. After all, this is an advantage of multimodal media: we can use the representational strengths of each mode together to "multiply meaning" (Lemke 1998) and create compelling messages. If a developer wants to create a sense that the player is performing heroic, epic actions, for instance, then the developer may have them slaying dragons (procedurality) while the controller rumbles with each sword thrust (haptics) within a fantastical environment (moving image) as an up-tempo orchestral score plays in the background (music). This is what we usually encounter, and it makes sense, as the various modes mutually reinforce one another and thereby provide a coherent overall experience.

Focusing on the procedural mode within an ensemble—and more specifically, a procedural-textual-visual-speech analysis—I will briefly examine one mechanic in the World War I (WWI)–themed FPS *Verdun* (BlackMill Games 2015). In this game, players from around the world battle to take, hold, and re-take virtual territory in the battlefields surrounding Verdun, France. Players may choose to fight for the Central Powers or Triple Entente and must attack enemy lines and repel enemy attacks in order to win the match. In many ways, *Verdun* is a typical online FPS: players must shoot at enemies, they can choose different classes and weapons, the game keeps track of kills and deaths, and so on. One rule in *Verdun* worth examining here pertains to *level parameters*, or the allowable area where players can travel. If a player goes outside the allowable zone, signified on screen by a "mini-map" in the bottom right-hand corner, the screen turns gray, and they are informed by an order delivered via text and speech that failure to return to the playable game area will result in execution. If the allotted time has passed and the player has not returned to their lines, they are "executed for desertion," which in turn reflects negatively on player statistics (e.g., kill:death ratio) and takes them out of play for a period of time.

In terms of the procedural mode, a conditional (if) statement checks whether or not the player is within the playable area after a given amount of time. If yes, then the player continues to live; otherwise (else if), the player is executed. Here we have multiple communicative modes—image, text, speech, and procedurality—all coming together to enforce the spatial parameters of the level. Each mode delivers the same message, that is, stay in the playable area. However, it is the procedural mode which is most impactful, as it is responsible for setting the parameters and for enacting the consequences. If the game only used visual, auditory, and textual modes to indicate a transgression, players likely would not comply as there would be no ludic consequences for disobedience. This strategy for constraining player movement works within the game because the designers have justified their constraints within the procedural mode with information from other modes that fit with the game's setting.

In contrast to modal consonance, modal *dissonance* occurs when two or more modes within an ensemble do not align but instead contradict or work against each other. Modal dissonance may occur intentionally or unintentionally and can produce a variety of aesthetic, rhetorical, and semiotic effects. Modal dissonance warrants a more detailed discussion than consonance since it is less common and is often viewed as an aesthetic failing (e.g., Hocking 2007). But perhaps more importantly, modal dissonance brings the interactions between semiotic modes to the fore. Like Heidegger's hammer, we become most aware of semiotic systems and their machinations when they break or refuse to act in a seamless, predictable way. Modal dissonance can cause positive or negative effects depending on the context, as I will discuss here. For the sake of clarity, it is perhaps best to first examine how modal dissonance works in other media before examining it within the context of the procedural mode.

Examples of modal dissonance are common in film and are often used to elicit humor, fear, or a sense of uneasiness. For example, the film *American Psycho* (2000) follows Patrick Bateman (played by Christian Bale), a serial killer whose day job is working in high finance. Bateman seems to be perfectly "normal" to the people around him. He is handsome, articulate, and likeable. The duality between Bateman's horrific crimes and his ability to fit in with those around him is a central theme of the film. The director, Mary Harron, uses multimodal dissonance to exemplify Bateman's duality in a particularly striking scene. One night, Bateman invites a work colleague to his apartment after a night of drinking. The colleague, played by Jared Leto, unknowingly bruised Bateman's ego earlier in front of a group of people at work. Once Bateman and his colleague get to the apartment, Bateman puts on Huey Lewis and the News' bubbly 1986 hit, *Hip to be Square*. The track is upbeat and rather silly at first blush. Yet as the music plays, Bateman viciously murders his colleague with an axe. The blood and sound of the axe hitting flesh is horrifying. However, what is truly unsettling about the scene is the juxtaposition between the visual content of a brutal murder and the auditory content of an upbeat, silly song. The tension created by this modal juxtaposition or irony reinforces the unsettling theme of Bateman's duality. He is a violent murderer who enjoys killing, but no-one would know it walking past him on the street.

Dissonance between component parts has also been examined in the context of videogames. Most notably, Clint Hocking (2007) coined the term "ludonarrative dissonance" to describe the frequent disconnect between gameplay (from the Latin for game, *ludus*) and story[6] in videogames. For instance, in BioWare's *Mass Effect* series (2007–2017), the player may be told during a cinematic cutscene—which conveys narrative—that a certain mission must be completed with the utmost urgency; any hesitation will result in the destruction of life in the galaxy. However, during the ensuing gameplay, the player may complete unrelated side-missions, romance a companion, aimlessly explore the gameworld, or engage in other mundane activities such as shopping, all without penalty. Here we have a disconnect between the information delivered in the narrative and the procedural rhetoric of the game, which encourages an open-world, exploratory style of play.

Building on Hocking's term, Lana Polansky (2015) proposes the terms "coherence" and "incoherence" to describe how the various parts of a videogame work together, that is, effectively or not. As Polansky points out, *dissonance* is not necessarily negative, though *incoherence* might be. However, while Hocking and Polansky describe the interactions between component parts and this is useful, the language of multimodality provides a more focused analytical framework for examining this phenomenon, as it allows us to pull apart the various semiotic components which may be coherent or incoherent. The dissonance or incoherence we see in games like *Mass Effect* is in reality a contradiction of semiotic modes. The informational content conveyed by the cutscene—a filmic/cinematic multimodal ensemble of moving image, speech, and text—does not align with the informational content conveyed by the procedural mode. Given the urgency of the mission briefing in *Mass Effect*, we might expect the procedure to look something like this:

(1)   receive mission details;

(2)   travel to appropriate destination;

(3)   complete task *as quickly as possible*.

In a linear game with limited player choice, this is precisely what we would see. Instead, the "sandbox" or free-form and exploratory style of gameplay in *Mass Effect* produces a different procedural message:

(1)   receive mission details;

(2)   choose to

a. complete main quest objective, or

b. explore the citadel (a primary hub), or

c. explore one of the planets, or

d. shop for better gear, or

e. pursue a romantic relationship, or

f. complete side quest i, or

g. complete side quest ii… and so on.

The procedural rhetoric of *Mass Effect* is designed to make players feel as though they are part of an expansive, living gameworld, and therefore, providing a high degree of player choice is important. However, this same procedural rhetoric is often in direct contradiction with information conveyed through the cutscenes, which rely upon the semiotic (filmic) modes of moving image, music, and speech.

Another example of multimodal dissonance involving the procedural mode can be found in examples involving unplayable areas. In the example of *Verdun* mentioned here, the designers took pains to explain the procedural constraints of the game's playable area. This is often not the case. In the open-world, FPS/RPG hybrid *Fallout 3* (Bethesda 2008), for instance, the player must traverse a hostile, postapocalyptic wasteland. Throughout the game, the player must battle enemies such as Raiders, Super Mutants, and radioactive insects. Players obtain many powerful weapons as they progress, such as automatic rifles and rocket-launchers. One of the core aspects of gameplay in *Fallout 3* is exploring bombed-out buildings and other ruins for loot such as food and weapons. In many instances, players will be obstructed from a viewable area in the building by a dilapidated, half-destroyed wooden door. In the context of the gameworld, this does not make sense, since it should be quite easy to destroy the door and enter the forbidden area. However, the designers make no attempt to justify this contradiction. What we have here is a contradiction between the mode of still image (the door asset) and the procedural mode (the constraint on player movement). For the first few encounters with these indestructible doors, there is a certain degree of frustration involved. Players, of course, recognize that *Fallout 3* is just a game, and so such instances of dissonance are quickly normalized as part of the gameplay; however, as the example of *Verdun* illustrates, it is possible to maintain constraints while also explaining them via non-procedural modes.

These examples illustrate how modal dissonance can be a detriment to a game's coherence and cohesion. In these cases, dissonance risks reducing player engagement or immersion and may appear as poor design. However, dissonance can also be used generatively. Games are ultimately about tension (Huizinga 1955), and so there is ample opportunity for designers and developers to juxtapose modes for the sake of generating all sorts of tension. One form of tension is comedic. Juxtaposition and irony are of course common comedic strategies. As an example, *Saints Row IV* (Deep Silver 2013) is an open-world action game that provides a good example of modal dissonance achieving a comedic effect. In the opening scene, the player is told via speech and text that they must stop a nuclear missile from launching and destroying Washington DC. This is, of course, a serious event, and it is treated as such in many other military-themed games. Visually, it is also quite serious, as a missile is indeed speeding toward Washington. However, the aural and procedural modes do not align with this sense of seriousness. Once the character reaches the rocket, Aerosmith's *Don't Wanna Miss a Thing* from the *Armageddon* soundtrack plays loudly. It is clearly out of place given the circumstances, and the juxtaposition creates a comedic effect.

The procedural mode is also dissonant in this scene. The interactive portions consist of some character movement and fairly easy QTE (quick-time-event) mechanics. The procedural and tactile (physiological) demands on the player are minor and so do not really align with the complexity of what one would associate with defusing a nuclear weapon. It never feels like anything significant is at stake, all in spite of the audio, textual, and visual signifiers saying otherwise. The ironic choice of music and simple game mechanics not only produce an amusing juxtaposition but will also shape how the player responds to the rest of the game. From the opening scene, modal dissonance tells the player what to expect in terms of tone, which might most closely resemble parody. If the musical score came from another game which treats the same scenario with gravity—e.g., *Call of Duty: Modern Warfare*—the scene would mean something very different than it does.

Comedy is one of the chief outcomes of modal dissonance; however, it can also be used for noncomedic purposes. In the 2016 WWI-themed FPS *Battlefield 1* (EA/DICE) there are very few moments of levity. The single-player campaign has players fight historic battles across Europe as a number of playable characters. The gameplay is typically intense, high-paced, and requires quick reaction times (see enemy, shoot enemy). There is one moment early in the campaign that goes against the grain, however, and it is quite memorable. During the mission "Through Mud and Blood," the playable character is a tank driver named Daniel Edwards during the battle of Cambrai in 1918. At one point during a particularly tense battle, Edwards' tank is damaged and unable to move. Enemy German soldiers surround the tank and seem ready to destroy the tank. Edwards' commander orders him to release a homing pigeon which carries orders to fire artillery on their own position, potentially destroying the tank crew along with the enemy forces.

As Edwards releases the pigeon, the game shifts perspective from Edwards to the pigeon itself, and it comes under player control. Once the player takes control of the pigeon, they get an overhead view of the battlefield, which is filled with smoke, explosions, flaming aircraft, and ash. The accompanying music, however, is antithetical to the chaotic scene below. It is essentially a down-tempo, light piano piece with some soft strings underneath.[7] But the most dissonant part of this multimodal ensemble is the procedural mode. The pigeon cannot be killed; the player can only gently control the flight path of the pigeon towards its objective. In direct contradiction to the mayhem communicated via the visual mode, the procedural mode communicates something akin to CALM or SERENE. Given the rest of the game and the genre as a whole, we might expect the pigeon scene to require the player to frantically dodge enemy fire. Instead, the musical and procedural modes work in tandem against the visual modes and produce a memorable moment, one that many critics point to as a highlight of the game (e.g., Famularo 2016). This scene is a strong example of the potential of multimodal irony as generative and affecting rather than incoherent.

6 Conclusions

In this chapter, I have attempted to outline the value of adopting the procedural mode into multimodal analysis generally. Procedurality is able to make meaning in its own particular way and can influence other components within a multimodal ensemble. As I described through numerous close-playings of videogames across multiple genres, procedurality can either align with (consonance) or be set against (dissonance) the informational content produced by other modes in an ensemble. In either case, it affects the overall meaning of an ensemble as well as the other modes within it. Procedurality, therefore, deserves to be included in the catalogue of modes within multimodal studies, as it provides an analytical framework for examining interactive artifacts like videogames and other computational media. Furthermore, while procedural expression has for some time been recognized as a communicative strategy within game studies, it is not typically used in the context of multimodal semiotics. One advantage of adopting a multimodal approach to videogame analysis is that it provides researchers and critics with an established toolkit for dissecting and understanding how meaning potentials are conveyed through games. Previous frameworks such as ludonarrative dissonance and coherence and incoherence were tremendously useful for identifying the dynamic relationship of a game's component parts; however, multimodality allows for a more focused, granular methodology for understanding the semiotic mechanics of that relationship and its impact on meaning.

My primary research area is videogames, and so these are the texts I have used as exemplars. The procedural mode can be useful outside of the fairly narrow band of videogame analysis, however. My hope is that researchers outside of game studies might adopt and apply the principles of the procedural mode in non-ludic texts, from participatory online media to interactive museum kiosks. These artifacts are highly multimodal and require procedural forms of expression too. Conducting multimodal analyses similar to the critiques described here may therefore be useful for understanding how these non-ludic artifacts function semiotically. A greater focus on modal affordance and dynamism might even be helpful outside of analysis. From a design perspective, understanding which modes communicate certain types of information best and how they interact with each other might lead to richer, more engaging instances of interactive media. Designers of games and other interactive media already intuit the importance of multimodality in their creative processes. Multimodal semiotics, however, provides a vocabulary for better understanding how these modes work together to achieve a particular expressive goal.

Notes

1 Portions of this chapter have previously been published in *Multimodal Semiotics and Rhetoric in Videogames* (Hawreliak 2018).

2 Bogost's earlier book *Unit Operations: An Approach to Videogame Criticism* (2006) also includes an extended look at procedurality and procedural criticism.

3 Matthew Payne's book *Playing War: Military Video Games after 9/11* (2016) on post-9/11 war gaming does not use the language of semiotics or modality, but offers an in-depth look at how games differ from other media in their representation of the War on Terror.

4 For a counterargument against procedurality, see Miguel Sicart's (2011) essay "Against Procedurality."

5 It is likely that multimodal analyses will often be "incomplete," and it must be acknowledged that choosing to focus on certain modes over others does not indicate one mode's inherent semiotic superiority over the others.

6 Although narrative is often considered a mode, this is typically in a discursive rather than semiotic sense. For my purposes, narrative is expressed *through* semiotic modes and is not a semiotic mode in itself.

7 This is a very common device in action films.

References

Aarseth, E. (2004), "Genre trouble," in N. Wardrip-Fruin and P. Harrigan (eds.), *First Person: New Media as Story, Performance, and Game*, 45–55, Cambridge: The MIT Press.
American Psycho (2000), [Film] Dir. M. Harron, USA: Lions Gate Films.
Bateman, J.A. and K. Schmidt (2012), *Multimodal Film Analysis: How Films Mean*, New York: Routledge.
Bogost, I. (2006), *Unit Operations: An Approach to Videogame Criticism*, Cambridge: The MIT Press.

Bogost, I. (2007), *Persuasive Games: The Expressive Power of Videogames*, Cambridge: The MIT Press.

Bogost, I. (2009), "Videogames are a mess," *Bogost.com*, September 3. Available online: http://bogost.com/writing/videogames_are_a_mess/ (accessed August 10, 2017).

Bolter, J.D. and R. Grusin (2000), *Remediation: Understanding New Media*, Cambridge: The MIT Press.

Burn, A. and G. Schott (2004), "Heavy hero or digital dummy: Multimodal player–avatar relations in *Final Fantasy VII*," *Visual Communication*, 3 (2): 213–33.

Caillois, R. (1961), *Man, Play, and Games*, trans. M. Barash, Champaign: University of Illinois Press.

Chion, M. (1994), *Audio-Vision: Sound on Screen*, trans. C. Gorbman, New York: Columbia University Press.

Crawford-Cammiciottoli, B. and I. Fortanet-Gomez (2015), "Introduction," in B. Crawford-Cammiciottoli and I. Fortanet-Gomez (eds.), *Multimodal Analysis in Academic Settings: From Research to Teaching*, 1–16, New York: Routledge.

Eisenstein, S. (1977), *Film Form: Essays in Film Theory*, trans. J. Leyda, New York and London: Harcourt, Brace & World Inc.

Ensslin, A. (2012), *The Language of Gaming*, New York: Palgrave Macmillan.

Ensslin, A. (2017), "Future modes: How 'new' media transforms communicative meaning and negotiates relationships," in C. Cotter and D. Perrin (eds.), *The Routledge Handbook of Language and Media*, 309–24, Abingdon: Routledge.

Famularo, J. (2016), "*Battlefield 1* reminds us of the cost of war with a pigeon," *Inverse*, October 26. Available online: https://www.inverse.com/article/22751-ba ttlefield-1-pigeon-scene-emotionally-devastating (accessed May 25, 2017).

Forceville, C. (2016), "Visual and multimodal metaphor in film: Charting the field," in K. Fahlenbrach (ed.), *Embodied Metaphors in Film, Television, and Video Games*, 17–32, New York: Routledge.

Flanagan, M. (2009), *Critical Play: Radical Game Design*, Cambridge: The MIT Press.

Gagnon, F. (2010), "Invading your hearts and minds: *Call of Duty* and the (re)writing of militarism in U.S. digital games and popular culture," *European Journal of American Studies*, 5 (3). Available online: https://journals.openedit ion.org/ejas/8831 (accessed August 24, 2018).

Galloway, A. (2006), *Gaming: Essays on Algorithmic Culture*, Minneapolis: University of Minnesota Press.

Gee, J.P. (2003), *What Video Games Have to Teach us About Learning and Literacy*, New York: Palgrave Macmillan.

Gee., J.P. (2013), "Proactive design theories of sign use: Reflections on Gunther Kress," in M. Bock and N. Pachler (eds.), *Multimodality and Social Semiosis: Communication, Meaning-Making, and Learning in the Work of Gunther Kress*, 43–53, New York: Routledge.

Gibbons, A. (2012), *Multimodality, Cognition, and Experimental Literature*, New York: Routledge.

Gray, K. (2015), *Race, Gender, and Deviance in Xbox Live: Theoretical Perspectives from the Virtual Margins*, Abingdon: Routledge.

Harper, T. (2011), "Rules, rhetoric and genre: Procedural rhetoric in *Persona 3*," *Games and Culture*, 6 (5): 395–413.

Hawreliak, J. (2018), *Multimodal Semiotics and Rhetoric in Videogames*, New York: Routledge.

Hiippala, T. (2015), *The Structure of Multimodal Documents: An Empirical Approach*, New York: Routledge.

Hocking, C. (2007), "Ludonarrative dissonance in *Bioshock*," *Click Nothing*, October 7. Available online: http://clicknothing.typepad.com/click_nothing /2007/10/ludonarrative-d.html (accessed February 5, 2017).

Huizinga, J. (1955), *Homo Ludens: A Study of the Play Element in Culture*, Boston: Beacon Press.

Jaimes, A. and N. Sebe (2005), "Multimodal human computer interaction: A survey," in *Proceedings from IEEE International Workshop on Human Computer Interaction in Conjunction with ICCV*, Beijing.

Jewitt, C. (2009a), "Introduction," in C. Jewitt (ed.), *The Routledge Handbook of Multimodal Analysis*, 1–13, New York: Routledge.

Jewitt, C. (2009b), "An Introduction to multimodality," in C. Jewitt (ed.), *The Routledge Handbook of Multimodal Analysis*, 14–27, New York: Routledge.

Kress, G. (1993), "Against arbitrariness: The social production of the sign as a foundational issue in Critical Discourse Analysis," *Discourse & Society*, 4 (2): 169–91.

Kress, G. (2009), *Multimodality: A Social Semiotic Approach to Contemporary Communication*, New York: Routledge.

Kress, G. and T. van Leeuwen (2001), *Multimodal Discourse: The Modes and Media of Contemporary Communication*, New York: Bloomsbury Academic.

Kromhout, R. and C. Forceville (2013), "Life is a journey: Source-path goal structure in the videogames *Half-Life 2*, *Heavy Rain* and *Grim Fandango*," *Metaphor and the Social World*, 3 (1): 100–16.

Lakoff, G. and M. Johnson (1980), *Metaphors We Live By*, Chicago: University of Chicago Press.

Layne, A. (2015), "Procedural ethics: Expanding the scope of procedures in games," *First Person Scholar*, January 7. Available online: http://www.firstpersonscholar. com/procedural-ethics/ (accessed June 21, 2018).

Lemke, J. (1998), "Multiplying meaning: Visual and verbal semiotics in scientific text," in J.R. Martin and R. Veel (eds.), *Reading Science: Critical and Functional Perspectives on Discourses of Science*, 87–113, London: Routledge.

Machin, D. and T. van Leeuwen (2007), *Global Media Discourse: A Critical Introduction*, New York: Routledge.

Manovich, L. (2001), *The Language of New Media*, Cambridge: The MIT Press.

Mateas, M. (2005), "Procedural literacy: Educating the new media practitioner," *On the Horizon*, Special Issue: *Future of Games, Simulations and Interactive Media in Learning Contexts*, 13 (1). Available online: http://homes.lmc.gatech. edu/~mateas/publications/MateasOTH2005.pdf (accessed March 15, 2017).

Murray, J. (1997), *Hamlet on the Holodeck: The Future of Narrative in Cyberspace*, Cambridge: The MIT Press.

Norris, S. (2004), *Analyzing Multimodal Interaction: A Methodological Framework*, New York: Routledge.

O'Halloran, K. (2004), "Introduction," in K. O'Halloran (ed.), *Multimodal Discourse Analysis: Systemic-Functional Perspectives*, 1–10, New York: Continuum.

Payne, M. (2016), *Playing War: Military Video Games after 9/11*, New York: NYU Press.

Polansky, L. (2015), "Coherence and dissonance," *Sufficiently Human*, August 31. Available online: http://sufficientlyhuman.com/archives/1006 (accessed August 30, 2017).

Rowsell, J. (2013), *Working with Multimodality: Rethinking Literacy in a Digital Age*, London: Routledge.

Sicart, M. (2011), "Against procedurality," *Game Studies: The International Journal of Computer Game Research*, 11 (3), December. Available online: http://gam estudies.org/1103/articles/sicart_ap/ (accessed June 30, 2017).

Sindoni, M.G. (2013), *Spoken and Written Discourse in Online Interactions: A Multimodal Approach*, New York: Routledge.

Stahl, R. (2006), "Have you played the War on Terror?" *Critical Studies in Media Communication*, 23 (2): 112–30.

Toh, W. (2015), *A Multimodal Discourse Analysis of Video Games: A Ludonarrative Model*, Singapore: National University of Singapore.

Van Leeuwen, T. (2005), *Introducing Social Semiotics*, London: Routledge.

Voorhees, G. (2009), "The character of difference: Procedurality, rhetoric, and roleplaying games," *Game Studies: The International Journal of Computer Game Research*, 9 (2), November. Available online: http://gamestudies.org/0902/ articles/voorhees (accessed July 12, 2017).

Walker, A. (2016), "Why *Mafia III* should tackle race head on," *Waypoint*, August 23. Available online: https://www.vice.com/en_ca/article/avav5j/why-mafia-iii-should-tackle-race-head-on-58477ff9ab4b6a022a05f752 (accessed August 14, 2017).

Videogames

Activision (2007), *Call of Duty 4: Modern Warfare*. M. Rubin [Producer].

Bethesda Softworks (2008), *Fallout 3*. G. Carter [Producer].

BioWare/Microsoft Game Studios (2007), *Mass Effect*. C. Hudson [Producer].

Blackmill Games (2015), *Verdun: 1914–1918*.

Deep Silver (2013), *Saints Row IV*. J. Boone [Producer].

Hangar 13 (2016), *Mafia III*. M. Orenich [Producer].

11

The Player Experience
of *BioShock*

A Theory of Ludonarrative
Relationships

Weimin Toh

1 Introduction

Player experience research (Sánchez et al. 2009; Chu et al. 2011; Drachen et al. 2018) is important and integral to the process of analyzing and developing games, as it provides us with a deeper understanding of *how* to design games to engage the players (Lindley 2004). During the playing process, the players may focus on the construction of mental or situation models in narrative comprehension (Zwaan et al. 1995; Arthur et al. 2002; Cardona-Rivera and Young 2012; McGloin et al. 2018) but they may switch to gameplay strategizing (Dor 2014, 2018) when they encounter the challenges in the game. A third perspective involves the narrative interpretation of the gameplay events or player character's actions as the player controls the character to interact with the gameworld (Larsen and Schoenau-Fog 2016; Sim and Mitchell 2017), which implies that players perceive the narrative and gameplay as a unified whole (Frasca 2003; Koenitz 2010, 2015).

This chapter aims to further our understanding of the structure of the videogame in the form of the *ludonarrative* (see Toh 2015, 2018) based on the lived experience of the players. The ludonarrative is defined as a frame for videogames which conceptualizes the gameplay, the narrative, and the player as a whole (Toh 2015, 2018). To date, very few studies have been conducted on how the players' understandings of ludonarrative relationships in a videogame contribute to their interactions with the videogames, their gameplay performances (see Nguyen 2016), and their narration-commentary (see Kerttula 2016) in gameplay videos. Drawing on Dena (2010), ludonarrative relationships are defined here as common semiotic principles, which not only serve to interlink the narrative and gameplay but also reflect the players' experiences because I have co-constructed them with the players during the interview portion of this research (Toh 2015, 2018). Examples of ludonarrative relationships include *ludonarrative dissonance* and *ludonarrative resonance*, types which are drawn from Hocking (2007) and Watssman (2012), respectively. Ludonarrative dissonance refers to the disjunction between the narrative and the gameplay (see also Hawreliak, this volume), and ludonarrative resonance refers to the congruence between the narrative and the gameplay (Watssman 2012). The contribution of the ludonarrative perspective to videogame discourse (analysis) is the foregrounding of the ways in which meanings are communicated by videogames and their makers to players via multimodality and narrative (see Ensslin 2012). It also helps players understand and communicate meaning of the ludonarrative relationships (Toh 2015, 2018) to the audience in gameplay videos, as discussed in this chapter.

Thus, the aims of this chapter are twofold. First, the discussion here aims to enable us to understand how the players' interactions and gameplay performances (see Nguyen 2016) in gameplay videos are actualized based on their understandings of the videogames' ludonarrative relationships (Toh 2015, 2018). Following Burnwell and Miller (2016), the player's understanding of the videogame's contents has been conceptualized as the process of meaning making within games, which is one of the two key characteristics of "Let's Play" practices, the other key characteristic being the player's mobilization of literacies associated with remix and appropriation. The remix literacy comprises the technical, aesthetic, and critical knowledge required to mix gameplay, commentary, and video into an effective Let's Play (Burnwell and Miller 2016). The appropriation literacy has been identified as the ability to meaningfully remix and sample media content (Jenkins et al. 2009). The player's meaning making of the ludonarrative relationships in a videogame can be derived from the narration-commentary in gameplay video recordings; in this chapter, the examination focuses on recordings collected by the researcher and on the empirical data from the open-ended qualitative interviews in Toh (2015, 2018).

Secondly, this chapter aims to analyze *how* the interaction of the player in the videogame with the player's simultaneous performance (see Nguyen 2016) of the gameplay in the video contributes to the actualization of the player's narration-commentary (see Kerttula 2016). The player's interaction,

Players' understanding of ludonarrative relationships (Toh 2015)

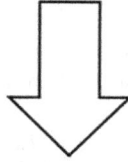

Players' interaction and performance of the gameplay in the video recordings

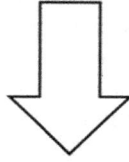

Players' narration-commentary in the video recordings

FIGURE 11.1 *Ludonarrative relationships and the players' experiences in gameplay videos.*

performance, and narration-commentary are all key elements in the video recordings first gathered in Toh (2015, 2018). The dual aims of this chapter are summarized in Figure 11.1. The implications of the analysis of the gameplay videos from the proposed model in this chapter will enable videogame developers and researchers to obtain a better understanding of how players make meaning and interact with the videogame content (ludonarrative) to produce their gameplay videos (Kerttula 2016).

2 Literature review

Think-aloud protocol (van Someren et al. 1994; Theodorou 2010; Tan et al. 2014; Eccles and Arsal 2017) or concurrent think-aloud protocol (Anders and Simon 1993; Barnum 2002; Woelke and Pelzer 2017) as an observational technique has for a long time been applied to study aggression (e.g., DiLiberto et al. 2002) and has also been used to study the structure of skill and mechanisms of skill acquisition in videogames (Boot et al. 2016). The thought processes of players, conveyed by the empirical data from my earlier study (Toh 2015, 2018) and foregrounded in players' narration-commentary, offer evidence of the relationship between players' meaning-making and their interactions, performances of the gameplay, and narration-commentary. Through the comparison of different (players') experiences which are elicited from their think-aloud protocol, we can obtain a more holistic understanding of what it means to play a videogame. Therefore, gameplay recordings provide an invaluable and indispensable supplement to videogame play experiential analysis.

A frame analysis has been proposed to divide the play phenomenon into different layers (Recktenwald 2014). The different layers are conceptualized as the *Let's Play Onion*, consisting of *Let's Player, game, YouTube*, and *Let's Play* conventions (Figure 11.2). Let's Play conventions refer to the conventions of creating Let's Play videos, in which the player's commentary is preplanned and monological, with smooth transitions and no turn-taking (Recktenwald 2014). From this conception, the analysis proposes a three-legged pattern of interaction between the Let's Player (metacommentary), game (synchronous interactions with nonplayer characters [NPCs]), and YouTube audience (and commenters). The Let's Player's commentary is the entry point for studying communication patterns with both the NPCs in the game, moment-by-moment, and the YouTube audience, asynchronously. In this chapter, I will focus on the two innermost layers of this model—*player* and *game*—because the focus is on the study of *how* the players make sense of the ludonarrative relationships in the game.

Radde-Antweiler et al. (2014) propose a three-level approach to analyze players' video recordings. Their framework conceptualizes plays in terms of the game level, the play level (comprising the players' performance, live comments, and self-recorded video images), and the comments on a play. Plays have also been conceptualized as both the videogame narrative, which is the result of an interaction between player and game, and the player's story

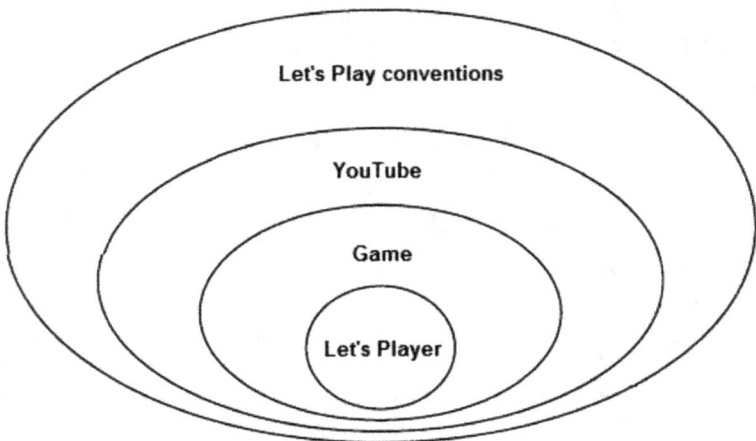

FIGURE 11.2 *The* Let's Play Onion *and its different layers (reproduced from Recktenwald 2014).*

(see Kerttula 2016) in the form of the player's narration (reaction to game events, explanation of choices, and general opinions on the game) (see de Rijk 2016). Smith et al. (2013) argue that in a play broadcast, the player/performer is simultaneously the commentator. In this way, the player is granted more control over the development of the narrative experience and can curate a play in ways beyond the scope of, for example, a live e-sports broadcast. From this angle, the player's experience of the videogame is revealed to the audience. Even though a nuanced analysis could be conducted by dividing the narration of a play into the different narrative elements and by explaining how these elements together form the player's story, the relationship between the player's story, the player's interaction and performance during gameplay, and the videogame's ludonarrative has not been analyzed. In this chapter, I will fill this research gap by using empirical data (gathered from audio interviews with players and from gameplay video recordings) to explain how the player's meaning making of the ludonarrative contributes to the interaction, performance, and narration-commentary in play videos.

3 The current study

The contribution of this study is the description of the manner in which players understand, perform, and communicate meaning to the audience in play videos from a ludonarrative perspective. The methodological approach detailing the think-aloud protocol in which the players' understanding of the ludonarrative relationships in videogames is elicited and recorded will be explained in the next section. The current study addresses the following three questions:

(1) How do players understand the ludonarrative relationships in the videogame? (discussed in Section 4.4)

(2) How do players' understandings of the ludonarrative relationships influence their interactions and performances of the gameplay in the play videos? (discussed in Section 5.1)

(3) How do the players' understandings of the ludonarrative relationships influence their narration-commentary in the play videos? (discussed in Section 5.2)

4 Methodology

The data reported in this chapter was originally gathered as part of my PhD dissertation research (Toh 2015, 2018), where a theoretical model is created based on the empirical study of player-participants. My main research aim is to explore players' understandings of how the videogame narrative and gameplay combine to create meaning. The contribution provided in this chapter is in the transfer and application of the ludonarrative relationships from the ludonarrative model to the analysis and description of the players' experience in their interaction, performance, and narration-commentary within the videos. The research findings will be discussed in Section 5.

4.1 Description of *BioShock* and rationale for choice of game

BioShock (2K Boston 2007) is a narrative-based, first-person shooter, which combines scripted narrative sequences with multilinear gameplay. The narrative is set in 1960 in the underwater city of Rapture. The player controls a blank-slate character (Jack) who does not speak and has to explore Rapture to uncover its secrets. *BioShock* incorporates role-playing elements, and the player is provided with different options to customize the Jack character, formulate different strategies to overcome the challenges, and choose whether to save or harvest the NPCs known as the Little Sisters— young girls in Rapture who have been genetically altered and mentally conditioned to gather resources. The player's choice of saving or harvesting the Little Sisters will contribute to one of the three narrative endings. Other than the scripted narrative, the embedded narrative, such as the audio recordings scattered throughout the gameworld, provides backstory for the player. Some of the audio recordings also complement the gameplay by providing hints to the player.

The main criterion for choosing *BioShock* was that the selected game had to provide both narrative and gameplay choices to the players, such that analyzing their choices would facilitate an understanding of how the players interpret the ludonarrative relationships. Related to this study, the term "ludonarrative" (Hocking 2007) originated from a critique of *BioShock* which gives the players' choices additional weight. The second criterion was that both the narrative and gameplay must have some relationship to each other, such that the players would not focus solely on either the narrative or the gameplay. Finally, *BioShock* was chosen because it is a popular commercial game with mass appeal; player-participants were not only familiar with the game, but also interested in playing it.

4.2 Participants

The four *BioShock* participants were recruited from the National University of Singapore. Snowball sampling was used, in which selected participants reached out to friends who fit the criteria of the study. Inclusion criteria included being aged between 17 and 35. The actual age range of the

TABLE 11.1 *Participants' profiles*

Participant no.	Age	Gender	Game experience
1	19	F	6–10 years
2	23	M	11–15 years
3	21	M	11–15 years
4	23	M	11–15 years

recruited *BioShock* participants was from 19 to 23 years old. In this chapter, I will use the data from four participants who completed the entire study, playing *BioShock* to demonstrate the proposed theoretical model. Their demographic profiles are provided in Table 11.1. During the first session, which was conducted in the lab, the researcher recorded the PC gameplay of the four participants where they play the game. For the remaining sessions, the participants recorded their plays while playing at home.

4.3 Procedure

4.3.1 *Participant observation and first session interviews*

In Toh (2015, 2018), I adopted a phenomenological and hermeneutic perspective, conceptualizing the videogame as a game object (see Vella 2015). Thus, instead of focusing on the formal analysis (Willumsen 2018) of the videogame in relation to the "functional characteristics and components of game objects, and the relations between them" (Aarseth 2014: 484), the emphasis is shifted toward "the game as played, as referring to the object of study for game studies from the player's perspective" (Leino 2010: 6). Therefore, my method involved the observation of the participants' gameplay in the lab, followed by the first of two open-ended qualitative interviews (Toh 2015, 2018). During the first session, the participants were asked to play *BioShock* for one hour in the lab. They started at the beginning of the game and ended at the part where the game offered them the choice to save or harvest the first Little Sister. I record their gameplay using *Fraps* (PC) for reference during interviews. The participants were just starting out in the study, and I recorded their gameplay so that they would be focused on playing the game and learning the mechanics. Most of them were playing *BioShock* for the first time (though one participant had played *BioShock* more than a few years ago). At the end of the first gameplay session, I asked them about their experience of the narrative and gameplay of the videogames they had played, using the first set of general, open-ended interview questions; interviews ended with more specific questions about the ludonarrative relationships, such as the congruence and disjunction between the narrative and the gameplay.

4.3.2 Subsequent session

In order to collect naturalistic data and to facilitate the participants' creation of their gameplay recordings with voiceovers, the subsequent session was conducted in their homes. They were asked to complete the entire game, playing for approximately 14 hours, and were taught to record their gameplay using *Fraps*. The participants recorded a modified form of play videos for the study. First, their target audience for the video recordings was the researcher, rather than a public audience, and their videos were not uploaded to YouTube for public consumption. Even though the participants were instructed to record their natural reactions and commentary, the weakness of this approach is that the participants were, to a large extent, self-conscious (and reflective) about what they said in the videos. Secondly, besides recording their natural reactions to the gameplay and providing narration-commentary and opinions, the participants were instructed to verbalize their gameplay experience using a think-aloud protocol (see van Someren et al. 1994; Theodorou 2010). Specifically, they were asked to explain *how* and *why* they made specific narrative and gameplay choices, based on their understanding of the ludonarrative relationships in the game. They were also asked to verbalize their interpretations of the videogame narrative in both cutscenes and gameplay.

4.3.3 Final session

The final session involved asking the participants more specific open-ended interview questions after review of the participants' video recordings. I conducted the retrospective protocol analysis (see Dorst and Dijkhuis 1995; Ericsson and Simon 1993) by talking with the participants about their play experiences (narrative, gameplay, and ludonarrative interpretations) and by reviewing the gameplay recordings together with them.

The postgame interview was used to clarify questions regarding the participants' narration-commentary in the play videos. I asked additional questions in order to further understand the players' use of both gameplay and narrative (ludonarrative) information for their decision-making during critical moments in the videogame. The final session lasted between 2 and 6 hours and was broken down into 2–3-hour sessions, depending on the participants. The "negotiation of interpretations" (see Toh 2015, 2018) was a core interaction dynamic during the final session interviews. This method is similar to the technique known as *member checks* in qualitative research, where the researcher obtains informant feedback to help improve the accuracy, credibility, validity, and transferability of a study (Yanow and

Schwartz-Shea 2006). During the retrospective protocol analysis, I clarified specific dialogue choices as gameplay option, narrative event, or both, with the participants. The "negotiation of interpretations" (Toh 2015, 2018) refers to the way in which the researcher explained the concepts in the interview questions to the participants and how this could have influenced the participants' responses. It also refers to the way the participants clarified the questions with the researcher to provide relevant answers.

4.3.4 Development of the ludonarrative model

The method used for collecting, interpreting, and analyzing the gameplay recordings is similar, in a general sense, to Kirschner and Williams' (2014) four-step analysis process of processual video data used for gameplay reviews with their participants. Step one of their method involves recording the gameplay video. Step two involves the researcher's observation of the video recordings with the participants to contextualize the record and interpret the *how* and *why* of actions. Step three involves conducting a gameplay review, wherein the researcher discusses the player's narration-commentary and interpretations. Step four analyzes the gameplay review and integrates data from multiple players.

When the researcher is able to involve the participants in feedback on their interpretations of the claims made, the validity of the researcher's claims is increased, as the researcher is able to obtain multiple players' interpretations of the *why* of specific gameplay actions. Finally, my original ludonarrative model (Toh 2015, 2018) was developed, refined, and validated using the data from the study's participants, including the retrospective protocol analysis (see Dorst and Dijkhuis 1995; Ericsson and Simon 1993) and the gameplay recordings. The research findings in Section 4.4 form another contribution in this chapter to provide the players' experience in the form of their understanding of the ludonarrative relationships to produce the play videos.

4.3.5 Ludonarrative relationships

The empirical analysis was conducted to (dis)prove these categories. The third main category of "ludonarrative irrelevance" (Toh 2015, 2018) was induced from the empirical data and refers to instances where the narrative and gameplay have a weak relationship with each other: neither conflicting, as in dissonance, nor harmonizing, as in resonance. The term *irrelevance* does not mean that the game is designed to offer irrelevant information, but rather refers to those moments when players fail to *perceive* relevance, meaning that the information has no impact on their gameplay. The ludonarrative dissonance subcategories of *contrast*, *metaphor*, and *demotivation*; the ludonarrative resonance subcategories of *motivation*,

TABLE 11.2 *Relevant subcategories of ludonarrative dissonance (Toh 2015, 2018)*

Ludonarrative dissonance subcategory	Description
Contrast	Players' narrative interpretations contrast with their gameplay actions.
Metaphor	Gameplay mechanics presented to the players in the form of fictional visual representation obstruct/delay their learning of the mechanics and vice versa.
Demotivation	Narrative demotivates the player from achieving the gameplay goals, choosing specific gameplay choices, or performing specific gameplay actions. The gameplay demotivates the player from progressing the narrative partly due to the lack of gameplay variety.

TABLE 11.3 *Relevant subcategories of ludonarrative resonance (Toh 2015, 2018)*

Ludonarrative resonance subcategory	Description
Motivation	Narrative motivates the player to achieve the gameplay goals and vice versa.
Semiotic metaphor	Metaphorical shifts (O'Halloran 2008) occur when the functional status of gameplay elements is not preserved as new narrative elements and gameplay mechanics are introduced.
Consequence/ contingency	Player's gameplay action creates a narrative outcome and vice versa, and the consequence can only be observed after some time.

semiotic metaphor, and *consequence/contingency*; and the ludonarrative irrelevance subcategories of *gameplay focus* and *guidance* were induced from the empirical data.

The relevant subcategories of ludonarrative dissonance, ludonarrative resonance and ludonarrative irrelevance discussed in this chapter are shown in Tables 11.2, 11.3, and 11.4, respectively.

In what follows, I shall provide further detail about how I arrived at these categories and how they relate to specific player-participant experiences.

TABLE 11.4 *Relevant subcategories of ludonarrative irrelevance (Toh 2015, 2018)*

Ludonarrative irrelevance subcategory	Description
Gameplay focus	Players focus more on the gameplay, pushing the narrative to the background.
Guidance	Guidance given by the narrative for the gameplay and vice versa is irrelevant to the players because the object is either obvious or implicit.

4.4 Analysis and results

4.4.1 Ludonarrative dissonance contrast

If the gameplay (outcome) prohibits the player from carrying out certain actions, but the narrative (interpretation) gives the player reason for doing them, this results in a ludonarrative dissonance contrast (see Toh 2015, 2018). In one example from *BioShock*, a narrative event occurs when the player flips the switch to open the door to a flooded cave/control room for the antagonist, Frank Fontaine/Atlas. A trap unleashes hordes of common enemies in the game, called *Ryan's Splicers*, to obstruct the player's progress. Participant 1 interpreted the narrative as wanting her to save Frank Fontaine/ Atlas (who has been guiding her in the game), but the gameplay did not allow her to reach him, only allowing her to fight the Splicers. Therefore, there was a ludonarrative dissonance contrast in this player's experience. The player had no choice but to perform the gameplay action of defeating Ryan's Splicers before the game could proceed. In her narration-commentary in her play video, the player highlighted her ludonarrative dissonance contrast thus:

> The narrative makes me like the character Atlas and now they don't let me go and save him and I feel very, very upset about it.

4.4.2 Ludonarrative dissonance metaphor

Sylvester (2013) defines metaphor in games as a concept "giving something new the appearance of something familiar to make it easier to understand" (220–21). Sylvester (2013) explains that metaphor is one of the most important concepts facilitating the player's learning of the gameplay mechanics by wrapping those mechanics in fiction and narrative elements to communicate information quickly. Participant 1 mentioned that she chose not to use some weapons, such as the crossbow, chemical thrower, and grenade launcher, based on their visual appearance of having a large physical size. She indicated that the huge physical size of the weapons obstructed her view; she chose not to use them, initially. She perceived them as hindering her gameplay and, by extension, her learning of the weapons' mechanics. It was only through discussion with participant 4 that she realized the potential functional usefulness of the weapons. Based on her self-generated interpretations of embodied experience (Howe 2017) and the lack of a mapping (see Lakoff and Johnson 1980) between the source domain (weapons' fictional visual representation) and the target domain (functional gameplay usefulness of the weapons), participant 1 experienced a ludonarrative dissonance metaphor when she initially picked up these weapons and, therefore, she chose not to interact and perform the gameplay with them. When she did not use them, she could not discover their gameplay function and usefulness. In her narration-commentary in the play video, she said:

> I think this weapon is quite hard to use as it covers half of my screen, so it will be very annoying to me because I cannot see the enemies when I am using it. If I have to walk around the whole map holding it covering half of the screen, it will make me very vulnerable to things that jump from my side, so I'd rather hold on to my wrench or something.

4.4.3 Ludonarrative dissonance demotivation

When the narrative demotivates the player from achieving the gameplay goals, choosing specific gameplay choices, or performing specific gameplay actions, ludonarrative dissonance demotivation is experienced (see Toh 2015, 2018). Ludonarrative dissonance demotivation can also refer to the gameplay demotivating the player from progressing the narrative partly due to the lack of gameplay variety. In *BioShock*, a narrative event occurs which requires the player to become a *Big Daddy*, a genetically enhanced human being whose main purpose is to protect the NPCs known as the Little Sisters. To do so, the player has to embark on the multipart goal of gathering the various parts of the Big Daddy suit, a gameplay object which, when completed, increases the player's resistance to damage toward the end part of the game. However, in participant 2's playing experience, he had become familiarized with the multipart goals given earlier in the game, such as gathering the components to create the Lazarus Vector (a chemical solution that would bring dead plants back to life) and gathering the various components to assemble the E.M.P. bomb (a device created to overload the energy core), to the extent that this event felt repetitive. Participant 2 mentioned during the interview that the multipart goal given by the narrative event was used to prolong the game's play time without offering new variety. The reward for becoming a Big Daddy is also not very significant, as he mentioned that, unexpectedly, he did not feel more powerful as a result. Therefore, he did not feel that the gameplay in the latter part of *BioShock* was fun; it felt like a chore to him. He merely went through the motions of playing the latter part of the game, as he wanted to finish the game as fast as possible. In his narration-commentary in the play video, he highlighted his ludonarrative dissonance demotivation:

> It's very boring this part. Too many parts to find and I have to keep walking around back and forth the place. I get lost here quite easily and for a long time. It is like dragging the game. Still have to find Big Daddy, as in become Big Daddy to get the Little Sisters to open the door. Why can't I have a potion that transforms me into a Little Sister? Then I can just crawl through.

4.4.4 Ludonarrative resonance motivation

A new narrative event may motivate the player's gameplay. In the latter half of the game, the antagonist Frank Fontaine initiates a new narrative event by telling the player in a scripted dialogue that the player's health will be reduced in intervals until he dies. Participant 3 mentioned that he was afraid that his character would die before he could complete the game. In his playing experience, there was ludonarrative resonance motivation. His subsequent interaction and performance of the gameplay in the play video was motivated by the new narrative event to find the antidote to remove the negative effects. In his narration-commentary, he indicated his ludonarrative resonance motivation thus:

> Frank Fontaine is using Code Yellow as a way to control me by reducing my maximum health at regular intervals until I find the antidote to reverse the negative effects of his control. I need to quickly find the antidote so that my life will not be lowered anymore until it becomes zero and my character dies.

4.4.5 Ludonarrative resonance semiotic metaphor

A *semiotic metaphor* is defined as a metaphorical shift that occurs when an element's functional status is not preserved and new elements are introduced (O'Halloran 2008). When applied to videogames, metaphorical shifts can occur when the gameplay elements' functional status is not preserved as new narrative elements and gameplay mechanics are introduced (Toh 2015, 2018). In *BioShock*, a narrative event occurs when the player character takes the first dose of Lot 192 in Olympus Heights (a location in *BioShock*), which successfully frees the player character from the antagonist Frank Fontaine's remaining influence. However, it has a side effect, which makes the player lose control of her active plasmid powers as they become randomized, as experienced by participant 1 during her gameplay. The narrative (the NPC Tenenbaum) explains to the player that the randomization is due to Lot 192 reorganizing the player character's entire plasmid structure. Therefore, the functional status of the plasmid powers is no longer preserved, and a new gameplay mechanic of randomized plasmid powers is introduced. During the interview, participant 1 mentioned that she was able to understand that the narrative event caused the randomization of her character's plasmid powers:

> Ya it's like they are trying to intersect the story and the gameplay, like trying to show how you need the freaking Lot 192 or something like that. It was so annoying. I was so angry with that part, but that part was really tough in my opinion. Mm. Because every time they changed, you would automatically switch to a plasmid thing which it might not be a fighting plasmid at that moment. So I keep attacking the Big Daddy [an in-game boss] with the stupid plasmid which doesn't help. So it was very annoying at that part. I guess they are just trying to link the gameplay and the narrative at that part. Make the link even clearer. But tension not so much, ah.

The shift in the functional status of the plasmid powers created a gameplay challenge for participant 1, as shown by her interaction and gameplay performance in the play video when she was fighting the boss called Big Daddy. She used the electric gel from the chemical-thrower weapon to stun and deplete the health bar of Big Daddy. However, when her plasmid power randomly shifted in the midst of the battle, the weapon could not be used simultaneously. Therefore, the functional status of participant 1's weapon was not preserved when the gameplay mechanics of the randomized power took control away from the player.

4.4.6 Ludonarrative resonance consequence/contingency

Participant 1 reflected that she understood the choice of saving or harvesting (killing) the Little Sisters to be both a narrative and gameplay (ludonarrative) choice with different narrative and gameplay (ludonarrative) consequences:

> The union between the gameplay and the narrative is very good. The narrative is brought to you by the text, the gameplay builds up the tension and shows you how twisted and ugly this Rapture is… so the combination of these two things, ya, that is why I chose to save the girl instead.

When the player encounters the first Little Sister, one NPC, Brigid Tenenbaum, will implore the player to save them and promise special gameplay rewards for the player. This narrative event is connected to all the situations when the player is provided with the choice to save or harvest the Little Sisters. For instance, choosing to save all of them will result in the consequence of the "good" ending for the player, whereas choosing to harvest all of them will result in the consequence of the "bad" ending. Choosing a combination of saving and harvesting will result in a less harsh version of the bad ending. The choice of saving or harvesting the Little Sisters is part of the gameplay, because the choice will provide the player with different rewards. One of the gameplay rewards consists of the ADAM reward system. In the narrative, the game explains that it is a substance harvested and processed from a type of Sea Slug and can cause cellular division of stem cells into any cell type. In the gameplay, ADAM is used for character customization and to give the player character new abilities or improve the ones already owned when they are spent at a Gatherer's Garden.

Participant 1 mentioned that she was influenced by both the narrative and the gameplay (ludonarrative) consequences when the game gave her the choice to save or harvest the Little Sisters. From a gameplay perspective, she mentioned that she chose to save all the Little Sisters because she wanted to collect all the special plasmid power rewards from the Little Sisters' presents. She chose to save them all after finding out she would be rewarded for every three Little Sisters she saved. She also mentioned that the game repaid the player with an ADAM gameplay reward, even though that reward was slightly less than if she had harvested the Little Sisters. The ADAM gameplay resource was not as important for her gameplay progress, though, as she mentioned that she did not perform a lot of upgrades.

From a narrative perspective, participant 1 mentioned during the interview that, because of Tenenbaum's promise, the narrative pushed her to save all the Little Sisters:

> I don't know maybe it's just me but I feel like, you know, you see how like the way Tenenbaum is, like, pleading you at that point. Give them a second life and everything like that and her own motherly change and everything like that. It just shows you have to save the Little Sisters... I mean if they wanted to push you to the evil side, maybe like they can have some maybe Atlas can keep telling you like, "Oh, maybe you should kill them to get more stuff."

In the play video, participant 1 reflected, during the narration-commentary, that the narrative caused her to save all the Little Sisters:

> Because if I harvest her, she will die, and I feel like from what the game has shown me, this world is very twisted, and it is mostly a result of

Andrew Ryan's and doctor Steinman. Like most of the other people, they do not really know what they are getting into. They don't know that the gene modifications and surgeries and things like that is so detrimental to them, so I didn't want to kill the child because I feel, from what I understand, she did not have a say either just like the rest of the Splicers. So that is why I choose to rescue her instead.

4.4.7 Ludonarrative irrelevance guidance and gameplay focus

The guidance provided by the narrative for the gameplay can be irrelevant to the player when it is embedded in a semiotic code that may foreground narrative yet be neglected or ignored by ludically immersed players (Bell et al. 2018; Toh 2015, 2018). The concept of immersion which I adopt in this chapter follows Thon's (2008) concept of the player experience of psychological immersion due to the player's shifting attention to and the construction of situation models of different parts of the game. The player can shift attention to the gameplay (ludic immersion), the narrative (immersion), or the ludonarrative (immersion). The use of irrelevance is related to players' construction of meaning, rather than game design. In *BioShock*, many of the door codes are given in the narrative of the audio logs. Players who prefer to overcome the gameplay challenges, rather than listen to the narrative in the audio logs, often miss the door code hints. One of the more important door codes was found in the "Paparazzi" audio log to unlock Frank Fontaine's penthouse door, as the door could not be hacked. Participant 3 did not note the door code in the audio log and ended up searching online for the code to unlock the door. In participant 3's experience, there was ludonarrative irrelevance guidance when he chose to background the information given to him in the audio to guide him in the gameplay, such that it became unimportant *to him*, even though the information was, in fact, highly important. The subcategory proposed here aligns with Bell et al.'s (2018) argument which highlights how different types of immersion might overrule each other. For example, ludic immersion might (temporarily) cancel out narrative/audio immersion. The absence of the player's narration-commentary in the play video when he was engaged in the mini game highlights his ludonarrative irrelevance gameplay focus, as clarified during the final interview. As we can see from the following quote, the player explained that because he was focused on the environmental interaction and gameplay challenges to find a means to unlock the penthouse door, he did not provide commentary during gameplay as it would distract him:

No, I didn't take notice. That's why, because you see I was focused on the gameplay, so I didn't take note that the code was actually playing in the background.

5 Discussion

5.1 Players' understandings of ludonarrative relationships and performance of gameplay

The players' understandings of the ludonarrative relationships will influence their performances of the gameplay. For instance, participant 2, who had gone through the multipart gameplay goal of the story twice, experienced ludonarrative dissonance demotivation when playing the game. His subsequent performance of the gameplay was rushed and less nuanced, as shown in his play video, where he frequently became lost as he tried to progress through the game as quickly as possible. Participant 1, after saving three Little Sisters, understood that saving all the Little Sisters would offer more attractive special rewards. Subsequently, participant 1 chose to save all the Little Sisters and to collect all the special rewards. Participant 1's experience of ludonarrative resonance semiotic metaphor also caused her to become more careful in the gameplay, especially when using the chemical thrower equipped with the electric gel to fight the Big Daddy boss. After her player character's near-death experience during the Big Daddy fight, she was encouraged in the gameplay to find the antidote to restore control to her character as soon as possible. Her ludonarrative experience of ludonarrative resonance semiotic metaphor therefore serves as a motivation for her gameplay.

5.2 Players' understandings of ludonarrative relationships and narration-commentary

The players' reflective narrations (Kerttula 2016) highlight their understandings of the ludonarrative relationships in the game. For instance, when participant 1 mentioned, in the video recording, that the narrative made her feel sympathy toward Fontaine but the gameplay prevented her from saving him, she felt very upset about the linearity of the game. The narration-commentary therefore highlights her experience of ludonarrative dissonance contrast. On the other hand, the absence of the player's narration-commentary in participant 4's play video highlights the dependence of players' experience of ludonarrative relationships on their different personality and playstyle preferences. For instance, participants who were more focused on overcoming the gameplay challenges tended not to narrate-comment, and their silence reflects the experience of ludonarrative irrelevance gameplay focus and ludonarrative irrelevance guidance.

6 Limitations

In this exploratory study, a qualitative approach in the form of interviews was adopted to provide empirical data to develop a theoretical model of ludonarrative relationships. As a result of the qualitative approach used and the small number of participants, the results obtained from the study cannot be generalizable. Another limitation is that the open-ended interview approach tends to be subjective and did not fully explain exactly why a specific participant understood a ludonarrative relationship or interpreted the videogame narrative in a specific manner. The cognition or emotions of the participants were not easily accessed in the interviews, as participants may choose to withhold information or certain responses.

The selection of the participants for the study is also a limitation. Some participants were recruited from the researcher's gamer friends, creating potential bias toward supporting the researcher's goals. Finally, some participants were obtained via snowball sampling, with current participants introducing friends to the study. Therefore, not all individuals fulfilled the study's criteria, which include a balanced gender ratio (instead of the 1:3 female to male ratio in the study), and players from diverse backgrounds (instead of having university undergraduates for the study). This means that the study's results are skewed toward male rather than female gamers and 100% college-aged gamers.

7 Conclusions and implications

This study has proposed a theoretical model for the ludonarrative relationships of *BioShock* based on the players' experience elicited from their play videos and interviews. Building on prior studies (such as Recktenwald 2014), it has focused on the two innermost layers of Recktenwald's (2014) Let's Play Onion to elaborate the relationship between the videogame's ludonarrative, the player's performance of the gameplay, and his/her narration-commentary in the videos. From a phenomenological perspective, it is important to understand how the players' performances of the gameplay and their narration-commentary are influenced by their experiences of the ludonarrative relationships of the videogame. In this chapter, I have shown how both the think-aloud protocol and the analysis of the gameplay recordings of the participants can be used to foreground the players' understandings of the ludonarrative relationships, thus demonstrating the transferability of the model in order to understand the players' experience in play videos.

Through study of players' recordings, game designers and researchers can better understand the internal workings (ludonarrative) of the videogame from the players' perspectives, because they have access to the players' experience (through gameplay performance and narration-commentary). Review of play videos offers opportunities for detailed study of moment-by-moment interactions and the specific strategies players utilize in response to the ludonarrative structures in videogames. By using the proposed theoretical model as a guideline, game designers might obtain a better idea of how to design the ludonarrative of computer and videogames in fully intentional ways to facilitate, disrupt, or confuse the player's performance of the gameplay and narration-commentary to produce a specific experience (see Kuznetsova 2017). A future study can be conducted on established Let's Players to enable our understanding of how ludonarrative relationships influence Let's Players' conception of the game structure and their Let's Play performance.

Acknowledgments

This chapter was partially funded by the National Youth Council's National Youth Fund (NYF) grant. The author would like to thank his supervisor, A/P Ismail Talib, for his guidance, and to thank the participants who took part in the study. The author would also like to thank the editors, the p2p reviewer, Jason Hawreliak, and proofreaders for suggesting improvements to earlier versions of the manuscript. This study was approved by the National University of Singapore's Institutional Review Board.

References

2K Boston (2007), *Bioshock* [PC Computer, PS3, Xbox 360, OS X, Cloud (OnLive), iOS] 2K Games. Quincy, Massachusetts, USA.

Aarseth, E. (2014), "Game ontology," in B. Perron and M.J.P. Wolf (eds.), *The Routledge Companion to Video Game Studies*, 484–92, New York: Routledge.

Arthur, C., A.C. Graesser, B. Olde, and B. Klettke (2002), "How does the mind construct and represent stories?" in M.C. Green, J.J. Strange, and T.C. Brock (eds.), *Narrative Impact: Social and Cognitive Foundations*, 229–62, Mahwah, NJ: Lawrence Erlbaum Associates.

Anders, E. and H.A. Simon (1993), *Protocol Analysis: Verbal Reports as Data*, rev. edn., Cambridge, MA: MIT Press.

Barnum, C. (2002), *Usability Testing and Research*, New York: Longman.

Bell, A., A. Ensslin, I. van der Bom, and J. Smith (2018), "Immersion in digital fiction: A cognitive, empirical approach," *International Journal of Literary Linguistics*, 7 (1): 1–22.

Boot, W.R., A. Sumner, T.J. Towne, P. Rodriguez, and K.A. Ericsson (2016), "Applying aspects of the expert performance approach to better understand the structure of skill and mechanisms of skill acquisition in video games," in W.D. Gray (ed.), *Special Issue: Topic Continuation: Visions of Cognitive Science—*

Game XP: Action Games as Experimental Paradigms for Cognitive Science, 9 (2): 1–24.

Burnwell, C. and T. Miller (2016), "Let's Play: Exploring literacy practices in an emerging videogame paratext," *E-Learning and Digital Media*, 13 (3–4): 109–25.

Cardona-Rivera, R.E. and R.M. Young (2012), "Characterizing gameplay in a player model of game story comprehension," in *FDG '12 Proceedings of the International Conference on the Foundations of Digital Games*, 204–11, May 29–June 1, Raleigh, NC: ACM Press.

Chu K., C.Y. Wong, and C.W. Khong (2011), "Methodologies for evaluating player experience in game play," in C. Stephanidis (ed.), *HCI International 2011— Posters' Extended Abstracts. HCI 2011. Communications in Computer and Information Science*, 173, Berlin, Heidelberg: Springer.

Dena, C. (2010), "Beyond multimedia, narrative, and game: The contributions of multimodality and polymorphic fictions," in P. Ruth (ed.), *New Perspectives on Narrative and Multimodality*, New York: Routledge.

De Rijk, B.J.S. (2016), "Watching the game: How we may understand Let's Play videos," Master's diss., Utrecht University, The Netherlands.

DiLiberto, L., C.K. Roger, L.B. Kenneth, and N.H. Gary. (2002), "Using *Articulated Thoughts in Simulated Situations* to assess cognitive activity in aggressive and nonaggressive adolescents," *Journal of Child and Family Studies*, 11 (2): 179–89.

Dor, S. (2014), "The heuristic circle of real-time strategy process: A *StarCraft— Brood War* case study," *Game Studies—The International Journal of Computer Game Research*, 14 (1). Available online: http://gamestudies.org/1401/articles/ dor (accessed August 9, 2018).

Dor, S. (2018), "Strategy in games or strategy games: Dictionary and encyclopedic definitions for game studies," *Game Studies—The International Journal of Computer Game Research*, 18 (1). Available online: http://gamestudies.org/18 01/articles/simon_dor (accessed August 9, 2018).

Dorst, K. and J. Dijkhuis (1995), "Comparing paradigms for describing design activity," *Design Studies*, 16: 261–74.

Drachen, A., P. Mirza-Babaei, and L. Nacke, eds. (2018), *Games User Research*, Oxford: Oxford University Press.

Eccles, D.W. and G. Arsal. (2017), "The think aloud method: What is it and how do I use it?" *Qualitative Research in Sport, Exercise and Health*, 9 (4): 514–31.

Ensslin, A. (2012), *The Language of Gaming*, New York: Palgrave Macmillan.

Ericsson, K.A. and H.A. Simon (1993), *Protocol Analysis Verbal Reports as Data*, Cambridge, MA: The MIT Press.

Frasca, G. (2003), "Simulation versus narrative: Introduction to ludology," in M.J.P. Wolf and B. Perron (eds.), *The Video Game Theory Reader*, New York: Routledge.

Hocking, C. (2007), "Ludonarrative dissonance in *Bioshock*—The problem of what the game is about." Available online: http://clicknothing.typepad.com/cli ck_nothing/2007/10/ludonarrative-d.html (accessed June 9, 2017).

Howe, L.A. (2017), "Ludonarrative dissonance and dominant narratives," *Journal of the Philosophy of Sport*, 44 (1): 44–54.

Jenkins, H., R. Purushotma, M. Weigel, K. Clinton, and A.J. Robison. (2009), *Confronting the Challenges of Participatory Culture: Media Education for the 21st Century*, Cambridge, MA: The MIT Press.

Kerttula, T. (2016), "'What an eccentric performance': Storytelling in online Let's Plays," *Games and Culture*. Available online http://journals.sagepub.com/doi/abs/10.1177/1555412016678724 (accessed January 2, 2018).

Kirschner, D. and P.J. Williams (2014), "Measuring video game engagement through the gameplay review method," *Simulation and Gaming*, 45 (4–5): 593–610.

Koenitz, H. (2010), "Towards a theoretical framework for interactive digital interactive storytelling," *Third Joint Conference on Interactive Digital Storytelling*, 176–85, Berlin, Heidelberg: Springer. http://doi.org/10.1007/978-3-642-16638-9_22.

Koenitz, H. (2015), "Towards a specific theory of interactive digital narrative," in H. Koenitz, G. Ferri, M. Haahr, D. Sezen, and T.I. Sezen (eds.), *Interactive Digital Narrative*, 91–105, New York: Routledge.

Kuznetsova, E. (2017), "Trauma in games: Narrativizing denied agency, ludonarrative dissonance and empathy play," Master's diss., University of Alberta, Edmonton.

Lakoff, G. and M. Johnson (1980), *Metaphors We Live By*, Chicago: University of Chicago Press.

Larsen, B.A. and H. Schoenau-Fog (2016), "The narrative quality of game mechanics," in F. Nack and A. Gordon (eds.), *Interactive Storytelling. ICIDS 2016. Lecture Notes in Computer Science*, vol. 10045, Cham: Springer.

Leino, O.T. (2010), "Emotions in play: On the constitution of emotion in solitary computer game play," PhD diss., IT University of Copenhagen, Denmark.

Lindley, C. (2004), "Ludic engagement and immersion as a generic paradigm for human-computer interaction design," in M. Rauterberg (ed.), *Entertainment Computing—ICEC 2004. Lecture Notes in Computer Science*, vol. 3166, Berlin, Heidelberg: Springer.

McGloin, R., J.A. Wasserman, and A. Boyan. (2018), "Model matching theory: A framework for examining the alignment between game mechanics and mental models," *Media and Communication*, 6 (2): 126–36.

Nguyen, J. (2016), "Performing as video game players in Let's Plays," *Praxis*, 22. Available online: http://journal.transformativeworks.org/index.php/twc/article/view/698/615 (accessed June 7, 2017).

O'Halloran, K.L. (2008), "Systemic functional-multimodal discourse analysis (SF-MDA): Constructing ideational meaning using language and visual imagery," *Visual Communication*, 7 (4): 443–75.

Radde-Antweiler, K., M. Waltmathe, and X. Zeiler (2014), "Video gaming, Let's Plays, and religion: The relevance of researching gamevironments," *Gamevironments*, 1: 1–36. Available online: http://elib.suub.uni-bremen.de/edocs/00104169-1.pdf (accessed June 8, 2017).

Recktenwald, D. (2014), "Interactional practices in Let's Play videos," Master's diss., Saarland University, Germany.

Sánchez, J.L.G., N.P. Zea, and F.L. Gutiérrez (2009). "From usability to playability: Introduction to player-centred video game development process," in *Proceedings of HCD 2009*, 65–74, Berlin, Heidelberg: Springer.

Sim, Y.T. and A. Mitchell (2017), "Wordless games: Gameplay as narrative technique," in N. Nunes, I. Oakley, and V. Nisi (eds.), *Interactive Storytelling. ICIDS 2017. Lecture Notes in Computer Science*, vol. 10690, 137–49, Cham: Springer.

Smith, T., M. Obrist, and P. Wright (2013), "Live-streaming changes the (video) game," in *EuroITV'13 Proceedings of the 11th European Conference on Interactive TV and Video*, 131–38, New York: ACM.

Sylvester, T. (2013), *Designing Games: A Guide to Engineering Experiences*, Sebastopol, CA: O'Reilly Media, Inc.

Tan, C.T., T.W. Leong, and S. Shen (2014), "Combining think-aloud and physiological data to understand video game experiences," in *CHI'14 Proceedings of the SIGCHI Conference on Human Factors in Computing Systems*, April 26–May 1, 381–90, Toronto, Ontario, Canada: ACM Press.

Theodorou, E. (2010), "Let the gamers do the talking: A comparative study of two usability testing methods for video games," Master diss., University College London, United Kingdom. Available online: https://uclic.ucl.ac.uk/conten t/2-study/4-current-taught-course/1-distinction-projects/8-10/theodoroue.pdf (accessed June 9, 2017).

Thon, J.N. (2008), "Immersion revisited. On the value of a contested concept," in O. Leino, H. Wirman, and A. Fernandez (eds.), *Extending Experiences: Structure, Analysis and Design of Computer Game Player Experience*, 29–43, Rovaniemi: Lapland University Press.

Toh, W. (2015), "A multimodal discourse analysis of video games: A ludonarrative model," PhD diss., National University of Singapore, Singapore.

Toh, W. (2018), *A Multimodal Approach to Video Games and the Player Experience*, New York: Routledge.

Van Someren, M.W., Y.F. Barnard, and J.A.C. Sandberg (1994), *The Think Aloud Method: A Practical Guide to Modelling Cognitive Processes*, London: Academic Press.

Vella, D. (2015), "No mastery without mystery: *Dark Souls* and the ludic sublime," *Game Studies*, 15 (1). Available online: http://gamestudies.org/1501/articles/vel la (accessed June 9, 2017).

Watssman, J. (2012), "Essay: Ludonarrative dissonance explained and expanded," *Escapistmagazine*. Available online http://www.escapistmagazine.com/forums/ read/9.389092-Essay-Ludonarrative-Dissonance-Explained-and-Expanded (accessed June 9, 2017).

Willumsen, E.C. (2018), "The form of game formalism," *Media and Communication*, 6 (2): 137–44.

Woelke, J. and E. Pelzer, (2017), "Cognitive assessment: Think-aloud and thought-listing technique," in J. Matthes, C. Davis, and R. Potter (eds.), *The International Encyclopedia of Communication Research Methods*, 1–9, New York, NY: Wiley-Blackwell. doi:10.1002/9781118901731.

Yanow, D. and S. Schwartz (2006), *Interpretation and Method: Empirical Research Methods and the Interpretive Turn*. Armonk, NY: M.E. Sharpe.

Zwaan, R.A., M.C. Langston, and A.C. Graesser (1995), "The construction of situation models in narrative comprehension: An event-indexing model," *Psychological Science*, 6 (5): 292–97.

12

Language Ideologies in Videogame Discourse

Forms of Sociophonetic Othering in Accented Character Speech

Tejasvi Goorimoorthee, Adrianna Csipo, Shelby Carleton, and Astrid Ensslin

1 Introduction

This chapter examines the distribution of speech accents among nonplayer characters (NPCs) of the fantasy role-playing game *Dragon Age: Origins* (BioWare 2009). The study is part of a larger project called "Speech Accents in Games" (SAG; funded by the Social Sciences and Humanities Research Council of Canada, through a 'Refiguring Innovation in Games' project subgrant) at the University of Alberta (2017–2018), which aims to investigate how selective uses and distributions of voiced-over linguistic accents in videogame characters may cast light on underlying language ideologies (e.g., Woolard 1998; Irvine and Gal 2000; Coupland 2007) and the ways in which they are negotiated in folk linguistic debate. It examines how games communicate and negotiate meanings within their specific communities of practice and seeks to identify game-specific patterns of representation and interaction. The key idea driving this project is to identify language ideological patterns underlying the use of specific accents of English by contemporary game designers, and to examine how these uses may contribute to perpetuating, functionalizing, or flouting extralinguistic meanings typically associated with certain accents. Attitudes toward speech accents can contribute to linguists' behavior and politics. Much like racism and sexism, *linguicism* (Skutnabb-Kangas and Phillipson 1995) refers to popularly held, stereotypical views about other people's use of language (in speech and writing). These views tend to be judgmental and can cause considerable harm to those they are directed against. For example, negative biases toward people's nonstandard, regional and/or nonnative accents may lead to exclusion, harassment, and other forms of direct and indirect discrimination.

Earlier, preparatory stages of this research (Ensslin 2010; 2011) looked at how conventional and unconventional oppositions are conflated multimodally in videogame character design, for example, by pairing moral binaries (good and bad) with artificial, sociophonetic opposites like received pronunciation (RP) vs. urban varieties of American English, and by looking at how *Pax Americana* (American hegemonic superiority; Bayard et al. 2001; Brice 2011) is embedded in and iconized by the voices of heroic characters (Irvine and Gal 2000; see also Lippi-Green 2011). This research is complemented by Brice (2011), who discusses how, in BioWare's *Dragon Age: Inquisition*, a multitude of characters speaking with standard North American accents surrounding the player help render the American accents invisible so the player may focus on other elements. A further study by Ensslin (2011) examined how speech accents can be used as tools of othering undesirable characters and as mnemonic devices, generating intertextual links to other elements of popular culture, thus increasing levels of immersion and entertainment. This early research was centered around a set of narrative 3D games, including *Black and White 2* (Lionhead Studios 2005), *Return to Castle Wolfenstein* (id Software and Grey Matter Interactive 2001), *Aion* (NCsoft 2008), *Fable* (Lionhead Studios 2004), *Wizard 101* (KingsIsle Entertainment 2008), and *Dragon Age: Inquisition* (BioWare 2014). It was grounded in mostly anecdotal evidence from personal gameplay and did not involve comprehensive and systematic data collection and analysis.

Our research is embedded in the sociolinguistic and social semiotic contexts of language ideologies (Joseph and Taylor 1990; Kroskrity et al. 1992; Schieffelin et al. 1998) and language ideological debates (e.g., Blommaert 1999), both of which have at their core an interest in the study of beliefs and belief systems relating to languages, their prescriptive rules and norms, and the ways in which these rules are used and deviated from. Language ideological debates in particular are debates "in which language is central as a topic, a motif, a target, and in which language ideologies are being articulated, formed, amended, enforced" (Blommaert 1999: 1). However, language ideologies are not just conveyed through explicit debate. Perhaps more subtly, they are also embedded in so-called implicit metalanguage (Preston 2004), which are language uses that convey, through implicature, attitudes and ideologies toward the ways in which people use language in speech and writing. Kathryn Woolard (1998) uses the term "implicit metapragmatics" in the sense of "linguistic signaling that is part of the stream of language use in process and that simultaneously indicates how to interpret that language-in-use" (9), meaning that the linguistic and paralinguistic choices we make systematically, if not always consciously, contribute to our personal image management: we want our communicative actions to shape other people's reactions to and views of us, and by selecting certain linguistic choices over others, we implicitly align ourselves with others whom we can relate to and identify with and, simultaneously, distance ourselves from the ways in which other people from whom we feel more detached might use language.

In a larger sociolinguistic framework, Judith Irvine and Susan Gal (2000) have identified three key semiotic processes that link linguistic forms and social phenomena such as class, race, and other simplifying generalizations and collectivizations. *Iconization* is the direct mapping of a linguistic feature onto a social image, for example, the process of inferring from repeated exposure that it is an inherently Canadian feature to say "hey" or "ey" at the end of a sentence to prompt agreement, and that, in a connotative process, this might be linked to sensations of affection or humor. Importantly, iconization implies that this associative link comes to be seen as inherent and socially representative. The second semiotic process they outline is *fractal recursivity*, where one semantic opposition is projected onto another. Fractal recursivity thus operates schismogenetically: it creates effects of othering between, but also within, groups and (language) communities (Andronis 2003: 264). For example, RP (received pronunciation, or Standard British English) speech is (still) commonly associated with "those in the upper reaches of the social scale, as measured by education, income, and profession, or title"—"at least in England" (Hughes et al. 2005: 3). In popular media representations, it is often contrasted with urban or rural vernaculars, which are inherently attributed a lower socioeconomic status, a lower level of education, and/or inclinations toward socially undesirable behavior or attitudes. *Erasure*, finally, denotes a process of simplification by erasing nuances of differentiation. It is "the process by which these distinctions are created and maintained. Erasure is integrally intertwined with both iconization and recursivity, as it is the erasure of any differentiation which is, according to the given ideology, inconsequential" (Andronis 2003: 264). It is this combination of representational essentialism and simplification that popular media are particularly prone to, and where videogames in particular can create or perpetuate socially and culturally damaging views because they often project holistic binary value systems (Ensslin 2010).

The use of accents in videogames can be just as ideologically charged as any other element of audiovisual representation, such as color of skin, type of clothing, and accompanying soundtrack. These semiotic choices can have the intent of being neutral, but, inevitably, they communicate certain assumptions and ideologies to the player—especially if neutral is seen as equivalent to normal. In games and other audiovisual media, the use of speech generates a matrix of predominant accent use, such as an RP matrix (if most characters speak with RP), or a Standard North American (SNAm) matrix. Accents that fall outside these matrices are, as a result, marked in that they represent deviations from the norm, and this gives rise to semiotic processes of *othering*: exposing specific individuals as different from the norm and therefore potentially threatening or suspicious. Players are subjected to the use of these semiotic signals coded into the user interface and are therefore led to adopt the often binary persuasive logic of the game mechanics. This again can lead to a perpetuation of stereotypical, hegemonic thinking about, as well as the naturalization of and willing submission to, "the dominant culture and prevailing power relationships" (Schniedewind and Davidson 2000).

In recent years, some narrative game developers have shown a tendency toward trying to paint a more diversified picture of the fictional societies they represent and how this is reflected by character accents. However, this diversification tends to follow fairly predictable, ideologized patterns, and it reconfirms a lot of the findings of accent attitudinal research. Lionhead's *Fable* series (2004–2017), for example, displays a variety of different British accents, mostly used for quest-giving villagers or, more generally, members of the "common" population of Albion (a fictional version of historical Anglo-Saxon Britain). They are generally portrayed as simple-minded, parochial, and often vulgar (often also morally corrupt), and they are iconically attributed a range of vernacular urban and rural accents (Northumbrian, Cockney, or Scouser), the kind of accents that "real" people often associate with either low prestige or low social attractiveness or both [Coupland 2007]).

The main goals of this chapter and of SAG more generally are, then, to examine, as systematically and comprehensively as resources allow, how native and foreign accents of English are distributed in contemporary narrative games, how they are functionalized politically in the gameworld, and what kind of language and specifically accent attitudinal ideologies they might perpetuate in this way. The main deliverables of the project are a database of speech accent samples (video and audio files), including metadata, and a combination of quantitative accent mappings and an understanding of accent social hierarchies and othering as relating to fractal recursivity. Our findings are expected to give a nuanced picture of how friend and foe, as well as more hybrid, dynamic, and rounded character roles, are framed phonetically through the allocation of linguistic accents. It may also tell us about the degrees to which accents are artificially constructed, or faked, by voice actors, either intentionally or unintentionally, to create othering within the social structures of the game. In this chapter, we showcase this agenda by examining, in detail, the speech accent distribution in BioWare's fantasy role-playing game *Dragon Age: Origins* (2009).

1.1 Game of choice: *Dragon Age: Origins*

Dragon Age: Origins (DAO) is a single-player fantasy game wherein the player character (PC) is a novice Grey Warden (a class of warriors) whose mission is to defeat the fifth Blight (the awakening of the Archdemon). Each choice the player makes has the ability to change the storyline of the game, persistently affecting the behavior of the NPCs and, thus, the player's experience. Choices arise through the elective options the player has within dialogues, where the choice of one line over another will lead to a different path of gameplay—for example, choosing to compromise with a character versus choosing to act violently toward them. *DAO* was chosen based on its extreme success and popularity upon its release, as well as its strong focus on narrative and character. *DAO* focuses on immersing the player in a complete world with class structures and races. The PC's journey is filled with hundreds of lines of dialogue from a variety of races, making it a valuable focus of study for accent distribution and the attitudinal ideologies they may perpetuate. Furthermore, as Bonnie Ruberg and Adrienne Shaw (2017) observe, "fantasy is always already political" (p. xxi) because, despite its typically superhuman and nonmimetic representations, it operates mimetically in relation to real-life hegemonic power assumptions. On the other hand, as Hanna Brady (2017) suggests, the fantasy genre—even in commercial game culture—offers worlds full of possibility that lend themselves to exploration of and experimentation with countercultural and counterhegemonial ways of seeing and "hearing" (as in our case).

There are multiple choices for the establishment of the player character (PC), including the class (Warrior, Rogue, Mage), and the race (Elf, Human, Dwarf) within six storylines: City Elf, Dalish Elf, Dwarf Commoner, Dwarf Noble, Human Noble, and Magi. The PC (male City Elf) was chosen because it stood, at the time of our decision-making process, at the top of the list for the most played character in this installment of the series (The Escapist 2013).

DAO is generally assumed to be a canonical text in the digital RPG genre (Henton 2012). It has been the focus of scholarly work, particularly in relation to its character design and development (Jørgensen 2010; Waern 2011). Annika Waern (2011), for example, considers the "bleed" effect in *DAO*, where the emotional investment of the player into the PC, as well as the PC's projection of the player's identity, creates a fuzzy boundary between the player and their PC. Furthermore, Waern conducted a case study, researching the possible romances the game allows in a single playthrough (a method we have emulated in this study), and their effect on the players. The results of her study point to players' inclination to convey their own identities, including their romantic tendencies, onto their PC. Moreover, there is substantial evidence that the "bleed" effect occurs in RPGs that place value on the construction and development of both PCs and NPCs, such as *DAO*.

While *DAO* has been debated extensively by ludo-narratologists and gender scholars (e.g., Greer 2013), it has not been examined at any length from the point of view of sound studies or sociolinguistics. We aim to fill this gap by looking at *DAO*'s distinctive sociophonetic soundscape and by contextualizing it with theories of language attitudes and ideologies, and accent attitudinal debates in particular.

2 Methodology and data

For this research project, we followed a quantitative approach for collecting and labelling character speech data to map the distribution of accents throughout a typical playthrough of *DAO*. We combined this with a qualitative approach to analyzing the multifaceted aspects of how the game is designed so as to link character accents to political and social roles and meanings. Anecdotal evidence had suggested to us that *DAO* reappropriates accent dynamics for the assumed North American player. Furthermore, we were aware that *DAO* embeds various European accents and attributes them to specific character races: Antivans have Spanish accents, and Orlesians have French accents, for example. City Elves, Dwarves, and the Qunari, in contrast, have American accents. The Humans, finally, have mixed British, yet mostly standard or even clipped, accents. Our quantitative data analysis enabled us to refine these anecdotal observations, as detailed here.

To obtain a comprehensive and representative sample of different accents in *DAO*, we played and video-recorded an entire playthrough of the game. The resulting video files (n = 476; length = max. 25 sec) were then fed into a database and subsequently labeled according to perceived accent types by one native speaker and three fully immersed, near-native speakers. The combined accent experience repository of all four coders comprised a wide range of North American, British, European, and Asia-Pacific accents of English. The characters whose accents/dialects were found to be ambiguous were taken into an in-depth analysis for linguistic variables that would unravel or explain both the ambiguity of their accent of origin and the reason behind it.

In our analysis, we will discuss how native and foreign accents of English are distributed in *DAO* by examining the accents that are most prevalent, the races that speak in those accents, and the intentional and unintentional othering that occurs based on accent distribution. We will then focus on how these accents function politically in *DAO* by examining how accents delineate class structure and social hierarchy among races. Finally, we will discuss the kind of language and accent attitudinal ideologies perpetuated throughout othering based on the PC's race and attributed accent.

3 Analysis

In *DAO*, a variety of accents are found throughout the game, with standard British and American forming a dominant double matrix. In addition to their accents, characters also have distinct races; humans, elves, and dwarves are the primary three. Each race belongs to a different social caste. City Elves are considered to be the lowest class, often referred to with racist slurs like "knife-ears." Human Nobles are considered to be the highest socioeconomic class, and Dwarves fall somewhere in the middle, often keeping themselves to themselves deep below the surface of the earth. Each race has a distinct accent assigned to it. As shown in Figure 12.1, Humans tend to speak primarily with British standard accents (RP), whereas Dwarves and Elves tend to speak with primarily North American standard accents (SNAm). Within this dominant RP-SNAm double matrix, other accents gain meaning through their contrast with standard varieties. For a general distribution of races encountered by the player-researcher (P-R) of this study, see Figure 12.2.

Out of 476 distinct character speech files in our dataset, three were inadmissible due to unintelligible speech;[1] the rest (*n* = 473) is analyzed as follows. While playing as a male City Elf, the player-researcher (P-R)

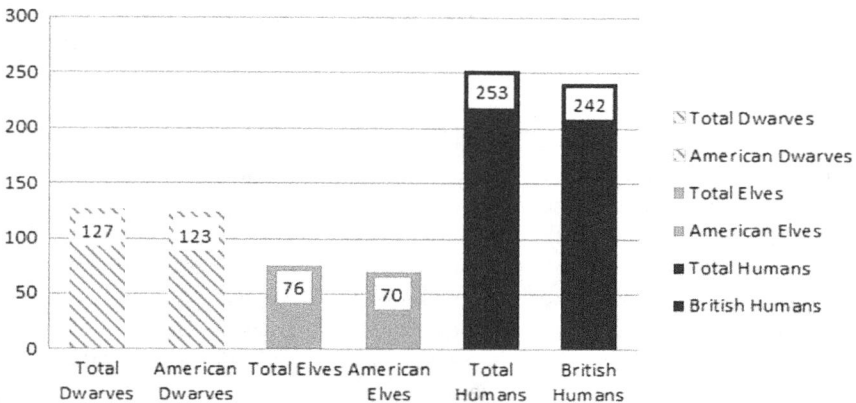

FIGURE 12.1 *Race and accent distribution in* Dragon Age: Origins.

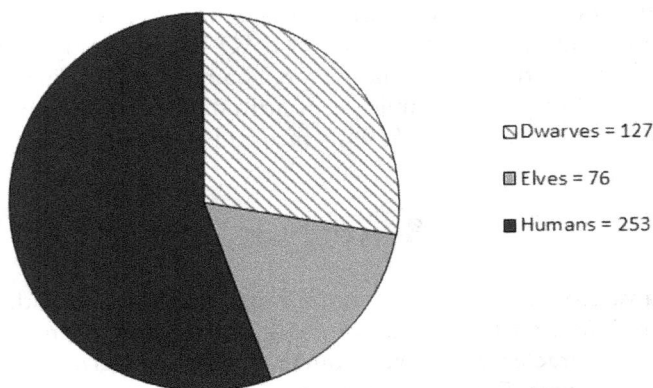

Dwarves = 127

Elves = 76

Humans = 253

FIGURE 12.2 *Race ratio in* Dragon Age: Origins.*

*This graph does not account for nonhuman characters, such as werewolves, demons, and animated trees.

interacted with (58–12.26%) female characters with a standard North American accent and (63–13.32%) female characters with a British accent (see Figure 12.3). The P-R also interacted with (157–33.19%) male characters with a SNAm accent and (185–39.11%) male characters with a British accent. Thus, a standard player is likely to encounter predominantly male RP speaking characters, developing an RP matrix, with male SNAm speaking characters closely behind. From the perspective of our PC (male City Elf), the RP accents form the dominant "other," as the Elves (i.e., the race to which the PC belongs) speak with a SNAm accent.

In addition to the dual-standard matrix, a small number of accents were noted that exhibited nonnative varieties (Spanish: *n* = 1 [Zevran]; French: *n* = 2 [Liselle and Erlina]). Other nonstandard accents were hybrids and hence difficult to classify with a single label: two characters (Master Ignacio and Cesar) exhibited a Spanish-Slavic hybrid accent; Riordan speaks with a mix of RP and Irish-accented English; Lady Vasilia shows elements of RP and SNAm; Barlin mixes RP with Cockney and SNAm; Isolde changes between Slavic, Germanic, and French accents; and Leliana's most mysterious accent shows elements of (clipped) RP and French, depending on narrative context. Hence, our results show that these intentional foreign accents tend to occur mixed in with the matrix accents. A plausible explanation for these hybrids is that they are voice-acting glitches, conveying a less than perfectionistic attitude on the part of the game's developers toward representations of foreign-accented speech, and general logistic-financial dictates surrounding voice-actor recruitment (i.e., preferring local, native speakers with put-on and often inconsistent foreign accents over more expensive, "imported"

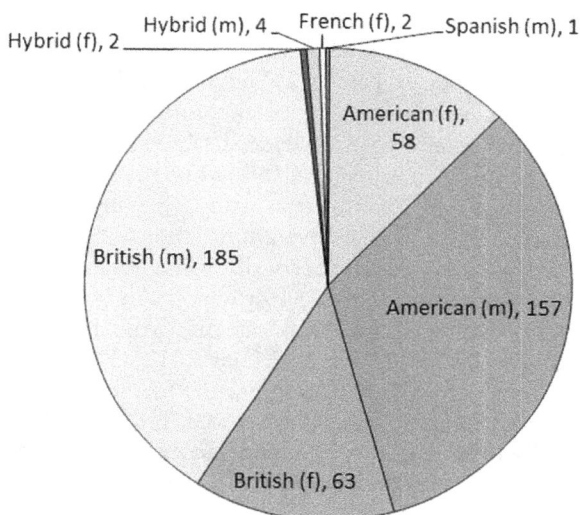

FIGURE 12.3 *Gender and accent distribution in* Dragon Age: Origins.

voice actors). In what follows, we will examine four of the aforementioned characters with hybrid accents in more detail.

3.1 Ignacio and Cesar

Master Ignacio has a peculiar accent, which mixes elements of Spanish and Slavic accented speech. The clip was listened to and labeled by five different coders (the regular four and a Spanish native speaker), who were, however, unable to arrive at a unanimous decision as to the exact accent provenance or type. Most of us agreed that there were elements of Spanish, but that there also some fairly noticeable, and perhaps unintended, Slavic traces. The intent for the accent was most likely Spanish, as the character comes from Antiva, whose residents are said to have a Spanish accent. Furthermore, his name alludes to a Spanish origin, which further intensifies the impression that the character was supposed to be framed as Spanish-Antivan. Regardless, he has the role of a tradesman which, in *DAO*, is a prominent line of work for characters with foreign accents. In other words, there are a number of working-class characters (especially merchants) who were assigned a foreign accent. That is not to say that there are no merchants who have RP/SNAm accents, but that, if there is a character who has a foreign accent, he/she is most likely a member of the working class.

A similar case appears with another character, Cesar, whose accent is a mixture of Spanish and a Slavic language (perhaps Russian or Ukrainian). He is also a working-class character, a merchant, and his name implies a

Spanish origin. It is near-impossible to pinpoint exactly which Slavic accent is present, as most Slavonic groups of languages share some of the same phonological features (Meier 2012: 318). This means that Slavic speakers, although admittedly an enormous group, encounter similar problems in their pronunciation of English words (Meier 2012: 318).

Within his speech, Cesar employs the following Russian features. First, the *face* lexical set is often pronounced as a long [ɛ:] or [e:], instead of a diphthong [ɛɪ] (Meier 2012: 319). For example, the character pronounces "trade" in a similar way as [trɛ:d]. Secondly, Cesar uses [u] in the *foot* lexical set, instead of [ʊ] (Meier 2012: 319), which can be heard when the character says "good" as [gud] and "too" as [tu]. Thirdly, quite common for Slavic languages is the trilled /r/ [r] (Meier 2012: 320), which can be heard only slightly in words such as "customers," "better," "trade," "north," "here," and "war" [kʌstəmərz, ˈbɛdr, treɪd, nɔr:θ, hir, vor]. The trilled /r/ is not as strong as within a natural Russian accent, presumably because the voice actor is a native speaker of English (possibly British English) and attempting to sound foreign and, thus, his /r/ sounds slightly odd, but not quite trilled. Overall, the voice actor employs lots of dentality, which can be detected in words such as "better," which is pronounced (ˈbɛdr). The speech also contains a fair amount of nasality and has a very peculiar intonation, which appears not to have a particular pattern and is thus quite chaotic. Although the target accent is Spanish, the execution of the same differs from it. The distinction between a Spanish and a Russian accent can be difficult to hear, especially to a native English speaker whose ear is not accustomed to hearing that specific accent often. These languages (Spanish and Russian) also contain some similar phonetic traits when pronouncing English words. For example, neither contains the phoneme [ɪ], and both are likely to use [i] instead (Meier 2012: 319, 332), in words such as "business" and "here" pronounced by our character as [biznis] instead of [bɪznɪs] and [hir] instead of [hɪə]. The features that could be Spanish are the ones that are in common with Russian, however, the more stereotypically Spanish delivery of the speech would employ different features; for example, the clichéd pronunciation of "yes" and similar words with [j] sounds would have been [dʒ] in Spanish. Incidentally, Cesar's Britishness also shows when the phonemes [p, t, k] are aspirated and slightly touching the alveolar ridge, instead of being unaspirated and a bit more dental—which would be the case in a Spanish speaker (Meier 2012: 335).

3.2 Barlin

Barlin is a character whose accent relates both to a voice-acting issue and to his socioeconomic class. His accent is a mixture of Cockney British and at times a type of Southern American. The character mostly has a Cockney accent, which can be seen in words such as "about" [əˈbæət]. Furthermore, in the *goat* set, the character uses a glide [ãə̃] when he says "suppose" [sə'pãə̃z] with a fair degree of nasality, which is typical of the local voice quality found in Cockney speakers (Wells 1982: 318), as opposed to RP speakers, who would use [əʊ] within that set (Meier 2012: 104). However, a North American accent, perhaps unintentionally, sneaks in within the *bath* lexical set, where the /a/ vowel is pronounced in a typically American way [æ:], as opposed to the British (and also Cockney) [ɑ] (Meier 2012: 185). This may be accounted for by the fact that Barlin's voice actor is from Ohio, the United States.

Cockney is often used in popular media for characters that aren't just lower class but in somewhat dubious professions (such as the trader in level 1 of Lionhead's *Fable 1*, see Ensslin 2010). Unsurprisingly, Barlin is a farmer and a merchant. The content of the conversation with Barlin also alludes to these characteristics, namely Barlin asking the P-R to get him some poison.

3.3 Isolde

Isolde is a character that originated from Orlais, a French-speaking region, and whose accent was problematic to pin down. In some respects, her accent implies Frenchness with the /th/ sounds being dentalized (s̪, z̪) as in her pronunciation of "this" [ts̪is]. Furthermore, her pronunciation of another character's name (Teagan) contains [i], which is one of the signature sounds in a French accent (Meier 2012: 274). Additionally, the stress patterns are often misplaced in her speech, which is a typical foreign trait, and specifically for this character, the attempt at French cannot be unheard. The pronunciation of the name Teagan has a stress pattern of **Tea**-gu-h-n, however this character pronounces it as Ti-g-**AN**, since in French the stress is on the final syllable in many words (Meier 2012: 277). Another signature French sound is the post-vocalic /l/, in RP velorized [ɫ], but here it is pronounced more lightly (Meier 2012: 276). We hear that in her pronunciation of "alone," "Alister," "all," and so on [ə'ləʊn, ælistr, ɔ:l]. Her use of the *bath* lexical set is, however, predominantly RP, which can be detected in words such as "castle" ['kɑ:sl], for example. The most famous French feature is the use of /r/, in this case a trilled uvular [ʀ], which can be a casual and fricative residue of that sound, at times voiceless (Meier 2012: 276). When pronouncing the word "here" [hiʀ] in particular, the character can be placed quite distinctly into that category, however, as some of her /r/ sounds in words such as "maker" and "return" [meɪkr, rɪ'trn] are less intensely trilled and more rolled. This can emerge from one of two reasons: either from so-called hypercorrectness (when a foreign speaker intentionally tries to imitate British pronunciation, and overdoes it in his/her attempt) or from the actress' native RP influence. The nonrhotic feature or RP is not the only signature sound of that accent that emerges in Isolde's speech, and she also employs [ɑ] of the *bath* lexical set.

4 Discussion

In the world of *DAO*, Dwarves and Elves speak primarily with an American accent, whereas Humans speak primarily with a British accent. With British and American accents being ubiquitously heard throughout the game, the player is signaled linguistically, through implicit metapragmatics (Woolard 1998), that these accents are most important as they define the world the PC inhabits. They form a matrix of normalcy against which all other, nonstandard accents are set off as marked and thus othered.

DAO draws attention to the impact of othering, both intentional and unintentional. Intentional othering derives from the player's chosen perspective. Depending on the race the player selects, other groups of characters will be othered by not sharing the PC's race, gender, or accent. For example, if a player is a Human, the American accent, which tends to be associated with lower socioeconomic classes in the gameworld, will be othered, as Human Nobles speak primarily with British accents. We refer to this effect as *intentional othering* because a conscious choice was made to separate character races and socioeconomic classes by iconizing them through a single accent. Intentional othering also takes the form of exoticized characters like Master Ignacio, a merchant who speaks with an unidentifiable Spanish accent. Spanish and other nonnative accents—in more or less hybrid form—are encountered by the player on very limited occasions throughout the game. This form of intentional othering is to make characters like Master Ignacio and Cesar appear as though they are from a different continent entirely, expanding the reaches of the player's imagination as to how big and how deep the lore of the *DAO* world really is. However, they also induce feelings of distrust or, at least, alienation, through processes of fractal recursivity vis-à-vis the normalized standard(s).

Unintentional othering happens primarily in the case of poorly performed or inconsistent voice acting of standard and nonstandard accents. The aforementioned Lady Vasilia is a prime example of unintentional othering. Her accent is difficult to discern due to incorrect pronunciations throughout the voice samples we were able to retrieve. The argument could be made that these hybrid accents might have the function of preventing effects of erasure (Irvine and Gal 2000), thus engrossing players into a more complex and possibly even more realistic-mimetic world of fantasy, where all manner of speech is possible; and/or to trigger debate among fans about the provenance of hybrid-accented characters. However, it is unconvincing to argue for a diverse fantasy world full of accidental accents that deviate from their standard British or American forms when the game itself seems to insist on an overall highly consistent dual-standard matrix, thus failing to include a breadth of accents across cultures of the world. The focus remains on hegemonializing British and American, and, in Lady Vasilia's case, the accent is just a *slightly different* form of British and American.

Humans in *DAO* speak with primarily British accents. In popular culture, the British accent has regularly been claimed as the accent of fantasy, forever immortalized within the characters of *The Lord of the Rings* and *Game of Thrones* franchises, which are all set in fantasy worlds. British accents dominate in these movies, thus forming an important counterpoint to another stereotypical use of British accents in popular culture—to designate the speech of villains (Poos 2013). Humans in *DAO* are considered the highest class, and it is not surprising that they are given the most prestigious accent derived from and reminiscent of British-Tolkienesque, medievalesque fantasy, which is common in popular film and media, to demonstrate, and thus iconize, their rank and power. In the fantasy realm, they could have any kind of accent, but British does seem to be the default. Even though the "Mid-Western accent is still what counts as 'normal' in the US-dominated entertainment industry, a British accent provides a 'splash of otherness' when set alongside it" (GenericGuy 2012).

This potentially leads to fractal recursivity, as the British accent appears to be at the top of the hierarchical chain, while the rest are experiencing "othering," ranking below the British in most social aspects. Humans, therefore, having the most prestigious accent, appear to be an isolated elite, essentially in binary opposition to the other races in the game. Captivating as it is, the use of language (predominantly the diversity of accents), seemingly already in the initial stages, creates the backdrop of the social structure within the game. The social structure is continually reinstated throughout the game, the only exception being that the PC (when played as a male City Elf) gets to change their social status and become a prominent member of the society. This, however, does not change the social status of their race. Elves remain to be mistreated, mostly by the use of profanity such as insults leading to disrespect or even racism, as some of the locutions are ethnophaulisms (e.g., racial slurs). The treatment of Dwarves is not much better as they are nearly ostracized, living below the surface of the earth. All of these aspects contribute and reinforce the established hierarchical system in the game, where British appear to be on the higher end of the social ladder, while SNAm and nonnative-accented characters are at the lower end.

Finally, we will discuss the kind of language and accent attitudinal ideologies perpetuated throughout othering based on the PC's race and attributed accent. The player begins the game with a choice of race for their character: Dwarf, Elf, or Human. Depending on the selection of race, like City Elf, the player will begin the game in an environment populated entirely by City Elves. In the game's opening moments, erasure (Gal and Woolard 2000) is a dominant force in eliminating any nuances of differentiation, as all City Elves speak with an American accent. This erasure leads to the generation of a SNAm matrix of predominant use, where the player becomes used to hearing a singular American accent in all of the City Elves, the group to which the PC belongs. The American accent then becomes the norm, and any accents falling outside of the SNAm matrix become deviations, thereby giving rise to the semiotic processes of fractal recursivity and othering. British accents encountered in Humans become the hegemonial other as Elves and Humans through their entirely predictable accents of SNAm versus RP.

However, dwarves also share the Standard American accent, so when a City Elf PC comes into contact with a Dwarf, they share the same matrix, as both races use the predominant accent. Hence, Dwarves and Elves have an important auditive component in common: through the similarity of their accented voices, which iconize their shared socioeconomic status, they can relate to one another in opposition to the social "other" represented by humans. When playing as a Dwarf or an Elf, humans are immediately set as the hegemonic other, but when playing as a Human, a standard RP matrix is formed with the British accent, thereby turning Dwarves and Elves into the other. Returning to Kathryn Woolard's concept of "implicit metapragmatics," by making a (conscious or subconscious) linguistic choice of American over British depending on the race of the PC, the player's views are shaped by character's reactions to their race and attributed accent. The player implicitly aligns themselves with those they can relate to, like speakers who fall within their predominant matrix of speech while simultaneously distancing themselves from other uses of language.

Thus, *DAO* does a superb job of perpetuating dominant accent attitudinal ideologies about the perceived elite status of RP (Hughes et al. 2005) and *Pax Americana* (Bayard et al. 2001). Although the phonetically augmented division of races can help in painting a clear picture of the gameworld and where allegiances lie, the division of accents between social groups from the player's framework of reference leads to stereotypical hegemonic thinking when it comes to power relationships. This design feature is perhaps surprising, given BioWare's widespread and long-standing reputation of seeking to diversify game culture (Rossignol 2009). However, as our research has shown, diversification of sociophonetic features, which are just as important as visual features such as color of skin and nonbinary, nonpolarized gender representations, can come across as parodic if, for instance, intentionally voiced foreign accents are unintentionally othered through strongly perceivable phonetic inconsistencies.[2]

5 Conclusions

The BioWare studio has long tried to be socially inclusive in their videogames, both in terms of gender and race, across their signature game series. However, our research has shown that *DAO* lacks cultural diversity in the speech accents assigned to its characters. Certain races are locked into specific accents, and the few accents found outside the typical British or American speech patterns, like French or Spanish, are found primarily in merchants as though they are "exotic" as they hail from far away. This depiction of accents in a fantasy world paints a highly limited society populated by primarily British and American speakers only. This is a dangerous perception to perpetuate as the ideas posited by videogames, as by any other media, can carry into the real world. This is especially true for normalized and idealized views of gender, race, class, voice, and language, whereby the latter two been largely neglected by critical game studies. Thus, the way a game represents its characters has the potential to affect real-world perceptions of people, and effects of underrepresentation and erasure are bound to carry over from fantasy world into real-world naturalization.[3]

DAO is guilty of portraying a "warped world" (Rosewater 2018), whose inhabitants are reduced to only two primary accents restricted by their race and socioeconomic status. In reality, of course, the world is populated with hundreds of different accents across a multitude of races, and the standard hegemony associated with RP in particular has lost significantly in social capital compared to even a few decades ago. These factors are further dependent on where someone was raised, whether they immigrated, or if they learned a new language along the way. The possibilities for human speech are extensive, and portraying only two possible accents of the world (which, furthermore, are the least naturally occurring) sends a message that these are the only accents worth portraying in an attempt to reach the player base. While the blockbuster videogame industry is hardly the only media industry to blame for popularizing linguistic reductionism, its voice-acting choices raise further issues in assumptions made about the player base, namely, that they are primarily American and British players, or players that willingly embrace popular culture as shaped by Disney and Hollywood (see Lippi-Green 2011). This assumption immediately alienates players who do not speak with a native or standard accent of English. Perceptions shape reality, and when only two accents are the primary focus, this perpetuates standard language ideology (Milroy 2006) by perpetuating the idea that standard varieties are more important, that they represent the "norm," and that any accents that deviate from that norm are othered as less desirable and socially less empowered.

Videogames are capable of constructing and reinforcing hegemonic political and attitudinal ideologies that can lead to or reconfirm harmful player perceptions outside of the gameworld. It thus becomes increasingly important for game designers to be aware of the ideologies they perpetuate through speech accents. To return to Ruberg and Shaw's (2017) important observation, "fantasy is always already political" (p. xxi), we would like to re-emphasize that the fantasy of videogames does not exist in a vacuum. It creates very real spaces for players to engage with, and because games act as lenses through which to understand society, they can also be seen as lenses through which to potentially unlearn deeply ingrained and naturalized views of hegemonial power relationships. These views must not stop at visual representations of nonhypersexualized or nonheteronormative bodies but need to include critical-creative engagement with commonly neglected yet nonetheless powerful and semiotically pervasive language ideologies.

Acknowledgments

Our thanks goes to the Artificial General Intelligence Lab at the University of Alberta, led by Dr Vadim Bulitko, for allowing us to use some of their facilities and resources for data collection and editing; and to Sergio Poo Hernandez for his support and advice throughout this project. We would also like to acknowledge the financial support of the Department of Digital Humanities and the Kule Institute of Advanced Studies at the University of Alberta, via Dr Michael Frishkopf's "Deep Learning for Sound Recognition" Cluster Grant.

Notes

1   Some of the minor characters whose speech we recorded had extremely short lines, at times restricted to a single word or a single phrase, which made it problematic, if not impossible, to determine a particular variety of British English or American English. Accent classification relies to a large degree on the usage of certain lexical sets, for example, words belonging to certain groups of words that share the same phoneme. If the usage of particular lexical sets that allow the apprehension of the dialect is absent, clear disambiguation is impeded considerably. In cases where it was impossible to distinguish an accent with any certainty, the data was omitted from the analysis.

2   The extent to which accents, (in)consistent, fake, authentic, or otherwise, are perceivable by native vs. nonnative speakers is a moot point. Some would argue that native speakers would pick up inconsistencies more readily than nonnative speakers and that therefore, to the latter, inconsistencies may not stand out in any noticeable way. This could be countered by the observation that nonnatives may be even more likely to detect the differences, especially European players, who are almost constantly surrounded by different English accents and are familiar with nonnative varieties and what they tend to sound like. Thus, if the player is, for example, French, and the character has a fake French accent that sounds far removed from French as it is known to Europeans and resembles more of a put-on accent by a native speaker of North American English, they

will notice and possibly take offense or at least be struck by the character's voice acting.

3 It is important to note here that *Dragon Age: Inquisition* (BioWare 2014), the third instalment in the series, features a significantly more varied set of speech accents. A comprehensive comparative analysis of this game in relation to *DAO* and the second instalment in the series is pending.

References

Andronis, M.A. (2003), "Iconization, fractal recursivity, and erasure: Linguistic ideologies and standardization in Quichua-speaking Ecuador," *Texas Linguistic Forum*, 47: 263–69.

Bayard, D., A. Weatherall, C. Gallois, and J. Pittam (2001), "Pax Americana? Accent attitudinal evaluations in New Zealand, Australia and America," *Journal of Sociolinguistics*, 5(1): 22–49.

BioWare (2009), *Dragon Age: Origins*, Redwood City: Electronic Arts.

BioWare (2009–14), *Dragon Age* series, Redwood City: Electronic Arts.

BioWare (2010), *Mass Effect 2*, Redwood City: Electronic Arts.

BioWare (2014a), *Dragon Age: Inquisition*, Redwood City: Electronic Arts.

Brady, H. (2017), "Building a queer mythology," in B. Ruberg and A. Shaw (eds.), *Queer Game Studies*, 63–68, Minneapolis: University of Minnesota Press.

Brice, M. (2011), "Speaking in accents and the American ethnocentrism in video games," *Popmatters*, November 15. Available online: http://www.popmatters.com/post/151275-speaking-in-accents-and-the-american-ethnocentrism-in-video-games/ (accessed May 20, 2018).

Coupland, N. (2007), *Style: Language Variation and Identity*, Cambridge: Cambridge University Press.

Ensslin, A. (2010), "'Black and White': Language ideologies in computer game discourse," in S. Johnson and T.M. Milani (eds.), *Language Ideologies and Media Discourse: Texts, Practices, Politics*, 205–22, London: Continuum.

Ensslin, A. (2011), "Recallin' Fagin: Linguistic accents, intertextuality and othering in narrative offline and online video games," in G. Crawford, V.K. Gosling, and B. Light (eds.), *Online Gaming in Context: The Social and Cultural Significance of Online Games*, 224–35, New York: Routledge.

GenericGuy (2012), "Why are fantasy world accents British?" *Giant in the Playground Forums*, April 2. Available online: http://www.giantitp.com/forums/archive/index.php/t-238364.html (accessed January 12, 2018).

Greer, S. (2013), "Playing queer: Affordances for sexuality in *Fable* and *Dragon Age*," *Journal of Gaming and Virtual Worlds*, 5 (1): 3–21.

Henton, A. (2012), "Game and narrative in *Dragon Age: Origins*: Playing the archive in digital RPGs," in G.A. Voorhees, J. Call, and K. Whitlock (eds.), *Dungeons, Dragons, and Digital Denizens: The Digital Role-Playing Game*, 66–87, New York: Bloomsbury.

Hughes, A., P. Trudgill, and D. Watt (2005), *English Accents and Dialects: An Introduction to Social and Regional Varieties of English in the British Isles*, 5th edn., New York: Hodder Arnold.

Irvine, J. and S. Gal (2000), "Language ideology and linguistic differentiation," in P.V. Kroskrity (ed.), *Regimes of Language: Ideologies, Polities and Identities*, 35–83, Oxford: James Currey.

Jørgensen, K. (2010), "Game characters as narrative devices. A comparative analysis of *Dragon Age: Origins* and *Mass Effect 2*," *Eludamos. Journal for Computer Game Culture*, 4 (2): 315–31.

Lionhead (2004–17), *Fable* series, Redmond, WA: Microsoft Game Studios.

Lippi-Green, R. (2012), *English with an Accent: Language, Ideology and Discrimination in the United States*, 2nd edn., New York: Routledge.

Meier, P. (2012), *Accents & Dialects for Stage and Screen: An Instruction Manual for 24 Accents and Dialects Commonly Used by English-speaking Actors*, Lawrence, Kan.: Paul Meier Dialect Services.

Milroy, J. (2006), "The ideology of the standard language," in C. Llamas, L. Mullany, and P. Stockwell (eds.), *The Routledge Companion to Sociolinguistics*, 133–39, London: Taylor & Francis Group.

Poos, D. (2013), "Why is the villain always British? An analysis of American attitudes towards British English and the influence of media, using the example of the movie 'Gladiator' (2000)," *GRIN*. Available online: http://www.grin.com/en/e-book/233422/why-is-the-villain-always-british (accessed February 15, 2018).

Preston, D. (2004), "Folk metalanguage," in A. Jaworski, N. Coupland, and D. Galasiński (eds.), *Metalanguage: Social and Ideological Perspectives*, 75–101, Berlin: Mouton de Gruyter.

Rosewater, M. (2018), "Why is wizards trying to fix something that isn't broken?" *Blogatog*, February 19. Available online: http://markrosewater.tumblr.com/post/171074786068/why-is-wizards-trying-to-fix-something-that-isnt#notes (accessed February 15, 2018).

Rossignol, J. (2009), "Bioware boss talks up PC diversity," *Rock Paper Shotgun*, January 19. Available online: https://www.rockpapershotgun.com/2009/01/19/bioware-boss-talks-up-pc-diversity/comment-page-2/ (accessed February 15, 2018).

Ruberg, B. and A. Shaw (2017), "Introduction: Imagining queer game studies," in B. Ruberg and A. Shaw (eds.), *Queer Game Studies*, ix–xxxiii, Minneapolis: University of Minnesota Press.

Schniedewind, N. and E. Davidson (2000), "Linguicism," in M. Adams, W.J. Blumenfeld, R. Castañeda, H.W. Hackman, M.L. Peters, and X. Zúñiga (eds.), *Readings for Social Diversity and Social Justice: An Anthology on Racism, Anti-Semitism, Sexism, Heterosexism, Ableism, and Classism*, 129–30, New York: Routledge.

Skutnabb-Kangas, T. and R. Phillipson (1995), "Linguicide and linguicism," *RoligPapir* 53, 83–91, Roskilde Universitetscenter, Denmark.

The Escapist (2013), "Which is your favourite origin story for the first Dragon Age game?" Available online: http://www.escapistmagazine.com/forums/read/9.404171-Poll-Dragon-Age-Best-origin-story (accessed February 15, 2018).

Waern, A. (2011), "'I'm in love with someone that doesn't exist!' Bleed in the context of a computer game," *Journal of Gaming and Virtual Worlds*, 3 (3): 239–57.

Wells, J.C. (1982), *Accents of English 2: The British Isles*, Cambridge: Cambridge University Press.

13

Playing It By the Book

Instructing and Constructing the Player in the Videogame Manual Paratext

Michael Hancock

1 Introduction

The inner cover of the instruction manual for the 1994 Super Nintendo game *Donkey Kong Country* (Rare 1994) begins in the same manner as most manuals from the period, serving a similar function to the copyright page of a novel. There is a warning to read the "consumer information and precautions booklet" that is also contained in the game box (Rare 1994: 2). Below the warning is the Official Nintendo Seal of Quality, which proves that Nintendo has reviewed the game and it has met their "standards for excellence in workmanship, reliability and entertainment value" (Rare 1994). A text box below the seal thanks the consumer for the purchase, cautions them to keep the manual close, and further warns them to avoid rapid switching of the POWER switch to avoid memory loss. A fourth and final box declares that the ESRB has rated the product. All of these elements are standard for Nintendo products, or, in the case of the ESRB notice, will become so, for as long as manuals are commonly printed. However, below the boxes at the bottom right, sandwiched between the ESRB notice and the copyright information, is a single panel depicting an elderly simian with a white beard and glasses, looking directly at the reader. The word bubble to its left reads "Look at the fancy box. Look at size of this instruction manual. You don't think they would have gone to all this trouble of [sic] the game was any good do you?!" (Rare 994: 2).

The character is identified as Cranky Kong in the following four-page introductory story, during which he mocks the game's protagonist, Donkey Kong, and makes veiled allusions to earlier times: "You'll never be as popular a character as I was! Why, in my heyday, kids lined up to play my games! The quarters were stacked on the machine as they waited for their turn!" (Rare 1994: 7). For younger player-readers who may not recognize Cranky's meta-reference, the manual eventually confirms what older players may already expect, that "Donkey Kong's grouchy pappy is actually the original Donkey Kong who starred in the many Donkey Kong classics of the eighties. He considers those games the pinnacle of game design and will have nothing to do with the newfangled graphics, sound and multiple-button controllers of today's Nintendo mega-hits" (Rare 1994: 27). Throughout the entire manual, the aged ape interjects another eight times, not counting his aforementioned comments in the manual's story, nor his page of advice near the manual's end. In each appearance, he heaps more scorn, either at the expense of the player ("I can't believe you're still reading this! What you need is a good thrashing!" [Rare 1994: 18]), the game ("I wouldn't believe a word of this! I've been everywhere and I found only two locations, bad ones at that!" [Rare 1994: 8]), and the very function of the manual itself: "Well, well, I've never seen so much rubbish. A good game shouldn't need any explanation!" (Rare 1994: 11).[1]

Why does the *Donkey Kong Country* manual feature Cranky instead of its own lead character? Why does an instruction manual that is meant to inform players about a game focus so heavily on a character who mocks the player for seeking that information, contradicts the manual's descriptions of the game, and opposes the very notion that the manual should exist? What messages about game culture is the *Donkey Kong Country* manual presenting to its players?

To answer these questions, we need to pull back to a broader question: what concepts and discourses are embedded in the videogame instruction manual? Ensslin (2012: 107) explores the discourses that reinforce gamer identity, pointing out that "gamers construct their identity as gamers by reinforcing discursively the social, phenomenological and ludic implications of gaming." My contention for this chapter is that in addition—and, sometimes, in opposition—to the purpose of instructing the player, the videogame manual is a means through which agents in the game industry— developers, publishers, and others—participate in these discourses. Through the manual, the game industry agents perpetuate their standards of what an ideal gamer (and an ideal consumer) should be, a task frequently supported by conflating a "discourse of 'cool'" (Ensslin 2012: 108–10) with a discourse of hypermediacy (Bolter and Grusin 2000). To illustrate these points, I will briefly discuss the history of the videogame instruction manual and the major theoretical touchstones for the chapter, including paratext and existing technical communication approaches to manuals that are relevant for game studies. This discussion provides the necessary background to explore both the *Donkey Kong Country* manual and the manual for *The Lord of the Rings Vol. 1* (Interplay 1994). In both cases, conceptualizing the videogame instruction manual as a paratext is the first step in understanding how the manual constructs the ideal gamer by pushing the player toward the new cool.

2 History, the instruction manual and paratext

The history of the videogame manual is frequently a reflection of the history of videogames. The amount of space devoted to explaining or highlighting novel game features in the manual demonstrates how much the designers wanted players to value those features. For example, the manuals for the earliest videogames for the Atari 2600, games that were part of the system's launch in 1977, emphasized for pages and pages the variety in different modes of play. These different modes of play included variations on speed and player numbers. There were 14 modes for *Surround* (Atari 1977) and 50 variant modes for *Video Olympics* (Atari 1977), including different events and different paddle responses. By 1982, in the manual for *Centipede*, Atari devoted about a quarter of a page to the mode description. However, the manual used an entire page to set up the background for the game, a complicated story revolving around a colony of friendly elves fighting off the giant pests that prey on their mushroom garden. The shift represented a transition in game design, a movement away from multiple forms of short play to longer engagements with structured levels. The change in game design eventually led to the idea of an overarching gameworld, which in turn puts greater emphasis on narrativity and continuity within the game. Likewise, later manuals emphasize other game features, such as the player character for franchise-based games (*Kirby's Adventure* 1993), 3D graphics (*Donkey Kong Country* 1994), and internet connectivity (*Metal Gear Solid 4* 2008).

However, while the changing sections of game manuals reflect the history of videogames in some aspects, there are also obvious limitations to conflating the two histories. A videogame instruction manual is, at best, an official document produced by a game publisher, presumably containing only the information they wished to include. If a game went through severe revisions, such information would be a part of its history but is unlikely to be something mentioned in the manual. Sometimes the publisher does not even produce the manual, as they may outsource the duties for composing the manual to third-party individuals who were not involved in creating the game. Furthermore, even in the years where hardcopy manuals were common, there were many games that never had an accompanying manual—arcade games had instructions displayed on the cabinet, if they had any at all. Unauthorized players surreptitiously distributed early computer games such as the 1976 *Colossal Cave Adventure* (Crowther and Woods) to run on university mainframe computers, with any instruction typically built directly into their code. Finally, the history of the videogame manual is at best an incomplete history of games at large, because the hardcopy manual itself has more or less ceased to exist. Since the late 2000s, game manuals have become increasingly shorter, and in 2010, Ubisoft, a massive game publisher that controls franchises such as *Splinter Cell*, *Rayman*, and *Assassin's Creed*, announced it would cease including paper manuals with its games. Ubisoft defended the move by citing environmental concerns and the chance to "offer the player easier and more intuitive access to game information" by placing the manual on disk (Graft 2010). A year later, another major publisher followed suit, as Electronic Arts (EA) announced that their sports line-up (which includes the series *FIFA*, *Madden*, *NHL*, and *NBA Live*) would no longer ship with paper manuals, again citing eco-related concerns (Good 2011). Other publishers have joined EA and Ubisoft, and the physical manual is increasingly rare (Totilo 2014).

While the game manual was a standard part of the overall game package for over 20 years, it is perhaps unsurprising that publishers would move to eliminate it; as Good (2011) editorializes in the article concerning EA's decision, the move may be more "green," but it also saves on printing costs. However, the comments below Good's article suggest a more mixed response to the removal among the game community. There, commenters ponder the ways in which games themselves are becoming increasingly digital, nostalgically recount stories about their favorite manuals, and debate about the movement toward games that increasingly rely on in-game tutorials for instruction. The consensus, however, is one of acceptance; as one commenter puts it, "I miss the days of large, color manuals with details and back story and whatnot, but I never really used them anyway, so I guess I'll probably get over it quickly." A responding comment notes that "most of the information you get from a manual, you can obtain it from the game itself … In fact, I have rather a good time finding out some thing [*sic*] on my own while playing."

A number of issues are at stake here: first and foremost, the value of the instruction manual, but also what instruction means in regards to a technological product designed for entertainment and what discourses a manual participates in beyond instruction of the game at hand—discourses on what it means to be a gamer and consumer. Before we can apprehend these forces and return to Cranky Kong, we must first turn to existing scholarship on videogame instructions, on instruction manuals in general, on paratexts, and the roles all three play in creating game communities.

Other scholars have studied in detail the connection between games and various forms of instruction (cf. Mayer 2014; Gee 2014; deWinter 2016; Ensslin 2011; Kocurek 2016). However, of that set, only Kocurek (2016: 63) focuses directly on manuals, technical ones in particular. Kocurek argues that the operation manuals for arcade machines, designed for the machines' owners rather than its players, deliberately deployed technical writing "as a means of shaping operator response and facilitating the diffusion of the new technology along established industrial distribution channels." Specifically, the maintenance manuals had to assuage operators' fears that the machines were not too complex to maintain while simultaneously promoting the complexity, novelty, and technology that drove the entertainment industry. The videogame manual, while aimed at a very different, frequently younger audience, must meet similar strictures by framing the game as complex enough to satisfy the gamer, but simple enough to be easily understood— and, on top of that, it must make the game appear fun. While few scholars (e.g., Flanagan 2016) write directly on the videogame manual, Kocurek's (2016) discussion of operating manuals comes closer than many through its discussion of how technical writing functioned to orient potential arcade operators toward a new technology.

Kocurek's (2016) study also touches on another area of scholarship relevant to videogame manuals: the study of instruction manuals in general. The first major hurdle to the typical manual—videogame or otherwise—is getting the user to acknowledge it at all; in this sense, the user's comment from the Good (2011) article that they never used their game manual is not by any means an uncommon sentiment. In a 2002 study, researchers interviewed subjects on how many of them read their automobile owner's manual, with about 60% reporting that they read any amount of it, and only 5.2% reporting they read 90% of it or more (Mehlenbacher et al 2002). If adult owners of a technological device that is capable of ending lives are not motivated to read the manual, then perhaps frequent younger users of an entertainment device should not be expected to do the same. However, the question as to why so few read them remains. In a 2006 study looking at instruction manuals for digital applications, Novick and Ward found that the most commonly cited problems were that the documentation is hard to navigate and is pitched to the wrong level of detail and expertise. However, they also noted that simply putting the manual into the program— the measure promised by Ubisoft and EA—failed to solve the problem and actually compounded it, as the manual was now competing with the program for screen space. Left to their own devices, users preferred to consult others or figure it out for themselves through trial and error (Novick and Ward 2006: 17)—just as the other commenter in Good (2011) did.

Concurrent with the 1980s and 1990s heyday of the videogame instruction manual, information scientist John M. Carroll was championing a new form of instruction manual, under the principle of minimalism. As Carroll (1990: 7) describes it, the three aspects of the minimalist manual are "(1) allowing learners to start immediately on meaningfully realistic tasks, (2) reducing the amount of reading and other passive activity in training, and (3) helping to make errors and error recovery less traumatic and more pedagogically productive." One relevant consequence of the minimalist manual design is that Carroll and his team deliberately omitted information they felt the user could discover through exploration (1990: 8). Summarizing the minimalist movement, Brockmann (1990: 118) emphasizes that point: "Minimalism gives the reader back their span of discretion by being intentionally incomplete and encouraging readers' active exploration." At the same time, however, Brockmann concludes with a list of the approach's potential faults: the possibility of creating gaps in the user's learning; the assumption of a motivated audience that wishes to actively develop the minimal instructions into full skills; the problem that if the user is free to choose goals, they may choose goals that are not possible or effective; the risk that conciseness can lead to being cryptic; and the fear that designers may take the easiest approach to minimalist design and just "cut words" without improving readability.

Parts of the *Donkey Kong Country* manual (Rare 1994) seem to invoke a minimalist design model—recall, for example, Cranky's entreaty that a good game does not need explanation. However, it is Brockmann's potential problems with minimalism that are particularly worth considering in terms of the videogame instruction manual, especially because the context of a videogame modifies and occasionally alleviates their effect. First, a game player is a motivated audience, to the point where many longstanding definitions of games maintain voluntary, free engagement as a necessary condition for games to exist (Caillois [1961] 2005; Suits [2001] 2005). In that sense, a minimalist manual can take advantage of a player's desire to play to fuel their own exploration of the rules. At the same time, it is certainly possible for a manual to be overly cryptic or withholding in its information; in that case, the players may then go so far as to accuse the developers or publishers of deliberately withholding information in order to boost the sales of strategy guides (cf. *Lunar: Eternal Blue*, Game Arts and Studio Alex 1995: 37). The risk of the player choosing the wrong goal is also a potential problem, if that goal is not clearly designated in the game as possible.[2] Perhaps the most obvious implication of the problem set Brockmann identifies is the concern that manual creators will read minimalist as "cut words," as later videogame manuals are cut to the bone, shrinking to a handful of pages that fail to go much beyond the system's basic set-up.

However, Cranky Kong's interjections cannot be explained purely through instruction, minimalist or otherwise. While some of his tips—notably, those attributed to the "Cranky's Advice" section—can be viewed as instructions designed to push the player toward self-discovery, others are comparatively pointless or even false. Instead, their inclusion is better explained through the concept of paratext, a concept that can be extended to the manual as a whole, but is particularly apt for the interjections. French literary theorist Gérard Genette adopted the term *paratext* to refer to the material relating to a text that "surround it and extend it, precisely in order to *present* it … to ensure the text's presence in the world, its 'reception' and consumption" ([1987] 1997: 1; emphasis in original); essentially, paratext includes anything an author might use to influence public interpretation regarding their book. Further, Genette divided paratext into two parts: the *peritext*, which was any paratextual element that was bound together or included within the text, from preface to afterword; and the *epitext*, which was any paratextual element that could be reasonably considered as being apart from the text, including publicity prior to the book launch and anything the author may say privately or publicly about the book at a later date. Any instruction manual, then, can be considered as paratext (and, generally, epitext) in the sense that its entire purpose is to make some product "present" in the world, to better enable a user to take advantage of its features.

At the same time, defining a manual as paratext goes beyond Genette's somewhat conservative application, as his object of investigation is print literature only, and paratext is restricted to what the publisher or author has to say about the text. Media scholar Jonathan Gray (2010) extends the definition of paratext considerably to include a wide variety of popular culture including film trailers, adaptations in other media, action figures, and more. Most relevant to the discussion at hand, Gray (2010: 35) coins another pair of terms that distinguish between types of paratexts: *entryway* paratexts that "control and determine our *entrance* to a text" and in medias res paratexts that "inflect or redirect the text following initial interaction." For instance, when design specialist Marc Rettig promotes "task-oriented" manuals that function like a cookbook, providing "recipes for all the things you might want to accomplish" (Rettig 1991), what he is calling for is a manual that functions as an in medias res paratext, helping the user clarify a specific function that a product they are already familiar with can perform.

Paratext has also been applied more directly to videogames, via Steven E. Jones (2008) and Mia Consalvo (2007), both of whom shift the focus from the creators of the main text to wider social networks. For Jones (2008: 9), that shift is a part of remembering that "playing [a game] is always in the social world, always a complicated mediated experience, never purely formal, any more than a text is purely a verbal construct." For Consalvo (2007: 18), paratext serves to explain how game-oriented magazines such as *Nintendo Power* function as paratext not just for individual games, but also for the game industry as whole. Further, such paratexts work toward shaping a concept of "gamer capital"—building from Bourdieu's notion of cultural capital, Consalvo (2007: 18) uses "gamer capital" to refer to the way players and other agents within the videogame industry establish themselves as knowledgeable or skillful regarding games. Together, these scholars take the original concept of paratext and transform it beyond a unilateral influence of author on audience into a constant recirculation of meaning and ideas.

The next step, then, is to consider videogame instruction manuals as paratexts that inform and influence the reception of videogame texts, illustrating a commonality of discursive structure and purpose among such manuals. In both of the following case studies, whatever their differences, the game designers shared a common goal: to use the paratextual influence of the manual to shape not just the players' conception of the game, but also their conception of the studio that made it, and their conception of what it meant to be a part of game culture through a combination of the discourses of "coolness" and hypermediality.

3 *Donkey Kong Country* and playing it cool

The *Donkey Kong Country* manual (Rare 1994) is, first and foremost, a paratext meant to explain and promote the game to its players, but both game and manual call on a web of other paratexts to convince the player that the game is the epitome of coolness and technology that only the publisher Nintendo and developing studio Rare can provide. The first and most obvious context is the original 1981 *Donkey Kong*, the arcade game that was Nintendo's first major success in the game industry.[3] While the game relies on the manual to make the connection explicit, it doesn't shy away from the association; after a brief display of Rare's and Nintendo's logos, the game begins with Cranky atop a pile of girders (scenery clearly meant to evoke the original game), listening on a gramophone to the opening music of the 1983 NES version of *Donkey Kong* (Nintendo). Suddenly, a boom box drops onto the gramophone, the music changes to a faster rock tempo, and Donkey Kong jumps down, dislodging Cranky as the girders change into jungle trees (Rare 1994). The supplanting of an antiquated media form by a newer one followed immediately by Cranky being supplanted by Donkey is a clear indicator that the two games should be interpreted in a similar manner: *Donkey Kong Country* is newer, more technologically sophisticated, and consequently better than the original *Donkey Kong*. While the reader who approaches the manual as an entryway paratext, studying it before actually playing the game, may not be entirely certain what to make of Cranky, a player who approaches it as an in medias res paratext after seeing the game's initial scene and others involving Cranky knows that he lives up to his name, as not just cranky but a crank, and much of his opinion on the game is not to be taken seriously.

Alongside Nintendo's history, the manual and game also come out of the culture of the contemporary game industry, particularly the rivalry between Sega and Nintendo and their respective consoles, the Sega Genesis and Super Nintendo. When Nintendo entered the North American console market in 1985 with its NES, it was the only dominant game console manufacturer, and it established itself by marketing to a young (generally male) child audience. Sega became Nintendo's main rival with the launch of the Sega Genesis in 1989 and the promotion of its mascot, Sonic the Hedgehog. Sega executives aimed the console at the now teen audience who had grown up with the NES, starting with its commercial slogan that attacked Nintendo directly: "Sega does what Nintendo'nt" (Donovan 2010). Writing for the games news and reviews site *Polygon*, journalist Mike Sholars argues that the difference between the two companies is that "Nintendo wants to be timeless, and Sega wants to be cool," where *cool* is defined as "[t]he act of emulating or creating something based on contemporary interests and arts." Sega's coolness is rooted in attempting to be at the forefront of technology and culture, as seen in acts such as allegedly bringing in Michael Jackson to consult for the soundtracks of *Sonic the Hedgehog 3*, and Nintendo aims for a broader, timeless appeal, drawing on stage and storybook aesthetics (Sholars 2017).[4]

Sholars' (2017) invocation of coolness draws his argument into larger discussions of the discourse of "cool." Ensslin (2012: 108–09) identifies several markers of the "cool" discourse when employed by gamers, including ownership and mastery of new technologies and the paradox of detached engagement. As Crabbe (2008: 28) describes it in the context of sport, the paradox of detached engagement is that cool is about detachment, but is expressed through being thoroughly engaged in play, through mastery in use. I would argue that one of the chief ways that game culture bypasses this paradox is through the discourse of *hypermediacy*. Following Bolter and Grusin (2000: 9), hypermediacy is a form of remediation, the ways in which new media visibly incorporates older media into itself in order to imply its superiority. The discourse of hypermediacy,[5] then, refers to the markers by which agents in the game industry incorporate older media and texts (games, in most, but not all, cases) to insinuate the superiority of the newer text. Thus, as Sholars (2017) frames it, Sega and its gamer fans could be enthusiastic about the company's pursuit of new technology and culture by performing how disengaged they were from Nintendo's approach.

Contrary to Sholars' claims, however, Nintendo has not always aimed for timelessness; in fact, *Donkey Kong Country* can be viewed an attempt to reinvent their first blockbuster videogame property as the epitome of cool. They did so by lending it to the development team Rare, who, starting in the 1990s, became known for a series of "cool" games, including the Ninja Turtle parody *Battletoads* (1991), the slick James Bond licensed shooter *GoldenEye* 007 (1997), and the deliberately crude and vulgar

Conker's Bad Fur Day (2001). *Donkey Kong Country* (1994)—and the *Donkey Kong Country* manual (1994)—does everything in its power to establish the same brand of contemporary coolness by making clear that the reader should not be laughing with Cranky but at him, belittling his claims about gameplay being more important than graphics with the implication that *Donkey Kong Country* can utilize both. The remediation of the gramophone with the boom box in the game's opening scene works alongside the hypermediacy of Cranky's interjections in the manual, prompting the player to signal their rejection of Cranky through play.

Near the end of the manual, there comes a page that makes even more explicit the link between discourses of coolness and hypermediacy. Under the heading "The Making of Donkey Kong Country," the manual states, "What makes the graphics of Donkey Kong Country so cool? Every image that you see on the screen was actually designed on Silicon Graphics Inc. workstations, the same powerful computers that were used to create computer animation in movies like Jurassic Park and Terminator 2: Judgment Day" (Rare 1994: 32). The game's graphics—and implicitly, the game itself—is cool in the contemporary sense that Sholars provides specifically because of its resemblance to contemporary films and the application of "the most realistic and three-dimensional graphics ever seen in video games," as demonstrated by the Donkey Kong wireframe screenshots accompanying the text. On the opposite page, Cranky appears a final time, mocking the notes section: "Waste of paper if you ask me!" (Rare 1994: 33). Despite his previously expressed attitude toward newfangled technology, Cranky has nothing to say here about the "use of computers to design graphics that will revolutionize the way games are made" (Rare 1994: 32), because doing so at that moment might risk players underappreciating the notion of gamer capital that Nintendo and Rare are trying to convey. Cranky Kong's interjections and "The Making of Donkey Kong Country" are two sides of the same coin; while the main purpose of the manual is to teach players how to play the game, its secondary purpose is to teach them how to value the game, through appreciation of the technology behind it and the discourse of hypermediacy via mockery of those who prefer the "old school" style. In doing so, the manual acts as paratext for game consumption as a whole, pushing players toward the newest technological innovation and the newest iteration of cool.

4 *The Lord of the Rings, Vol. 1* and playing with story

In the remainder of this chapter, I wish to establish that the *Donkey Kong Country* manual is not simply a flash in the pan, and that the game manual frequently performed this dual role of instructing and enculturating players in gamer discourses, albeit some more overtly than others. To that end, the computer game manual for Interplay's *The Lord of the Rings Vol.1* (*LotRV1*) (Interplay 1990) is particularly worth investigating for the ways it overtly encourages its players to extend their notion of play beyond the game at hand.

Starting with the postapocalypse-themed 1988 game *Wasteland* (Interplay), a series of home computer games included a "paragraph section" in their manuals. When the player reached an appropriate point in the game, the game would tell them to consult a numbered paragraph from the manual in order to continue. To avoid the player reading ahead of their position in the game, the paragraphs were not only randomized, but a number of them—sometimes a majority of them—were also false, describing events that the player could not actually encounter. The intent behind these manuals seems to be mixed, addressing two different technology-derived problems. First, in an era where game text was often blocky and displayed in a handful of lines, the paragraphs offered a more elegant way delivering description and exposition. Second, in an era where it was easier to copy a disk than copy and distribute a manual, the paragraphs served as a form of copyright protection, in that they typically included codewords that the player would come across and need over the course of the game (cf. Hancock 2016: 278–81). Arguably, the paragraphs turned the manual from epitext to peritext, ensuring that they were not just influencing reception of the game, but actively a part of it.

The manual for Interplay's 1990 *The Lord of the Rings Vol. 1* demonstrates not just how the paragraphs section worked to shape a notion of an ideal player, but also how the manual as a whole works to shape what it means to play an adaptation of a nongame text. In discussing the 2007 *The Lord of the Rings Online*, Randall and Murphy (2012: 121) argue that it constitutes a "comprehensive expansion" of the original novels, generally being faithful to the books but depicting its locations, characters, cultures, and general environments in much greater detail. *LotRV1* (Interplay 1990) takes a similar approach, adding new encounters and characters, but it faces a slightly different problem in terms of fidelity. Anyone playing is likely to be familiar with the film series or books or both, but in 1990, long before Jackson began adapting Tolkien, it is possible that the players may be experts on Tolkien's books, but also possible they have not heard of them at all. The game's manual addresses this issue with an opening section with two subsections. "If You Are Not Familiar with Tolkien's Books…," tells the player that "we [Interplay] are honored to be your introduction to one of the greatest works of imaginative literature ever written" (Interplay 1990: 3), establishing it as a gateway paratext to both the game and the book, and a source of authentic knowledge of what the books contain. The other subsection, "If You Are Familiar with Tolkien's Books…", functions as a gateway paratext to the game but an in medias res paratext for the books themselves; it acknowledges that the Tolkien-familiar player is "going to be our toughest critic" and offers both a defense and apologia that the game is different from the books:

> While this is one of the largest computer games ever created, we couldn't fit every place in Middle-earth into this game. At the same time, we didn't want to clone Tolkien's World directly into the game and have anyone who knows the book be able to easily solve the game. You'll find plenty of new encounters, new characters, and even a plot twist or two, that are not included in Tolkien's epic fantasy. The reason we did this was not to "improve" Tolkien's work (this would be extremely arrogant and stupid of us to say), but to challenge the computer gamer who is familiar with Tolkien's work. Expect to be surprised. (Interplay 1990: 4)

Through this introduction, Interplay not only establishes itself as an authentic authority on Tolkien (the manual also concludes with a brief biography) and the books, but also establishes that videogames based on print properties can and should deviate from the source material. Essentially, it offers an argument in favor of transmedia storytelling, whereby "integral elements of a fiction get dispersed systematically across multiple delivery channels for the purpose of creating a unified and coordinated entertainment experience" (Jenkins 2007). However, in doing so, it is still participating in a discourse of hypermediacy, implicitly arguing that a player should value novelty and challenge in a game more than fidelity to any original source.

It may seem that the manual is violating the discourses of "cool" and hypermediacy by emphasizing the developers' respect for the original source. The accompanying paragraphs immediately dispel such fears. As is typical for these paragraph sections,[6] the first entry is a fake one that breaks diegesis in order to remind the player how these passages work. In the scene, the player gazes into a palantir (a stone that lets people communicate with those who possess other such stones) and sees an unusual sight:

> The glow gives way to a misty red-tinged vision of a dark figure sitting upon a ceramic stool, reading a scroll.
> Suddenly, the Dark One looks up, his single flaming red eye glaring with malice. "Ssssss," he hisses. "Read NOT those paragraphs for which you have been given no instructions. There is a special place in Mordor for the likes of you!"
> And with that, the vision disappears. Yet even as it fades, you hear a muttered, "You'd think being a Dark Lord would grant you some privacy, but NOOOO!" (*Lord of the Rings Vol. 1*, Interplay 1990: 49)

In *Cheating*, Consalvo (2007) raises the contradiction of game guides and magazines of the 1980s and 1990s—that they simultaneously encouraged players to consult them while creating the notion that the skilled player did not need any help, theirs included (39); any acknowledgment of the player's own failure to master the technology is a step away from the discourse of "cool." A similar contradiction arises here; the paragraphs section exists in part as an attempt to keep players from "cheating" by making sure they purchase the game. The passage here is chastising players specifically for cheating, in the form of reading a passage the game did not instruct them to read. At the same time, however, it is rewarding the player by giving them a comic scene they would not witness if they had followed the rules. Further, while not as overtly technologically focused as the *Donkey Kong Country* manual (Rare 1994), this manual too implies disengagement with the original model by slyly presenting a scene of Sauron on the toilet.

In order to parse how the manual is attempting to discursively shape the player, Gray's (2010) concept of entryway and in medias res paratext is again useful, along with his discussion of *The Lord of the Rings* films. Referring to the extra paratextual material included with the films' DVD special editions, Gray argues that the DVDs don't just demonstrate the gallery-worthy art of the film but also provide the production literacy for appreciating it: "The DVDs work to give us the information and teach us to appreciate the work" (2010: 98). As I have argued throughout, this teaching is the work that game manuals have always done, and the *LotRV.1* manual (Interplay 1990) is no different; it teaches the players to appreciate the original text, and teaches them, through the paragraphs section, to appreciate how the game plays with the original text. A player using the manual as an entryway paratext for the game will be encouraged to play the game for themselves and discover which paragraphs are true. The player who returns to it as in medias res paratext, especially after finishing the game, is now in on the joke; they can re-read the entire set of paragraphs and knowingly recognize entries involving a Ringwraith Bilbo (Interplay 1990: 79) or a vampire Strider (Interplay 1990: 53) as fake. At 88 pages, the *LotRV.1* manual is a far cry from the minimalist design of the instruction manual, but by inverting the minimalist formula— by adding false information rather than leaving gaps—it encourages the same sort of user-motivated exploration, a mastery of paragraphs by being able to distinguish which ones are fictionally "real" within the context of the game. In short, it teaches players to respect the affordances videogames offer and to treat all stories, even its own, as mutable and as sources of play.

5 Conclusions: Paratext as playful tech

As videogames themselves are increasingly becoming nonphysical, downloadable goods purchased online rather than a box from a retail store, it is no wonder that the print game manual has become a rarity. The rhetorical functions it used to serve as paratext, however, are as relevant as ever. Some, such as the need to provide entryway instruction, have shifted into a closer state of peritext, embedded, as De Winter (2016) describes, as in-game tutorials. More specialized, in medias res information shifts outwards into epitext, into Wikis, playthroughs, and walkthroughs, as does the information regarding what it means to value games, into game sites, YouTube videos, and Twitch streams. The game manuals of the 1980s and 1990s offer snapshots of game history, examples of how their publishers used them as paratext to instruct and influence how players played their games using discourses of "cool" and hypermediacy. The *Donkey Kong Country* manual uses the exaggerated claims of an old ape to teach players to value graphical superiority and technological prowess; *The Lord of the Rings* (Vol. 1) manual uses sharp deviations from its source material to encourage the player to recognize the flexibility of the videogame narrative. Both manuals push players to take their knowledge and consider it in a playful context, to treat instruction itself as something potentially playful. As such, they illustrate the part game manuals played in shaping what it means to play and consume videogames.

Notes

1 Perhaps Cranky has forgotten in the intervening 13 years, but the original *Donkey Kong* arcade cabinet did have an instruction set, consisting of six rules, a "scoring value" section, and two short notes on points. The essential rule, rule 4, however, was marked out with a distinct yellow font for emphasis, and is admittedly tautological enough to enforce Cranky's argument: "Jump button makes Jumpman jump" ("Donkey Kong Instruction Cards" 2012).

2 A variation of this problem is that the player becomes too effective, to the detriment of play, and discovers a single approach in a game so effective that they avoid all other alternatives and end up playing in a monotonous, repetitive manner. Referring to this issue in regards to the videogame *BioShock*, critic Kieron Gillen concludes that "most players would rather be efficient than have fun." This tendency points to a difference between the technical manual and the videogame manual: the fear for the technical manual is that a minimalist approach will have users rely on a single kludge more complicated than other approaches; the fear with the game manual (or too-open game) is that the player will rely on a method too effective and reliable and thus get bored.

3 Of course, *Donkey Kong* wasn't without its own paratexts; as the name and the plot of a giant ape kidnapping a woman suggests, the game is clearly drawing on the *King Kong* film franchise—while being legally distinct, as the ensuing Universal Studios lawsuit eventually determined (*Universal City Studios v. Nintendo Co.* 1983).

4 Sholars is here referring to the North American *Super Mario Bros. 2* and *Yoshi's Island* and *Yoshi's Story*, respectively.

5 Arguably, *hypertextuality* or *intertextuality* may be more appropriate terms than *hypermediacy*, as the texts being compared under this discourse generally both belong to the medium of videogames. A full comparison of the merits of each term is beyond the scope of this paper; I chose *hypermediacy* to maintain the link to Bolter and Grusin (2000) and to reinforce how both of my chief examples involve at least tangentially a comparison of mediums (the gramophone and stereo in the case of *Donkey Kong Country* [1992] and the original novel and the game in the case of *The Lord of the Rings, Vol. 1* [1990]).

6 In fact, the *Wasteland* manual (1988), the first manual to contain such a section, uses its first paragraph to describe the player coming across a nude woman about to bathe, who shoots them for reading the wrong passage (1), meaning that both this passage and the one in *LotRV1* frame the "cheating" player as a juvenile voyeur.

References

Bolter, J.D. and R. Grusin (2000), *Remediation*, Cambridge: The MIT Press.

Brockmann, R.J. (1990), "The why, where and how of minimalism," in *SIGDOC '90: Proceedings of the 8th Annual International Conference on Systems Documentation*, 111–19.

Caillois, R. ([1961] 2005), "The definition of play," in K. Salen and E. Zimmerman (eds.), *The Game Design Reader: A Rules of Play Anthology*, 123–28, Cambridge: The MIT Press.

Carroll, J. (1990), *The Nurnberg Funnel: Designing Minimalist Instruction for Practical Computer Skill*, Cambridge: The MIT Press.

Centipede (1982), Atari [Atari 2600 manual].

Consalvo, M. (2007), *Cheating: Gaining Advantage in Video Games*, Cambridge: The MIT Press.

Crabbe, T. (2008), "Avoiding the numbers game: Social theory, policy and sport's role in the art of relationship building," in M. Nicholson and R. Hoye (eds.), *Sport and Social Capital*, 21–38, New York: Routledge.

DeWinter, J. (2016), "Just playing around: From procedural manual to in-game training," in J. de Winter and R.M. Moeller (eds.), *Computer Games and Technical Communication: Critical Methods & Applications at the Intersections*, 69–86, New York: Routledge.

Donkey Kong Country (1994), Rare, published by Nintendo [Super Nintendo game].

Donkey Kong Country Instructional Booklet (1994), Rare, published by Nintendo [Super Nintendo manual].

"Donkey Kong Instruction Cards" (2012), *The Unpaidgamers*, September 2. Available online: https://theunpaidgamers.wordpress.com/2012/09/02/lets-g ive-a-little-love-to-the-instructions/donkey-kong-instruction-cards/ (accessed September 25, 2017).

Donovan, T. (2010), *Replay: The History of Video Games*, Great Britain: Yellow Ant.

Ensslin, A. (2011), "'Recallin' Fagin: Linguistic accents, intertextuality and othering in narrative offline and online video games," in G. Crawford, V.K. Gosling, and B. Light (eds.), *Online Gaming in Context: The Social and Cultural Significance of Online Games*, 224–35, New York: Routledge.

Ensslin, A. (2012), *The Language of Gaming*, New York: Palgrave Macmillan.

Flanagan, K.M. (2016), "Introduction—Videogame adaptation: Some experiments in method," *Wide Screen* 6 (1): 1–18.

Gee, J.P. (2014), *What Video Games Have to Teach Us about Learning and Literacy*, 2nd edn., rev. and updated edn., Canada: Macmillan.

Genette, G. ([1987] 1997), *Paratexts: Thresholds of Interpretation*, trans. J.E. Lewin, Cambridge: Cambridge UP.

Gillen, K. (2007), "*BioShock*: A defence," *Eurogamer*, December 6. Available online: http://www.eurogamer.net/articles/bioshock-a-defence-article (accessed September 29, 2017).

Good, O. (2011), "No more manuals as EA Sports goes green," *Kotaku*, March 19. Available online: https://kotaku.com/5783650/no-more-manuals-as-ea-sports -goes-green (accessed September 22, 2017).

Graft, K. (2010), "Ubisoft adopts environment-friendly game packaging," *Gamasutra: The Art & Business of Making Games*, April 19. Available online: https://www.gamasutra.com/view/news/119094/Ubisoft_Adopts_Environment Friendly_Game_Packaging.php (accessed September 22, 2017).

Gray, J. (2010), *Show Sold Separately: Promos, Spoilers, and Other Media Paratexts*, New York: New York University Press.

Hancock, M.J. (2016), "Games with words: Textual representation in the wake of graphical realism in video games," PhD diss., University of Waterloo, Canada.

Jenkins, H. (2007), "Transmedia storytelling 101," *Confessions of an Aca-Fan*, March 21. Available online: http://henryjenkins.org/blog/2007/03/transmedia_ storytelling_101.html (accessed February 24, 2018).

Jones, S.E. (2008), *The Meaning of Video Games: Gaming and Textual Strategies*, New York: Routledge.

Kocurek, C.A. (2016), "Rendering novelty mundane: Technical manuals in the golden age of coin-op computer games," in J. deWinter and R.M. Moeller (eds.), *Computer Games and Technical Communication: Critical Methods & Applications at the Intersections*, 55–68, New York: Routledge.

Lunar: Eternal Blue (1995), Game Arts and Studio Alex [Sega CD manual].

Mayer, R. (2014), *Computer Games for Learning: An Evidence-Based Approach*, Boston: The MIT Press.

Mehlenbacher, B., M.S. Wogalter, and K.R. Laughery (2002), "On the reading of product owner's manuals: Perceptions and product reality," in *Proceedings of the Human Factors and Ergonomics Society's 46 Annual Meeting*, 730–34.

Metal Gear Solid 4: Tactical Espionage Action – Guns of the Patriots (2008), Konami [PlayStation 2 manual].

Novick, D.G. and K. Ward (2006), "Why don't people read the manual?" in *SIGDOC '06: Proceedings of the 24th annual ACM International Conference on Design of Communication*, 11–18.

Randall, N. and K. Murphy (2012), "*The Lord of the Rings Online*: Issues in the adaptation of MMORPGs," in G.A. Voorhees, J. Call, and K. Whitlock (eds.),

Dungeons, Dragons, and Digital Denizens: The Digital Role-Playing Game, 113–31, New York: Continuum.

Rettig, M. (1991), "Nobody reads the documentation," *Communications of the ACM*, 34 (7): 19–24.

Sholars, M. (2017), "Nintendo vs. Sega: The battle over being cool," *Polygon*, August 22. Available online: https://www.polygon.com/2017/8/22/16179048/ni ntendo-vs-sega-the-battle-over-being-cool (accessed October 1, 2017).

Suits, B. ([2001] 2005), "Construction of a definition," in K. Salen and E. Zimmerman (eds.), *The Game Design Reader: A Rules of Play Anthology*, 173–90, Cambridge: The MIT Press.

Surround: Game Program Instructions (1977), Atari [Atari 2600 manual].

The Lord of the Rings, Vol. I (1990), Interplay [PC instruction manual].

Totilo, S. (2014), "Nintendo is slowing reinventing the video game instruction manual," *Kotaku*, February 5. Available online: https://kotaku.com/nintendo-is -slowly-reinventing-the-video-game-instructi-1515814941 (accessed June 23, 2018).

Universal City Studios v. Nintendo Co, 578 F. Supp. 911 (S.D.N.Y 1983).

Video Olympics: Game Program Instructions (1977), Atari [Atari 2600 manual].

Wasteland Paragraphs (1988), Interplay, published by Electronic Arts [PC manual].

Afterword

James Paul Gee

The book you have just finished successfully defines a new field of inquiry. This new field is discourse analysis of gaming as a domain of human activity. *Discourse analysis* can mean different things, but, for the most part, it means the analysis of (oral and written) language in use. Analyzing language in use always involves analyzing more than language, since it must deal with *context* (specific situations in which language is used, and all the bodies, minds, things, actions, interactions, beliefs, and values that compose those situations).

The papers in this book tell us how language is used to make meaning when people play games. The papers look at the talk and texts around gaming as social practices and ways of enacting and recognizing different sorts of gamer identities. As such, the papers are a model for how discourse analysts can analyze the language foundations of different distinctive domains of human social activity. Each such activity has different "ways with words": different vocabulary, different ways of adapting the meaning of (old and new) words to specific contexts, different ways of phrasing things, different ways of recruiting grammar for meaning making, and different conventions of what counts as "acceptable" language and interactions.

Furthermore, gaming as a domain of human activity is, like so many other such domains today, global and cross-cultural, but yet still a shared set of ways with words inside social practices. So, the papers study, as well, how translation in games works across different types of words and across different cultures and how a game can be made to make sense to speakers of different languages in different cultures all of whom are gamers gaming.

Any specific domain of human activity has "strange" properties in comparison to other domains, especially domains we have come to take for granted. So we see here, among other "strange" things, the "odd" practices of insulting people in real-time interaction who cannot hear or respond to you (your competitors in multiplayer games where you are not using voice chat or messaging) and the practice of using hostile language to people with whom you are collaborating to solve problems. Seeing language and interaction in new ways in new settings is a key way for us to see old ways with words and social conventions we have come to take for granted as new and "strange" again, thereby attaining new metaknowledge about ourselves and our institutions and social and cultural groups (Gee 2014).

There are a billion things in the world we linguists could describe, so why gaming? Well, in part, as we have just said, gaming is a new and distinctive domain in which we can develop methods for describing a great many other domains and, in the act, sketch out the linguistic social geography of humans in our global world. But gaming is also an interesting and important case in its own right. Games are simulations in which the player is "half inside" and "half outside" the simulated world, a doubled actor. This is why gamers often say "I died" when their avatar dies in a game.

Imaginatively, books can work in a similar way, but in games, a player's choices make real differences, differences the player has to think reflectively and strategically about. Just as books have served humanity as (relatively passive) vicarious experiences from which we humans can learn, thereby greatly supplementing our everyday experiences, so, too, games can greatly supplement our everyday experiences via agent- and choice-driven experiences in virtual worlds that bear deep resemblance to our embodied experiences in the "real world."

Since humans learn from experiences and plan their futures through using those experiences in imagination, games may well change the way humans think and act and who they are, as have books and media generally (Gee 2017b). Thus, the papers in this book start the deeply important study of how language helps form, and in turn is changed by, gaming as a distinctive human activity.

This book is a deep dive into the distinctive language practices that occur in and around videogames with crucial implications for the future of how we study gaming. But, papers like these can form, also, an opening into yet a broader sense of discourse analysis, the application of tools that were built by linguists to study language in use to the wider domain of nonlanguage modes of meaning making and to multimodal modes.

Of course, the semiotic study of multimodality has already well begun (e.g., Kress 2010; Kress and van Leeuwen 2001). Nonetheless, we still know little about how to build and use trustworthy methods of analysis here. We know much less about how to take discourse analysis as a distinctive set of methods grounded in linguistics and extend those methods fruitfully to multimodality. Videogames are, however, a great place to begin, since they are a superb example of multiple moods integrated into a system (i.e., a game).

There are a great many different approaches to the analysis of meaning in oral and written language, not all of them connected to linguistics or linguistically based forms of discourse analysis. For me, any linguistically based form of discourse analysis must be connected to grammar (Gee 2017a).

Grammar is the set of principles—stored in the mind/brain—that tell speakers two things: (1) what count as syntactically well-formed phrases, clauses, and sentences and (2) what the basic (general, literal) meanings of words, phrases, clauses, and sentences are.

Let me give you a simple example. The grammar of English tells us that a sentence like "The coffee is missing" is grammatical, while "Coffee the missing is" is not. The grammar of English tells us also that the basic meaning of the grammatical sentence (its "literal" or "type" meaning) is that COFFEE (whatever constitutes our concept of coffee) SPILLED and YOU are BEING TOLD TO GET A MOP (where all capital-lettered words represent concepts and not words).

Concepts are generalities. The concept of a bird, connected to the word *bird*, covers a whole range of things that all count as birds and share some features (even of these are only family resemblances). So, that's grammar (syntax and basic meaning). Simple, yet important. It is important because grammar allows us, in actual situations of language use, to "riff" on general meanings and make them specific to the situation. So, if I say "The coffee spilled, go get a mop" in an actual situation, you know I do not mean coffee in general, but coffee as a liquid. If I say, "The coffee spilled, stack it again" in an actual situation, you know I mean in this case coffee as tins. Note other ways the meaning of the word/concept *coffee* can vary in actual situations of use:

(1) The coffee spilled, go get a broom.

(2) I need more coffee pickers for the harvest.

(3) Coffee prices are rising on the stock market.

(4) Big Coffee is as bad as Big Oil.

Note, also, in the case of "The coffee is missing," only the actual situation in which the sentence is uttered will tell us what *coffee* means here specifically. The meaning we humans make in actual situations, based on the general grammatical structures and meanings that anchor those specific meanings, we can call *situational meanings*. Grammatical structures, not just words, have situational meanings. The sentence form "X gives Y to Z" means that something was transferred from Y to Z, and that is the basic meaning of a sentence like "I gave Mary the virus." But, in different situations of use, "I gave Mary the virus" can mean I made her sick; I handed over a vial of the virus to her in the lab; I ceded her the research rights to the virus; and thanks to the ways the word *virus* can vary in meaning in different situations of use, it could even mean "I turned Mary onto a meme."

Situational meanings are construed, constructed, negotiated, transformed, and even invented by people in interaction within contexts, specific practices, and social and cultural groups. To study them, we need to study contexts, practices, social groups, and cultures in history, within institutions, and in local and more global contexts. While our ways with words like *coffee* may not be all that important in the larger scheme of things, our ways with words like *democracy, literacy, diversity, race, marriage*, and many others are far more consequential.

Grammar (syntax and basic meaning) not only sets the schema which anchors our situational-meaning work but also sets up the choices available to a speaker as to how to say what she wants to say. Grammar is anchor and choice maker. For example, consider the sorts of choices below:

1a. Mistakes were made.

1b. The company's president made a mistake.

1c. To err is human.

1d. Unforeseen circumstances intervened.

1e. Mistakes happen.

1f. It was a real blunder by the boss.

2a. Hornworms sure vary a lot in how well they grow.

2b. Hornworm growth exhibits a significant amount of variation.

2c. Hornworms come in lots of different sizes.

2d. *Manduca sexta larva* grow up to 70 millimeters in length, but can vary significantly.

3a. Could you please help me?

3b. I need help.

3c. I hate to ask, but could you possibly help me?

3d. Get a move on and help me.

4a. They are freedom fighters.

(said of people who use terror to attack our enemies)

4b. They are terrorists.

(said of people who use terror to attack us or our allies)

4c. They are guerillas engaged in guerilla warfare.

4d. They are mujahideen engaged in jihad.

Imagine that in the case of the utterances in (1) a company spokesperson has been asked why something bad has happened. The spokesperson must choose among all the alternative choices the grammar of his or her language makes available. The available choices are determined by grammar (a few of which are listed in 1). The actual choice made is that language in action—discourse—is determined by a human being in real time. Saying "mistakes were made" allows the spokesperson to leave out the person or people responsible for the mistakes. Saying (1f) might be a good way for the spokesperson to get fired.

Choices have meaning not just by themselves, but also in relation to all the other choices that were available, but excluded, once a choice was made (Saussure 1986). If all neckties were black, wearing a black tie would just mean you chose to wear a tie. If there are many colors of ties, then wearing a black one means you did not choose to wear other colors (e.g., brighter ones) for the occasion. And we can then ask why you didn't. So, too, with language.

Choices can allow us to try to capture the truth as we see it, to lie effectively, or to shape how people think without directly lying to them. They allow us to express what we want to say in ways that can reach people's emotions and minds and even encourage them to act.

So, at a very basic level, linguistic discourse analysis is the study of how grammar leads to situational meanings and to choices that, in context, constitute what the speaker (or writer) is trying to both say and do (accomplish). Now we can immediately see what a large task would lay ahead of us if we sought a discourse analysis of games. We would have to discover the grammar of various modes beyond language (like images, sounds, music) in their own right and then study how the basic meanings these grammars give rise to are situated (modified, adapted, made specific) in actual situations and practices.

We would have to discover as well, for each mode, what choices it made available for meaning making (at different levels) and then study what choices players made in given situations and why. Then we would have to the same things, at a higher level, for the multimodal system itself, the game that combines and integrates different modes and has properties (its own higher grammar) beyond its parts. This task has not even begun and would require the integration of different disciplines and tools.

But, what would be the point? What is the point of any linguistic discourse analysis? The point is to illuminate how humans make meaning in interactions in specific contexts and practices. And why do this? Because such studies, done right, illuminate the nature of human beings as certain sorts of creatures and the distinctive ways they bring help and harm to themselves, each other, and the world.

As we said above, videogames are deeply unique technology. Humans learn and think through experience (Gee 2017b). They store their experiences, of the real world and experiences they have had in media, in their long-term memories. In turn, they use these experiences to form simulations in the mind (a type of mental role play) in order to plan before they act, make sense of things, reflect on the past, and imagine better futures.

Videogames are an externalization of just what we do in our minds. In a videogame, we players stand both outside the game and inside it (with an avatar or a god's eye view we can often manipulate) and work out actions, interactions, and possible problem solutions in the service of sense-making. What was once private in the internal theaters of our minds is rendered public and very often social. We face a technology with the potential to be a new form of public and shared imagination. So this book could be the beginning of something big.

References

Gee, J. P. (2014), *An Introduction to Discourse Analysis: Theory and Method*, 4th edn., London: Routledge.

Gee, J. P. (2017a), *Introducing Discourse Analysis: From Grammar to Society*, London: Routledge.

Gee, J. P. (2017b), *Teaching, Learning, Literacy in Our High-risk High-tech World: A Framework for Becoming Human*, New York: Teachers College Press.

Kress, G. (2010), *Multimodality: A Social Semiotic Approach to Contemporary Communication*, London: Routledge.

Kress G. and T. van Leeuwen (2001), *Multi-Modal Discourse: The Modes and Media of Contemporary Communication*, London: Arnold.

Saussure, F. de (1986), *Course in General Linguistics*, (originally published in French in 1916), Chicago: Open Court.

CONTRIBUTORS

Carola Álvarez-Bolado is a lecturer at Universidad Politécnica de Madrid, Spain, in the Department of Linguistics applied to Science and Technology. She also lectures at the Games UPM Master on Videogame Development. Her main research interests are in the field of Applied Linguistics, with a special focus on videogame terminology in Spanish.

Inmaculada Álvarez de Mon is a professor at Universidad Politécnica de Madrid, Spain, in the Department of Linguistics applied to Science and Technology. Her research focuses on semantics, translation, and contrastive discourse analysis in the domain of Information and Communication Technology (ICT). She has both authored and co-authored articles on cohesion in English and Spanish, linguistic annotation for the semantic web, specialized translation, and the lexis of ICT.

Isabel Balteiro is a senior lecturer in English Lexicology at the University of Alicante, Spain. Her teaching and main research interests focus on lexis and word-formation mechanisms in English and in specialized languages. She has published quite widely on these fields; worth mentioning are internationally recognized publications such as two authored books, book chapters, and articles in journals such as *Lingua, Journal of Pragmatics, International Journal of Lexicography, English Studies*, and *English Today*.

Miguel Ángel Campos-Pardillos is a senior lecturer at the Department of English Studies at the University of Alicante (Spain), where he teaches legal English and translation in postgraduate programmes. He has written extensively on legal English and translation and contrastive lexicology. His research interests currently focus on the language of contracts, legal stylistics (specifically, metaphor in legal language), and legal English as a lingua franca.

Shelby Carleton is an undergraduate student at the University of Alberta, Canada. She studies English and game design, focusing on the narratives of games and what they can communicate to a player base. She has a strong interest in sociolinguistics and most recently was a co-author of a paper on "Deep Learning for Speech Accent Detection" (2017).

Adrianna Csipo is an MA student of linguistics at Ludwig Maximilian University, Munich, Germany. Adrianna is currently finishing her thesis focusing on metaphorical expressions from speeches given in the months leading up the 2016 US election. Her interests cover cognitive linguistics, morphology, and phonology, as well as computational linguistics.

Scott Dutt is a PhD candidate in Applied Linguistics at the University of Memphis, USA. His research interests include multimodal digital interactions, gaming, and impoliteness. His dissertation investigates informal learning communities in digital third spaces.

Astrid Ensslin is a professor in Digital Humanities and Game Studies at the University of Alberta, Canada. Her main publications include *Small Screen Fictions* (Paradoxa, 2017), *Literary Gaming* (MIT Press, 2014), *Analyzing Digital Fiction* (Routledge, 2013), *The Language of Gaming* (Palgrave, 2011), *Creating Second Lives: Community, Identity and Spatiality as Constructions of the Virtual* (Routledge, 2011), *Canonizing Hypertext: Explorations and Constructions* (Bloomsbury, 2007), and *Language in the Media: Representations, Identity, Ideology* (Bloomsbury, 2007). Ensslin is secretary and director of the Electronic Literature Organization, founding editor of *Journal of Gaming and Virtual Worlds*, a review board member of *Game Studies*, and an editorial board member of *Discourse, Context & Media* (Elsevier) and *Digital Culture & Society* (transcript). She has led externally funded research projects on videogames across cultures; reading, curating, and analyzing digital fiction; speech accents and language ideologies in videogames; and specialized language corpora.

John Finnegan is a lecturer in Screenwriting at Falmouth University, the U.K. His research interests focus on spectatorship studies, historical studies of screenwriting, and the relationship between technology and the screenplay format. He has published on the subject of digital screenwriting practices and is a representative for early career researchers as a member of the Screenwriting Research Network.

Christopher Gledhill has held posts at the universities of Aston, St Andrews, Strasbourg, and Lille. Since 2011, he has been Full Professor of English linguistics at Paris Diderot, France, where he is currently director of LANSAD (Languages for Specialists of other Disciplines) and co-coordinator of a research master's in Languages for Specific Purposes, Corpus Linguistics and Translation Studies (LSCT). His teaching and research interests include English as a lingua franca, text and discourse analysis, interlinguistics, specialized translation, and phraseology.

Tejasvi Goorimoorthee is a research assistant in the Humanities Computing Department at the University of Alberta, Canada. She received her degree in Mass Communications from Sunway University, Malaysia. She has a background in mass media and communications, media technology, and public relations, and her main research interests lie in digital media, international communication and media studies. She has also worked on various projects, namely "Speech Accents as Language Ideologies in Video Games" and "Videogames as Life Formation Narratives." Goorimoorthee is currently working on a thesis paper about decentralized and Twitter networks.

Dr. Sage L. Graham is an Associate Professor of Linguistics in the Department of English at the University of Memphis in Memphis, TN, USA. Using language as an analytical lens, her research explores misunderstanding, conflict, and impoliteness in digital, medical, and religious contexts. Her recent research addresses multimedia multimodality in online gaming.

Michael Hancock teaches technical writing and popular culture in the English department at the University of Waterloo, Canada. He has published on gothic choice and doubling in videogames *Planescape: Torment* and *Catherine*; his recent work emphasizes the connection between superhero adaptations, games, and aesthetics of cuteness. His research interests include branching narratives in 1980s gamebooks, paratextual game elements in title screens, and transmedia textualities.

Jason Hawreliak is an assistant professor of Game Studies at Brock University's Centre for Digital Humanities in Ontario, Canada. His research examines the semiotics and rhetoric of videogames and digital media with a focus on multimodality. He is the author of *Multimodal Semiotics and Rhetoric in Videogames* (2018) and a co-founder of the online game studies periodical *First Person Scholar*.

Elisavet Kiourti (PhD) is a sociolinguist lecturing at University of Nicosia, Cyprus. Her recent publications focus on gaming and longevity (Elsevier 2018, Frontiers 2018), literacy education in prison (*Research Papers in Education*, 2018), and lexicography (*Proceedings ICGL12*, 2017). Her research interests focus on language and literacy in digital contexts and videogames, social identity, and nonformal education. She has contributed to several research programs such as "Promoting Authentic Language Acquisition in Multilingual Contexts" (http://dev.palm-edu.eu/), the Digital Cypriot Slang Dictionary "Cyslang" (http://www.cyslang.com/), and "The Revitalization of Cypriot Maronite Arabic," contributing in writing books for teaching at A1 and A2 level.

Alice Ray is a translation PhD student in the English Department at Université d'Orléans, France. Her research interests focus on the translation of science fiction in French and, more particularly, the invented words found in science-fiction stories. She has published some articles on retranslation and is a member of the editorial board of the academic journal *ReS Futurae*. She also works as a scientific and literary translator.

Luke A. Rudge is Senior Lecturer in Languages and Linguistics at the University of the West of England, Bristol, U.K. His teaching and research interests include multimodal communication, intercultural communication, the intersection of language and technology, and functional approaches to understanding language in action. He has recently completed his doctoral research, presenting the first empirical analysis of British Sign Language via Systemic Functional Linguistics, and is expanding on this work while extending its application toward the visual-spatial elements found across a range of communicative environments.

Weimin Toh (PhD) teaches in the Department of English Language and Literature at the National University of Singapore. His main research areas are social semiotics, multimodal discourse analysis/multimodality, game studies, and narratology. He is also interested in other research topics such as child language development. He has published on anime in *Social Semiotics* and has a book chapter on weapon manipulation in *Emotions, Technology, and Digital Games*. His monograph *A Multimodal Approach to Video Games and the Player Experience* was published by Routledge in 2018.

INDEX

7 Days To Die   120, 122, 134, 136

abbreviation   4, 49, 52, 54, 62
accent distribution   8, 273, 275, 277
acronym   4, 45, 46, 54, 168
affective discourse   142, 154
affix   45, 47, 91, 100, 104
Alien: Isolation   5, 88, 109
appraisal   64
Atari   290

backformation   45, 47
Bad Language Expression (BLE)   6,
    139, 143–8, 150–4
ban   1, 205, 206, 208, 213–20
Battlefield 1   241
BioShock   7, 252, 253, 257–9, 262,
    264
blend   45, 48, 51, 143, 148
boilerplate language   127
borrowing   3, 4, 45, 49

chat   2, 7, 39, 41, 52, 197, 202,
    205–7, 210–15, 217, 218,
    220, 305
chatroom post   205, 206, 211, 212,
    214
clause
    abandoned   190–92, 197, 198
    declarative   188–9
    elliptical   6, 189, 190, 192, 197
    full   189, 189, 193, 197
    imperative   189
    interrogative   188–92
clipped compound   45, 46, 48
clipping   45, 46, 49, 126
code of conduct   202, 204

collaboration   179–81, 184, 185, 197,
    203, 208, 220
collocation   15, 16, 21, 24, 26, 28, 59,
    62, 83, 144
colloquialism   5, 125, 129
common semiotic principle   248
community of practice   6, 141, 148,
    152–4, 161, 215
compound/-ing   45–9, 70, 91, 92,
    94, 96, 97, 99–101, 103, 106,
    108–10, 126
conditional advice   70
context   1, 3, 5, 14–16, 22, 31, 36, 42,
    43, 48, 49, 60, 70, 72, 73, 75–7,
    79, 92, 94, 97, 98, 100, 101,
    103, 107–9, 130, 141, 142, 144,
    147, 149–51, 154, 158, 159–65,
    168, 170, 171, 174, 178, 179,
    183, 184, 190, 201, 203–5, 208,
    209, 211, 212, 217, 228, 229,
    232–4, 236–40, 242, 270, 276,
    293, 295, 296, 300, 301, 305,
    307–9
conversion   45, 49, 62
conversational discourse   53, 125,
    129
conversation analysis   6, 178, 182,
    183
cool   6, 8, 52, 139, 142, 143, 146,
    147, 158, 168, 212, 290, 296,
    297, 299–301
coolness   295–7
cooperation   6, 9, 139
copyright   117, 121, 132, 195, 288,
    289, 298
corpora   16, 17, 31, 34, 60, 63, 66,
    67, 69, 70, 73, 75, 78, 79, 97

corpus
    analysis   21, 61, 63, 65, 152
    -driven analysis   4–6, 35
co-situated gaming   2, 153
CS:GO   158, 162, 164, 165, 167,
    168, 170–4

deontic modal/-ity   65, 75, 78
derivation   45, 48, 49, 91, 92, 103
dialogue   2, 3, 125, 155, 189, 255,
    259, 273
directed imperatives   65, 78
discourse
    of cool   142
    of fun   142
    function   60, 79
    oral   41, 64, 129
    referents   5, 78
Donkey Kong Country   8, 288–90,
    293, 295–7, 300, 301
Dragon Age: Origins / DAO   8, 269,
    273–5, 277, 280–3
dysphemism   6, 160–2, 166, 168–70,
    173, 174

EULA   3, 5, 116, 117, 119, 121–5,
    127, 129–34
euphemistic expression   52, 53
evaluation   16, 61, 64, 65, 70, 77,
    201, 285
expletive   6, 143–50, 152–4, 158,
    166, 169, 171, 174

face-saving   139, 160, 174
FAQ   202, 204
fictive word   92, 99, 109
first-person shooter   6, 50, 157,
    162, 164, 205, 230, 232,
    237, 252
flaming   171, 202, 204, 242, 299
forum/fora   3, 4, 28, 39, 40–5,
    47, 48, 50–4, 120, 124, 127,
    140, 205
    asynchronous   40, 41
    language   50, 52
fractal recursivity   271, 272,
    280–2
frame analysis   161, 162, 250

fun   6, 35, 52, 72, 120, 129, 139, 142,
    143, 151, 154, 163, 168, 170,
    211, 212, 215, 259, 292
functional shift   49

gameplay
    experience   90, 254
    strategizing   247
    video   7, 248, 249, 251, 255
game-specific   43, 91, 269
gender   144, 155, 201, 207–9, 213,
    216, 217, 221, 264, 274, 280,
    282, 283
genre   2, 4, 5, 8, 14, 16, 17, 20,
    21, 23, 26–9, 36, 40, 43–5,
    52, 53, 59–63, 68, 69, 73,
    75, 78–80, 88–91, 108, 132,
    140, 142, 150, 153, 233,
    242, 273
grammatical item   5, 59, 66–9, 71,
    75, 78

Halo 4   144, 150
harassment   207–9, 216, 270
Helldivers   144, 147, 151
homophobia/homophobic   125, 144,
    145, 205, 207–9
hypermediacy   290, 296, 297,
    299, 301

identity   6, 41, 150, 158, 159,
    161–3, 171, 173, 174, 207, 209,
    273, 289
immersion   6, 87, 109, 147, 148, 240,
    262, 270
impersonal expression   62
impoliteness   6, 159, 160, 202, 203,
    204, 220
"Indie" game   120, 133
in-group code   43
initialism   4, 45, 54
instruction manual   3, 8, 62, 120,
    288–95, 300
insult   161, 162, 208, 210, 215–17,
    220, 281, 305
interaction markers   64
interactivity   62, 65, 90, 99, 109, 231
interplay   229, 290, 298–300

interruption   179, 185, 188–91, 196
    196, 197
interview   6, 139, 140, 144, 147–52,
    154, 163, 164, 170, 227,
    248, 251, 253–5, 258, 260–2,
    264, 292
invented word   88, 89, 92

jargon   3–5, 40, 42–4, 50, 54,
    118, 159

language
    creativity   40
    ideologies   269, 270, 284
League of Legends   157, 203,
    205, 206
legal action   125
legalese   5, 125–7, 133, 134
lexical
    choice   15, 39, 40, 43, 44
    creativity   40, 88, 89
    expansion   64, 65
    reduction   64
    set   278, 279
lexico-grammatical pattern   4, 59,
    60, 69
linguistic strategy   158, 168, 170,
    173, 174
loanword   49
localization   3–5, 16, 21, 23, 25,
    87–90, 92, 99, 109
The Lord of the Rings, Vol. 1   290,
    297–9, 301
ludolect   3, 43
ludonarrative
    dissonance   7, 239, 242, 248,
        255–9, 263
    irrelevance   255, 256, 262, 263
    relationship   7, 248, 250, 251, 252,
        254, 255, 263–5
    resonance   248, 255, 256, 259,
        263

Mafia III   227, 228, 234
Mass Effect   46, 239, 240
meaning-making   140, 229, 248,
    251, 306
microlinguistic choice   42

Minecraft   5, 117, 119–34
MOBA (Multiplayer Online Battle
    Arena)   205
modal
    consonance   237, 238
    dissonance   238, 240, 241
mode   3, 7, 15, 28, 46, 47, 50, 131,
    143, 164, 167, 188, 201, 221,
    228–43, 290, 306, 309
moderator   7, 126, 206, 210, 213–15
morphological choice   41, 44
multimodal/-ity   2, 3, 7, 43, 139, 145,
    158, 162, 164, 165, 202, 205,
    228, 230, 231, 234–40, 242,
    243, 248, 270, 306, 309
multiplayer   6, 20, 23, 28, 45, 48, 91,
    120, 122, 157, 164, 178, 184,
    198, 203, 205, 305
multiple embedding   63

narration-commentary   7, 248, 249,
    251, 252, 254, 255, 257–9,
    261–5
narrative comprehension   247
neologism   4, 14, 15, 22, 29, 30, 34,
    36, 70, 92, 108
new
    creation   45
    formation   4, 45
    word   14, 42, 45, 47–50, 88–91,
        103, 106, 109, 305
Nintendo   42, 46, 50, 150, 288, 289,
    294–7

online
    communication   39, 41, 43,
        51, 54
    interaction   41, 42, 202
oral language   41, 52
othering
    sociophonetic   8
    unintentional   8, 280
Overwatch   203, 205, 206

paragame genre   80
paratext/-ual/-ity   2, 3, 5, 7, 8, 62,
    80, 139–43, 154, 290, 292, 294,
    295, 297–301

pause 148, 159, 165, 188, 191, 194, 195, 197
performative action 158, 165, 168–71, 173, 174
personal pronoun 52, 125, 130
phraseology 4, 42, 58–61, 71, 78, 80
plain language 117–19, 133
player experience 7, 140, 247, 262
prefix 29, 48, 100, 104, 105
prefixation 48
procedurality 2, 7, 228–34, 236–8, 242
procedural rhetoric 229, 230, 239, 240

racism 205, 208, 209, 228, 234, 270, 281
racist 125, 126, 144, 145, 207, 208, 228, 275
rapid written conversation 53
readability 119, 122, 123, 130, 133, 134, 293
received pronunciation (RP) 270–2, 275–7, 279, 282, 283
reformulation 70, 77
register 4, 41, 53, 54, 61, 68, 69, 71, 78, 79, 89, 121, 125, 129
rephrasing 127
rhyming slang 52
rule 7, 43, 117, 119, 141, 148, 205–7, 210, 212–14, 219, 220, 227–9, 237

Saints Row IV 241
science fiction 5, 88, 89, 91, 92, 105, 106, 108, 109
Sega 46, 296
semantic
  neology 15
  shift 4, 36, 49, 92, 93, 108
sexism/sexist 144–6, 152, 205, 207–9, 216, 221, 270
silence 191, 194–8, 263
situational meaning 8, 307, 308, 309
slang
  expression 52, 54
  oral language 52
  word 43, 51, 53, 54

sociophonetic othering 8
spam 7, 206, 207, 211–13, 217, 219, 220
spatial extent 70
specialized
  collocation 62
  languages 14
speech accent 8, 7, 269, 270, 272, 273, 283
spoken language 41, 53, 190, 197
Standard North American (SNAm) 270, 272, 275–7, 281, 282
Starbound 120, 122, 123, 126, 128, 131, 132
Starcraft II 5, 88, 109
stream 209–16, 271
streaming 2, 205, 213, 215
Streetfighter X Tekken 144
stress 34, 158, 161, 169, 174, 180, 182, 185, 191, 195, 197, 279
subordination 63, 65
suffixation 48
swearing 2, 6, 142, 143, 147–50, 152, 154, 158, 161, 162, 166, 168, 169, 174
systemic functional linguistics 6, 178

taboo expression 53
task completion 6, 182, 185, 195, 195, 197
technical
  discourse analysis 60
  term 50, 64
technicity 50, 62
techspeak 40, 43
terminological network 62
terminology 4, 13, 14, 40, 43, 54, 58, 60–2, 65, 66, 71, 80, 104, 107, 119, 235
terms of abuse 6, 144–6
think-aloud protocol 249–51, 254, 264
timeout 213–18
transitional location 70
translation
  pattern 89, 94–6, 98, 99, 101, 103, 108
  strategy 107

trolling   51, 125, 126, 202, 204,
     209, 210
Twitch   2, 7, 171, 205–7, 209, 211,
     213, 217, 219–21, 301

Verdun   237, 240
videogame
   manual   8, 289–92
   review   16, 36
   tutorial   5, 59, 68

walkthrough   62–5
word-formation   4, 40, 44–6, 50,
     54, 108
World of Warcraft   2, 46, 47, 117,
     122, 124, 157

Yoshi's Woolly World   144,
     150, 151

zero-derivation   49

www.ingramcontent.com/pod-product-compliance
Lightning Source LLC
Chambersburg PA
CBHW060144280326
41932CB00012B/1634